Praise for AGAINST THE TIDE

"A very fine and engaging book."

— **Bill McKibben, author of *The End of Nature***

"An eloquent and thoughtful elegy which anticipates the loss of yet another fishing community along our coasts."

— **Peter Matthiessen**

"This is a book concerned in equal measure with endangered fishermen and endangered fish, both of which are in a deep sense beautiful and worth preserving. Policy analysts tend to look over the heads of human beings. Good writers, like Richard Carey, look them in the eye. Mr. Carey brings particular people and places to life in these pages, and he makes us share his worry and his love for them." — **Tracy Kidder**

"*Against the Tide* is deep economic journalism at its best, an effective and compassionate chronicle of a threatened way of life, and a worthy successor to such classical portraits of American fishermen as William W. Warner's *Beautiful Swimmers* and Peter Matthiessen's *Men's Lives*."

— ***New York Times Book Review***

"The author has sailed and fished with the draggers and longliners and lobstermen of Cape Cod, and in this deeply observant and well-written book he portrays the complexity of their work while delineating the web of tortuous economic and environmental regulations in which they're enmeshed."

— ***The New Yorker***

Books by Richard Adams Carey

RAVEN'S CHILDREN

An Alaskan Culture at Twilight

AGAINST THE TIDE

The Fate of the New England Fisherman

Against the Tide

THE FATE OF THE NEW ENGLAND FISHERMAN

Richard Adams Carey

A MARINER BOOK
HOUGHTON MIFFLIN COMPANY
Boston • New York

First Mariner Books edition 2000

Visit our Web site: www.hmco.com/trade.

Library of Congress Cataloging-in-Publication Data

Carey, Richard Adams.
Against the tide : the fate of the New England fisherman / Richard Adams Carey.
p. cm.
Includes bibliographical references (p.).
ISBN 0-395-76530-7
ISBN 0-618-05698-X (pbk.)
1. Fishers — Massachusetts — Cape Cod. I. Title.
HD8039.F66U474 1999
338.3'727'0974492 — dc21 99-18146 CIP

Book design by Robert Overholtzer

Printed in the United States of America

QUM 10 9 8 7 6 5 4 3 2 1

FOR

MARY JANE CAREY MINER
1920–1998

whose days were radiant with
courage, grace, and love

For an animal its natural environment and habitat are a given; for a man . . . reality is not a given: it has to be continually sought out, held — I am tempted to say salvaged.

— JOHN BERGER, "The Production of the World"

Contents

Against the Tide

Prologue

I GREW UP in central Connecticut, but in my family's more fortunate days we rented a summer cottage on the Connecticut side of Long Island Sound. I shared a bunk bed with my older brother, and on certain mornings, before Jack or anyone else, I woke there to the sort of stillness that I imagined reigned on Sabbath days in heaven. Black Point was always a quiet place: no TV, no radio, no hi-fi record player, and certainly no traffic. There was, however, almost always a low-level background hum (something like the refrigerator at home), Black Point's own sort of traffic: a musical susurrus spun from the murmur of the wind, the slap of the waves on the beach, the keening of the seagulls. But on these special mornings even that was hushed. I lay in the early morning chill, in the thin half-light, and heard absolutely nothing except Jack's even breathing from the bunk above me and, at unpredictable intervals, the basso profundo of foghorns out on the sound.

I imagined the fog as thick as a snowbank on the water, except for that peculiar island of clarity that an observer occupies. I felt sorry for the sailors, who had to maneuver their boats and ships through the fog, but I believed that their horns kept them reasonably safe, and I thrilled to the sound of those horns. Their blasts ran throbbing through my bones. They set me all in motion, set my bones to rocking somehow, as though my bunk were a cradle set loose on the sea. The horns spoke of vast distances, of great errands, of hidden portents, of inarticulate sorrows. Outside the cottage all creation had halted in order to listen to them. I rocked in my bones between sleep and wakefulness. The sea was a theater of dreams.

Black Point, and later neighboring Giant's Neck, where for a time we actually owned a cottage, were always much more vivid places to me than suburban Hartford. Both Giant's Neck and Black Point were once ruled by the Niantic chief Mamarakagurana, also known as the Giant. Some hold that Giant's Neck took its name from that imposing chief, while others say the name came from the events surrounding a siege once laid on the Niantics by the larger and fiercer Pequot tribe. The Niantics, indisposed to pay tribute to the Pequots, had holed up in a fort at Black Point and were near starvation. Finally, on the same foggy day that the Pequots chose to launch an all-out attack by canoe from Groton, a party of white settlers friendly to the Niantics loaded their own boats and canoes with supplies and set off toward Black Point. The fog lay particularly thick about Millstone Point, and the Pequots were holding close to shore. As they rounded Wigwam Rock, the attackers were astonished to find an opposing armada bearing down on them out of the mist, a fleet led by a man of extraordinary size. (This giant, a white man named Lester who was said to be big anyway, happened to be standing on a crate of supplies.) The Pequots panicked and fled, the siege was lifted, and the rest is history — maybe.

I always believed in that story myself, but mostly for sentimental reasons. I liked its elements of chance and inadvertency, not to say its whiff of comedy. I liked its peaceful alliance between white man and Native American, though it was unfortunate — and ominous — that the alliance was at the expense of another Indian tribe. I liked its specificity in place and landmark (I knew of no place in West Hartford that, like Wigwam Rock, had a story to tell). And I liked the role of fog, which I considered supernatural stuff, at least in the vicinity of Long Island Sound.

Down there fog arrived with meanings and implications that the fogs of West Hartford entirely lacked. In its power to still the traffic of both wind and man across the water, to blanket the horizons, to set the foghorns ringing, to simultaneously rock me in my dreams and reveal the world's capacity for catastrophe, the summer fog on the sound was a captivating communication from a world more complex, intriguing, and dangerous than that of the suburbs. During the other three seasons, in West Hartford, I stared blankly at the weather reports on the evening news, mystified by their notices of high and low temperatures and barometric readings, their probabilities of rain, their gradient maps with swirling lines of teeth, maps that looked like battle plans. I didn't see what it all meant, couldn't understand why anybody cared, since — with the notable exception of a possible snow day, home from school — we were all going to do precisely what we intended to do regardless of the weather. I supposed at last that there just wasn't enough news to occupy a whole half-hour and that TV stations needed something to fill in the time until the first of the evening quiz shows and sitcoms. In assuring us that something wasn't going to sneak up on us out of the sky, these weather reports harmonized, I supposed, with the neat lawns and manicured bushes of West Hartford: they suggested that everything was under control.

Of course, it wasn't just the fog that caught me, or even its opposite, the sound and fury of the occasional hurricane that blew through. It was also the mussels we scoured off rocks, the crabs we caught with them (green crabs, ghastly-looking spider crabs, ferocious blues) after we cracked the mussels' shells and hung them on strings from docks. It was the starfish we harvested from the bottom and baked on the concrete of our front steps into pale odorless facsimiles of living starfish, with most of their wonderful tangerine color sizzled away. It was what we saw when we went diving for those starfish: the eerie ripple of an eel's tail as the beast snaked its way through a bed of eelgrass; the eye-blink, take-your-breath-away speed of a sand shark vacating its former plot of sand, leaving nothing but a cloud of silt and pebbles behind.

These were all part of a world of astonishing color and shape and texture, of surprise and a perceptible knife-edge of menace. I hunted and fished along its margins, burned by the sun, stung by sea nettles, bitten by greenheads. Over its deeps, meanwhile, enterprising men made a living of what I did for fun. Their boats — dwindling in numbers even then, in the early 1960s — went out of Stonington, Mystic, Groton, and Old Saybrook. I couldn't say at the time what it was they caught, but I took note of them as they plied the horizon far beyond Snake and Griswold Islands, off Giant's Neck.

I knew that the boats' occupants lived in the weather, that the disposition of the tide and the wind and the barometer made an enormous difference in what they did each day. I knew that the dramas of life and death that they engaged in as a matter of course were conducted on a scale and with a gravity of purpose that revealed my own pursuits as frivolous. I conceived of them as astronauts, orbiting far beyond the physics that bound me to the shore. If they were not themselves the sources of those booming horns on a foggy morning, the audial equivalent of quasars at the edge of the universe, then at least they bobbed in their eggshells among the tankers that were. I tried to imagine what they knew, and could not. As I grew older, and as the girls I once played with — the daughters of welders or machinists at Electric Boat in New London, where nuclear submarines were made, or else of vacationing bankers or insurance salesmen — put on bikinis and stretched themselves on towels spread on the sand, I took less notice of the boats out to sea.

When I was in my early teens my family's fortunes changed. My brother, Jack, twelve years older than me, had been gone for some time anyway. We were forced to sell the cottage at Giant's Neck, and I returned there only occasionally, sometimes in winter, to stay with friends or sleep on the beach. A pretty girl who had grown up within earshot of the surf at Giant's Neck, whose hair, even when perfumed, always smelled to me like the morning breeze off the sound in summer, came away with me to college in

Boston. That romance ended sadly just before our graduation. Jack died suddenly of a brain tumor one night in his apartment in Los Angeles, and was not found until several days later. I thought it puzzling that his breath could just stop unnoticed, that I didn't rise from my bed in alarm, stunned by that most absolute silence, even though I was several thousand miles away. Grief stole upon me unawares. In my homes out of the weather, the circumstances of catastrophe — a lover's recrimination, a brother's death, even an unpaid mortgage — were prepared in steps as abstract as those that built the gradient maps on TV. There seemed to be no logic to their onset, no preparatory shift in wind direction or telltale accretion of fog.

I went west, not to Los Angeles but to Alaska. I taught school in an Eskimo village on the Bering Sea. The people I met there lived far beyond the pale of the catastrophes that had overtaken both the bellicose Pequots and the friendly Niantics in King Philip's War and its aftermath, beyond the pale of three hundred years of manifest destiny, the U.S. Cavalry, the Homestead Act; a people who in the 1980s possessed telephones, TVs, and videocassette recorders in every house and yet still lived primarily by hunting and fishing. There was no beach; the tundra simply dissolved by degrees into the muddy water of Kuskokwim Bay. But the land was rich in precisely the way wealth had been defined for most of human history. I saw split salmon hanging as thick and red as autumn leaves from driftwood drying racks, Canada geese and pintail ducks heaped in full plumage into stand-alone freezers, ringed seals looking in their sleek silver coats like extra fuel tanks as they lay dead beneath the thwarts of a skiff.

These were people to whom beef or chicken or vegetarianism was not an option, who could not afford to sentimentalize their wildlife, as other Americans did, but who loved their wildlife far better. They ranged across the tundra and the Bering Sea, through rain and wind and Plutonian cold, often at peril of their lives, with a *joie de vivre* that made me not at all offended when they described my own kind of work as "a jail sentence." Sometimes I hunted and fished at their side, though my own skills were comic compared to theirs. They knew the answers to questions I didn't even know how to ask.

They also knew about the political controversy surrounding subsistence hunting and fishing in Alaska, and they knew their American history as well, if not back to the Pequots and the Niantics, at least back to the Sioux and the Nez Perce. They were beginning to understand also the titanic, black-hole-like gravity of the cash economy, something that allowed no turning back once a person approached and made that first transaction. They went about their vocation, and avocation, with a tragic sensibility, an underlying and pervasive fatalism. I wrote *Raven's Children: An Alaskan Culture at Twilight* not so much as an effort to preserve and explain that world, which was beyond my power in any respect, but as an effort to understand my own, which by comparison seemed thin, attenuated, and smugly ignorant of alternative economies, patterns of living, and terms of engagement with the land and sea.

I left that village, finally, and others in which I taught after that, entirely out of respect for such an alternative. I was afraid that if my own children grew up in that corner of the bush, they would find life outside the bush — in the world of commerce and weather maps, in the midst of what Brian Gibbons in this story describes as "the golden-arches, white-lines, and car-payment culture" — entirely too much like a jail sentence. Once back east, however, and especially after a summer of commercial fishing with Oscar Active, the protagonist of *Raven's Children,* I was reminded that (to paraphrase Bob Dylan) you don't have to be an Eskimo to know which way the wind blows. I remembered those boats out of Stonington and Mystic, and those portentous foghorns on the sound, when I read such newspaper stories as one in the *Boston Globe* in 1990 titled "The End of the Line," which described how one of Gloucester's proudest fishing families was selling off its boats.

The newspapers, and many observers, were describing the end

of New England family fishing as a morality play. In act one, factory trawlers from the Soviet Union and most everywhere else in the world scooped up tons of cod from the Georges Bank fishing grounds. In act two, Congress passed the Magnuson Fishery Conservation and Management Act, banishing the factory trawlers and reserving the wealth of Georges Bank for American fishermen. In act three, American fishermen made lots of money, some of which they invested in drugs and parties, the rest in bigger boats and more sophisticated gear. In act four, those fishermen howled in protest whenever fisheries managers or conservationists or sport fishermen suggested they might cut back a little. In act five, the cod disappeared and commercial fishermen received the wages of greed. In many stories they were described as our last cowboys, the implication being that it was high time to fence in the range. Better they should get regular jobs, like the rest of us.

My years in Alaska had led me to suspect that commercial fishermen loved the cod and other New England species even more than urban conservationists did. But a fisherman's conservationism is necessarily of a more complicated sort than a city dweller's, influenced in its germ by the need to keep doing, in some manner, what a fisherman does. Others condemned this as an entirely spurious conservationism. And frankly, they saw little need for fishermen. So the Brancaleones in Gloucester were selling their boats? Well, too bad, but you can't stop progress. A few generations ago most of us were farmers, they argued; now a few agribusinesses accomplish industrial-scale farming for us, and while it's sad that the modern family farm exists only in TV commercials, we console ourselves at the market with greater efficiencies and lower prices. We ourselves lead lives less strenuous than a farmer's. Agribusiness-style industrial fishing and/or aquaculture — strictly regulated, of course — is simply the next logical step up on the car-payment culture's evolutionary ladder. Let them become appliance repairmen and have their weekends off.

During the foreclosures of the 1970s and 1980s, I was always impressed by the tears and rage that accompanied a farmer's abandonment of his strenuous lifestyle. I also worried about the progressive loss of knowledge that accompanied those foreclosures, a process analogous to the death of a generation of elders in an Eskimo village. More than the rest of us — more than anybody except a fisherman or an experienced subsistence hunter — a farmer has a command of what the writer Bill McKibben terms "fundamental information." This in part is information specific to farming and general to most of recent human history — how to clear land, how to build a house and barn, how to raise crops and cattle, how to thresh grain, how to bake bread, how to kill and butcher an animal, how to forecast the weather — and in part is that philosophical information natural to anyone who lives in deep engagement with either the land or the sea: the double helix of life and death; the resourcefulness and intelligence of other animals; the manifest triviality of a human being in the great scheme of things; a keen apprehension of limits.

Even the dullest farmer quickly learns about limits, says McKibben:

> You can't harvest crops successfully until you understand how much can be grown without exhausting the soil, how much rest the land requires, which fields can be safely plowed and which are so erosion-prone they're best left to some other purpose. This sense of the limits of one particular place grants you some sense that the world as a whole has limits, a piece of information we've largely forgotten, in part because being a successful businessperson involves constantly breaking through limits.

A fisherman possesses fundamental information as well, and precisely that same sense of limits, though knowing where to draw the line necessarily becomes a trickier proposition when dealing with fish and fishing grounds than with crops and fields. When those fishing grounds are held in common, and when a businessperson's limits-be-damned bias is mixed in, then perhaps inevitably a fisherman becomes a suspect in his own demise. But I wondered if the fishing crisis of the 1990s was in a certain way like the death of my brother or other personal losses: a catastrophe not so easily predicted as hindsight might suggest, a misfortune whose elements rise above blame or innocence and plumb far below the ordinary gestures of a workaday life. In any event, I saw no blush of progress, no evolutionary advance, in the foreclosure of another fishing boat. And I was frightened by the evident willingness of many outside the fishing industry to dispense with the services of small-boat family fishermen and to consign them and all that they knew to the dustbin of history.

In 1994 I drove to another Stonington, this one in Maine, at the northern mouth of Penobscot Bay. I ate lunch there with Robin Alden, who has since served a term as Maine's commissioner of marine resources but at that time was the publisher and executive editor of the *Commercial Fisheries News,* the newspaper she had founded in 1973. Her paper offered the best and most comprehensive coverage of the New England fishing industry, and I went to her for both her knowledge of the business and her contacts. I didn't really know at that time what I was interested in, specifically, but I knew who: someone who was the owner-operator of a small family boat and therefore working inshore waters; someone who was in some way imperiled by the events in the industry; ideally, someone whose family had been fishing for several generations.

Robin suggested the names of several fishermen in Maine and Massachusetts, and talks with these men produced other names. These boiled down to a short list of attractive and willing candidates, men who were content to share a calendar stretch of their everyday lives with me. One had been on Robin's original list: a lobsterman from Cape Cod whom Robin didn't know personally but whom she had heard speak at a New England Fisheries Management Council meeting on lobster industry issues. Brian Gibbons was not in any way the product of a fishing family, but he knew his business, loved books, and in his public service work as a representative of Outer Cape lobstermen had become as familiar with the political behavior of *Homo sapiens* as he was with the no less scrofulous feeding behavior of *Homarus americanus*.

Perhaps too familiar: Brian was interested in my project, but not interested in taking on the entire burden of this sort of representation. For my part, I was more interested in the spectacularly troubled groundfish industry — and the sacred cod — than I was in lobstering. Eventually a compromise was struck that gave my project added breadth. Brian approached several other owner-operators living in his home town of Orleans, each of them in a different sector of the Outer Cape's very diversified small-boat fishery. We were joined at last by Carl Johnston, a mate and sometime skipper on a Chatham dragger; Dan Howes, who dredges quahogs out of Cape Cod Bay; and Mike Russo, who runs a long-liner out of Chatham. This is the story of their various engagements during 1995–96, not only with the species for which their boats were geared, not only with the winds and tides that toyed with those boats, but also with the currents of history and the squalls of politics and economics that were as inescapable for these men as weather systems. There was nothing in their lives, I discovered, that was as pat as a suburban lawn, as seemingly specious as a Hartford weather report.

In the end I was glad that I had been sent to Cape Cod. Down around Long Island Sound, the battle between the fishermen and the gentry for land on and near the water had already been lost by the former. But on Cape Cod such places as Orleans and Chatham and Provincetown were still as much fishing towns as they were resort communities, and as such their inhabitants had more than a sentimental connection to their history; they believed — or at least hoped — that their history had something to do with their future. And the literature of Cape Cod afforded me the company of such graceful historians as William Bradford, Samuel Eliot Morison, and Henry C. Kittredge, such naturalists and philosophers, such petitioners of fundamental information, as Henry David Thoreau and Henry Beston.

In fact, if Alaska's Kuskokwim Bay might be the place where America ends (or possibly has yet to arrive), then Cape Cod is arguably the place where America began. The straight line that joins the United States Congress to the Mayflower Compact, signed in Provincetown Bay in 1620, extends as well to the New England Fisheries Management Council and all its angry and uneasy constituents. On Cape Cod, America has returned to its cradle to wrestle democratically with one of the unresolved issues of its maturity: how to tap this continent's wealth without plundering and despoiling it; how to reconcile our hard-wired demand for growth and consumption with a husbandman's concern for sustainability; how to mark our limits and resolutely stay within them.

In my dreams I still hear those foghorns ringing. The fishermen ply their nets somewhere on the other side of grief, and the fish rise shining from the sea.

≈ 1

Siegfried's Fabulous Horde

BRIAN GIBBONS likes jazz. He likes Miles Davis, John Coltrane, Pharoah Sanders, Thelonious Monk. He likes music that's tough and sinewy and inventive, that sings against the mortal tenor of its own heartbeat, that blue, throbbing pulse of fatal and proximate sadness. He doesn't care so much for Bob Dylan, the troubadour of his generation, but he knows Dylan well enough to parody him: "We're artists and we don't look back."

He stands on Snow Shore, gazing across Nauset Inlet and out to sea. At his back is history: a historical marker saying that the French explorer Samuel de Champlain anchored near this spot in 1605 and that Champlain's ship's carpenter, an unfortunate named Malouin, from St. Malo in Brittany, was killed on this beach by Nauset Indians in a dispute over an iron kettle. It also notes that this was the landing for the undersea telegraphic cable that from 1898 to 1959 stretched three thousand miles between Brest, France, and Orleans, Massachusetts, and carried the first word to America of the success of Lindbergh's solo flight. It notes as well that this was the home port of the early Nauset fishing fleet on Cape Cod.

At his back is history: the site of the little two-room camp that Dr. Ralph Wiggin, a Boston urologist and surgeon — Brian's grandfather on his mother's side — bought in rural Orleans in 1910 for his hunting and fishing pleasure, his family's rest and solace; the spot on a curving stretch of the Southeast Expressway near Braintree where Brian's friend Eddie's MG sports car went out of control in 1968 as it was carrying Brian to his induction into the army and from there possibly to Vietnam; the cove in Pleasant Bay where Brian, by then a husband and breadwinner, was pleased to catch his three-bushel limit on his first day of bullraking littleneck clams in 1972; all the newspaper racks and magazine stands, the television sets and radios, that in 1976 trumpeted the news on the Cape and along the Gulf of Maine that President Gerald Ford had signed into law the Magnuson Fishery Conservation and Management Act, which promised a prosperous new era for hard-pressed New England fishermen; the dock at Wychmere Harbor in Harwichport, where the famous Harry Hunt, who taught Brian how to catch lobsters, was wont to pull his lips tight against his teeth and remind his young sternman, "The wind, the tide, the weather, and *every man* is against ya."

At his back is a rampart of glacial till cocked like John L. Sullivan's left fist and thrust thirty miles out into the Atlantic; at his feet is one of the Atlantic's most notorious graveyards. Since the English ship *Sparrowhawk* grounded against a sandbar off Nauset Beach in 1627, more than three thousand ships have met a proximate sadness off Cape Cod, and uncounted sailors have laid their bones near the carpenter Malouin's.

But that sort of catastrophe seems impossible now. The terns — common, roseate, and least — have returned to Nauset like the swallows to Capistrano and are gathering into colonies to mate and nest. The striped bass are back too, swimming north from their spawning grounds in Chesapeake Bay and ravenously following the herring into the bays and inlets. Whales — finback, humpback, the nearly extinct right whale — are feeding off the beach on krill or bait fish schooled near sandbars such as the one that doomed the *Sparrowhawk*. The migrating whales occasionally break water like sandbars themselves, then settle, submerge, dissipate into spring's seething broth of blossoming diatoms and dinoflagellates, opalescent moon jellies and scarflike nudibranchs, a profligate and proliferous brew comprehending a million animals in a single quart, as many as three thousand different species among them, each tiny creature as distinct an organism as a sixty-foot finback, each as eloquent as that whale of creation's wealth and invention.

The rim of the harbor is layered with pollen from Cape Cod's ubiquitous pitch pines, as were the hood and bed of Brian's Ford pickup as it sat in his driveway twenty minutes ago. Broad magenta splashes of salt-spray rose fleck the gritty scrub behind the beach. The scent of honeysuckle seeps down from the heights behind us, where great and fine houses of glass and cedar shingle keep watch.

But Brian Gibbons does not look back. He holds to the harbor, where a small fleet of lobster boats, including his own, the *Cap'n Toby,* rests placidly at anchor. Waves roll under the yellow pollen like wrinkles being smoothed from a carpet. The horizon has a telescopic clarity, the boats a cardboard-cutout inertness. Offshore, unseen, scattered pods of *Homarus americanus,* the American lobster, are moving into shoal waters to their summer feeding grounds.

Brian lingers at the door of the pickup, drinking the morning in, joyful just to be here, content in this truce he has struck, at least for the moment, with the wind and the tide and the weather, if not necessarily with other men or the history that in recent years has cast such a shadow across his and others' lives.

"Once a week, for thirty seconds or so," he says, "you get to just love it."

At thirty seconds or so past seven, on this morning of May 20, 1995, Brian lifts the engine cover, amidships on the *Cap'n Toby,* and opens up the lobster boat's seacocks. The raised cover reveals a 135-horsepower, 4-cylinder, turbocharged Volvo marine diesel. The seacocks allow water from the harbor to be pumped into the engine's cooling jackets. Brian kneels at the engine hatch and checks his other fluids: coolant, crankcase and transmission oil. He switches on the battery, jabs an extra shot of fuel into the cylinders, and turns the engine over.

Brian doesn't love the cut-down plastic milk jug that hangs jury-rigged on a breather line from his crankcase, which shivers like a nest of angry wasps as the motor kicks into life and sucks seawater into its cooling jackets. He says that Volvos are notoriously hard on their o-rings and seals, and that each day now his own Volvo is leaving a quart of engine oil in the bilge. The milk jug on the breather line slows that loss, collecting oil vented out of the crankcase and allowing it to drain back into the engine. "Where it can then be leaked into the bilge," Brian adds, smiling.

The only thing he can do with his bilge water is pump it into the sea. If a quart of oil is mixed into that water, then Brian, according to the federal Water Pollution Control Act, the terms of which are tacked beneath his cabin's port window, is subject to as much as a $5000 fine from the Coast Guard. This happens to equal the amount of money he spent in March at Nauset Marine in an unsuccessful attempt to stop the oil loss. Those two amounts combined equal the price of a new engine for the *Cap'n Toby.*

Brian stays on the right side of the feds, and keeps the inlet clean for lobsters and striped bass and littlenecks with an oil-absorbent cloth inside the milk jug and absorbent pads packed into a fine-mesh bait bag in the bilge. Every week or so he fishes this bag out with a gaff and changes it. He actually enjoys the jiggling milk jug's combination of grade-school science project tackiness and genuine efficiency, but he doesn't like the fact that it's necessary, nor the extra housekeeping it requires. He is also increasingly fearful that the money he has poured into this engine so far might as well have been leaked into the bilge or pissed over the side — especially in light of what has been so far an inexplicably poor lobster season on Cape Cod.

The *Cap'n Toby* is moored several hundred yards off Snow Shore. The beach gives way to a level mud plain only one to three feet under water at low tide, which requires boat captains in Nauset Inlet to reach their vessels on the other side of the plain through a three-stage shuttle system. Moments ago Brian rowed a dinghy just bigger than a bathtub out to a sixteen-foot wooden skiff moored halfway between the beach and the *Cap'n Toby.* This was slightly comic, with Brian's big arms and jackknifed legs squeezed into that dinghy, his baseball cap askew. He looked like a grown Huck Finn, only now getting around to lighting out for the territories, putting on the lineaments of boyhood again as he did so.

He tied the dinghy to the skiff's mooring and then motored the skiff into shore, beaching it behind the Ford pickup, which he had parked at the water's edge. Then we loaded the skiff with items from the pickup's bed: twelve wooden lobster traps to be added to the gear already offshore, an equal number of buoys and lengths of creosote-dipped rope (what Brian calls warp; that is, rope for hauling), two coffee-table-sized plastic totes of mackerel and squid for bait, two six-gallon diesel fuel canisters, a cooler containing peanut butter sandwiches and Diet Pepsi, and a bucket of such odds and ends as fresh rubber gloves and a handheld VHF radio. "In case of in case of," Brian said, nodding at the radio, repeating the cautionary words of Olav Tveit, the captain of a ninety-four-foot scalloper out of New Bedford on which Brian once served. When the skiff was loaded, we pushed it off the beach and pointed for the *Toby.*

Now Brian arranges gear in the cabin of the twenty-five-foot boat while the engine idles. The wooden traps are piled in stacks of three in the stern, and the spindly black-pennanted buoys, constructed so their flags will bob several feet above the waves, are jammed upright through their slats. The totes of bait are stacked amidships behind the engine cover, which has been put back in place. The cooler and the plastic bucket are stowed in the wheelhouse. The fuel canisters are for topping the boat up at the end of the day, and they remain in the skiff, which has been tied to the *Toby*'s mooring. When the engine's temperature gauge moves off dead cold and there is that first blush of heat to absorb as the seawater swirls through the housings encasing the coolant lines, Brian scrambles over the starboard gunwale to the foredeck, unties the mooring, and swings back behind the wheel.

Spitting oil and breathing water, flying as many flags as a unit of hussar cavalry, the *Cap'n Toby* slips into gear and points east to the Atlantic.

Brian remembers reading somewhere in the knotty science fiction of Philip K. Dick a passage to the effect that paranoia is a fine thing, actually, a disorder that earns its keep as a sort of preemptive survival mechanism. Just being paranoid, this thinking goes, doesn't necessarily mean that everybody's not out to get you.

But Brian doesn't like paranoia. He described it to me once as "the most self-serving of mental disorders, and beyond a certain point its value in survival is lost as it starts to erode the mind and health of its host." He also warned me that I would see a lot of beyond-a-certain-point paranoia in the New England fishing industry, that it didn't necessarily take hard times for its discordant song to be heard, that its thematic variations are as subtle and far-ranging as any Coltrane riff. "It runs the gamut from the madman who accuses everybody on the ocean of hauling his traps to the quiet fellow who always knows damned well that if you said *this,* then you must be thinking *that.*"

These thoughts were part of a letter Brian wrote me in which he sought to explain how a man born to a pair of journalists in Delaware in 1950 came to be chasing lobsters, owning a boat and a house, raising a family, and taking the measure of his occupation's psychological rip tides in Orleans, Massachusetts, in 1995. To all this Brian attaches a certain onus of fate. "Shortly after Hagen tricked Kriemhilde into relinquishing Siegfried's fabulous horde of gold," he began, "my great-great-great-great . . . Perhaps I shouldn't begin so far back."

He didn't. He began with his grandfather, Dr. Ralph Wiggin, who bought a small camp in Orleans. "This camp eventually served as a retreat and center for low-cost existence for his children (and grandchildren). As an example, while he was stationed in Fort Dix during World War I and then a field hospital in France following the Armistice, my grandmother closed their

Cambridge home and set up house with their two young children at the Orleans camp. A kerosene heater, a hand pump, and an outhouse were the amenities provided at the two-room camp. We still used the outhouse until I was four or five."

Brian's grandmother died when his mother, Marian, was approaching adolescence, and his grandfather remarried two or three years later. "Within a few years his health began to fail to arteriosclerosis, and his fortunes declined from sundry reasons spawned by the Depression and poor health. Hence, for my mom, the Cape became a place of physical retreat and also spiritual serenity."

Marian graduated from Colby Junior College in New Hampshire and returned to Orleans, getting by, or nearly so, on the three or four dollars a week she earned writing book reviews for the *Boston Evening Transcript.* Eventually she moved to Boston to work full-time at the newspaper. There she met young John Gibbons, who worked at the paper with his father and commuted with him to Boston from their native New Jersey every week. "They met; they courted; their first date was the night of the 1938 hurricane; the rest is history. (Historians have overlooked the fact that my parents were climbing Mount Monadnock the day Hitler invaded Poland.)"

John Gibbons never went to war, and Brian isn't sure whether it was his age, his occupation, or his children that kept him ahead of the draft. During the war he left the *Transcript* and landed at a small paper in Wilmington, Delaware, where Brian and his two brothers and sister were raised. Marian Gibbons, however, still relished the solace of the Orleans camp, and she adjourned there each summer with the children.

It was a life of many pleasures and few embellishments. "Believe me, it was the budget plan all the way. The summers on the Cape were a nearly cashless enterprise, the main activities being baseball, swimming, picking berries, endless combat in the beach plum entanglements while garbed in German World War I gear gathered in France by my grandfather, and a vigilant monitoring of the comings and goings of the boats in nearby Rock Harbor. Thus, the brood became Cape Codders of a sort."

Dr. Wiggin, the first Cape Codder of a sort, lived in an age when medicine was not so remunerative as it is now. Neither did John Gibbons ever make a lot of money in journalism. In the early 1960s he lost his job at that Wilmington newspaper, but he was able to parlay his knowledge of the ins and outs of the Delaware statehouse into some catch-as-catch-can political public relations work. Brian described his grandparents' hold on middle-class status as "tenuous." He concedes that the slipperiness of that hold has shadowed his family for as long as he has lived.

For the moment the *Cap'n Toby* lies at rest outside the harbor, and the twelve-inch mackerels that Brian and I are cutting up for lobster bait are ripe, their bellies gravid with either blood-red roe or milt the color of sailors' bones. Brian slices them behind the gills; I put the heads and tails into nylon bait bags the size of a small purse. The bags' quarter-inch mesh openings are convenient for a lobster's claws, but not for most larger scavengers. The mackerels' flesh is dark and firm and scentless, their skin a steel blue along the spine and cut with jagged zebra stripes of indigo. These in turn are limned in iridescent tones of silver and brass and copper.

The squid in the tote next to the mackerels have neither color nor form. They lie like gobs of phlegm in the May light, their collapsed tentacles defined only by the purple pinpricks of their suckers, their glabrous heads by the wet inkdrops of their eyes. Usually Brian sticks to mackerel and herring for bait at this time of year; these squid are here on a trial basis. Brian told me yesterday that sometimes "oddball bait" works well with lobsters, something different from their customary fare, a little more piquant, perhaps. He recalled a brief infatuation among Orleans lobstermen with squirrel and rabbit after an immigrant from Provincetown said that road kill worked pretty well up there. But when a rumor circulated that some Provincetown lobstermen were going so far as to kill dogs for bait, this practice was abandoned. Brian isn't sure about these squid, but it's been, after all, the sort of spring to drive a man to opossum.

His main VHF hangs from a mount in the pilothouse above his Si-Tex electronic fish-finder. The radio crackles with the chatter of long-liners — fishermen pursuing cod and haddock with lines of baited hooks anchored to the bottom of the sea — who complain that as many as three fourths of their hooks today are being taken by dogs, that is, spiny dogfish. The dogfish are small but voracious sharks, two to three feet long, whose numbers have ballooned on Georges Bank and the waters immediately off the Cape as populations of cod and other groundfish, or bottom-feeders, have precipitously declined. "I love hearing that," Brian says, noting this first untoward report in his truce with the elements. "Dogs just love mackerel. They'll bite holes in these traps to get at them, and then the traps won't hold lobsters."

The traps that Brian means to haul from the bottom today were baited with mackerel and herring three days ago. The two hundred or so traps that he has gotten into the water so far this spring are arranged in north–south lines of ten to twelve, called strings, at points two to three miles off the beach. The precise locations of the first and last trap in each string are entered into a logbook kept next to the wheel, and the direction in which Brian hauls — whether he works from north to south or vice versa — depends on the direction of the tide. Today he'll start at the south end of each string, steaming against the tide to create slack in the ropes as he picks up his buoys and runs the lines through his hauling equipment. We head up the beach toward Wellfleet, cruising at ten knots and navigating between corridors marked by the flags of other lobstermen's buoys. "By June it's going to look like a goddamned miniature golf course out here," Brian says. "From Truro to Chatham, you'll be able to walk on the buoys."

The narrow northern finger of Nauset Beach, alternately split open and stitched together again by nor'easters, yields within two miles to Eastham and the neat red-and-white buildings of the Nauset Coast Guard station and lighthouse. Above the lighthouse a line of gritty cliffs rears up behind the beach, running north to Truro and climbing as it goes. Within five miles the beach shrinks to a thin blank strip footing the cliffs. The sand becomes a specimen of moat, the cliffs military ramparts raised against a besieging sea that, at its current rate of increase, will entirely engulf Cape Cod within six thousand years, drowning municipal miniature golf courses and lending an element of prophecy to the Puritan cleric Cotton Mather's observation that this land would be known as Cape Cod until "shoales of coddefishe bee seene swimming on its highest hills."

Opposite the cliffs of Wellfleet, with the tide running south down the beach, the *Cap'n Toby* slows and tiptoes up to the southern end of Brian's first string. This stretch of bottom is known to lobstermen as the Can, because of a large buoy that once floated here. Brian throws the motor out of gear, leans over the starboard gunwale, nabs the bobbing lobster buoy with a boathook, and then runs its rope, its warp, through a block suspended from a four-by-four oak beam bolted across the wheelhouse roof. He drops the buoy to the cabin floor and turns a handle that activates a hydraulic winch, the trap hauler, which is bolted into the cabin rail at his feet. Finally he lassos the warp around the turning wheel of the Hydro-Slave hauler. This takes but a few seconds and is accomplished with an athlete's economy of motion. The quarter-inch warp shivers and throws off water as it runs into the sheave of the hauler, then frees itself into black loops on the cabin floor. The hauler sings with a frog's mad chortle.

The first trap is coming 120 feet up from level, sandy bottom. I peer over the gunwale into green water of deceptive clarity, which seems wholly without secrets but which frays and dissolves the straining warp into milky nothingness within a fathom of the surface. The trap is just a mote in an emerald void, then a gradually spreading cloud. In an eyeblink the cloud crimps and hardens into a geometry of wire and twine barreling toward the surface like an oncoming truck. The trap foams from the water and for an instant swings lengthwise, green and dripping beneath the block, its cargo scuttling within, the whole apparatus like a core sample torn from the bowels of a wreck, now glinting like a trophy in the sun.

Brian throws the hauler into neutral with one hand and with the other pulls the trap over the gunwale, where a touch of the Hydro-Slave's reverse gear allows him to lay it gently upright with its gate on top. Brian makes most of his traps himself out of oak lathing, but the ones in this string are factory-made, welded together out of a square-mesh vinyl-coated wire that lasts longer under water than the oak. There are three lobsters in this trap, their claws jabbed forward in rage, their tails snapping backward in panic. Brian keeps clear of the claws as he works them out of the trap.

One lobster is plainly a short — undersized — and Brian sends it pinwheeling back into the water. The others he puts on the pegging board, a chessboard-sized piece of plywood laid on top of the engine cover and partitioned into compartments for such items as a handful of thumbnail-length wooden pegs, dozens of yellow rubber bands as small as wedding rings but as thick as bracelets, a scissorslike banding tool for stretching the heavy bands over the lobsters' claws, a gauge for measuring the length of a lobster's carapace from the thorax to the eye socket to determine the legal minimum of three and a quarter inches, and corral space for one or two free-ranging lobsters. The pegging board's name comes from lobstermen's former practice of pushing the small wooden pegs into the lower hinges of a lobster's claws to keep them from opening. Nowadays this is more easily done with rubber bands, but Brian always keeps some pegs on hand for that rare lobster whose claws are too big for his bands.

These two lobsters, both in the neighborhood of minimum size, are something different from the mackerels and squid, another sort of invention entirely. They claim in their coloring, perhaps, some degree of the mackerels' designer beauty. Their shells are green and black and olive, prettily mottled in aquamarine and dusky orange, spiked and tuberculated in red. Otherwise they occupy a point not even on the scale between the mackerels' wind-tunnel symmetry and the squid's broken-egg shapelessness. Our familiarity with the lobsters on our dinner plates, motionless and with all their pigments boiled away except that well-known mineral red, robs them of their real strangeness. To observe their chitinous, appendage-laden skittering on Brian's pegging board is to go far toward restoring it.

They look like nothing so much as Swiss Army knives brought to life, given limbs and difficult personalities on the day Hieronymus Bosch was hired as a Disney animator. The eight broomstraw legs, tipped with pincers, are all out of proportion to the armored plugs that are their bodies. So too, in an opposite sense, are the claws, which even on these small specimens, one or two pounds, look so great and weighty as to be pushed like millstones ahead of them. But somehow the lobsters dance about the pegging board with disturbing agility. Somehow they hold their big claws aloft and wide apart. The claws' snaggled forceps gape. Deftly and with a sneaky quickness, the lobsters parry the threatening movements of my hands. They rotate like monstrous mantises, their stalked eyes and pronged snouts kept square to me, their claws hair-triggered like leg-hold traps. These are good special effects, I think to myself.

Eventually I find ways for my hands to move over and behind the lobsters' air defenses. Their eyes drop like periscopes into their sockets at the approach of the carapace gauge. One of the lobsters is just barely a short, and I pitch it back in over the port rail. The other is a keeper, a chick, and with the banding tool I slip a rubber band over the smaller cutting claw, the quicker of the two, and then another over the more powerful crusher claw. Meanwhile Brian removes from the trap the old bait bag, containing now only an assortment of clean bones, and replaces it with a fresh bag of mackerel. He drops the old bag onto the pegging board for me to clean after I place the banded lobster into an empty tote at my feet.

Brian shuts the gate on the trap. He works the last of the warp free of the hauler's sheave, puts the *Cap'n Toby* in gear, and lets the trap slip clear of the rail as the boat starts to move again. The trap settles comfortably into the water, like a cat into a pillow, and slips wholly under the surface as Brian looks to be sure that its warp pays out tangle-free over the boat's transom. At last he throws the buoy back overboard and throttles north to the second trap in the string.

Brian has begun at the northernmost of the strings he is hauling today, and slowly works toward home as the day wears on. There is talk. He apologizes for the occasional saltiness of his language, and the saltiness of fishermen's talk in general, with a story of that talk's sense of borders: "I was fishing on the *Bell*, that big scalloper that Olav Tveit ran, and Olav was bringing his sixteen-year-old son out with him for the first time. It was the old cook — cooks frequently being the mentoring types — who took the kid aside and told him that he was going to hear a lot of hard language out there, language that he wouldn't necessarily use in front of his mother, and that he should be damned well sure that he didn't. I thought that was nice. It was something that Olav couldn't say to him. Olav was a Lutheran, but he could turn the air blue with the best of them. He claimed *fuck* and *shit pile* had no equivalent in Norwegian or any other language."

Concern for his own children was one of the factors that compelled Brian finally to abandon the *Bell* and other big New Bedford boats, with their ten-day voyages out to the rim of the continental shelf, out to Georges Bank and the Cultivator Shoals, in favor of lobstering. The wisdom of hindsight might call that an economically savvy move as well, now that New England groundfish stocks on Georges Bank have collapsed, now that the scallop harvest is in a tailspin, now that the docks of New Bedford and Gloucester have gone idle and their slips are full of ships for sale or in receivership. New England lobster harvests, meanwhile, have climbed steadily throughout the 1990s, so much so that this little spot off Wellfleet — which once very few people fished, coming here under cover of fog lest other lobstermen start working this rich bottom — now has that proto miniature-golf-course look.

Brian himself isn't so sure. Again, he likes to take the long view, suggesting that the fate that made him a lobsterman has more to do with history and circumstance than market forecasts or biomass analysis, albeit he — like many others, both fishermen and scientists — saw the industry's current crisis coming a long way off. But not so far back as the early 1960s, when Brian was still summering on the Cape with his mother and siblings and when the vigilantly monitored boats sailing out of Orleans's Rock Harbor, a nick on the western or bay side of the Cape, were transforming themselves from a fleet of small commercial quahog draggers to charter boats catering to tourists and sport fishermen out for striped bass. Meanwhile, out on Georges Bank, 150 miles east of Cape Cod, the first great ships of the international distant-water fleet were making their appearance.

"In my early teen years, the occasional odd job of helping to scrape and paint the bottom of one of the old draggers was sometimes available to otherwise listless wharf rats," Brian wrote. "Sometimes the wharf rat might even be pressed into service out on Cape Cod Bay, picking piles and bagging up quahogs." During his later teenage years, he worked — "slaved," Brian corrected me — on one of the Rock Harbor charter boats, the *Empress,* under a skipper known as one of the harshest taskmasters in the fleet. The not-so-patient Stu Finlay, Brian told me, "would be driven into apoplexy by my wooden-headed adolescent stupidity."

In 1969, however, Brian was drafted into the army. He hadn't gone to college after graduating from high school in Delaware, though at that time college would have provided him with a deferment. He says he was only an indifferent student, and he was put off by the campus upheavals of the late 1960s. "I was, shall we say, ambivalent about the politics of the day," he says, "and it seemed to me that you could party and carouse just as well without the formality of attending a university." After receiving his greetings from Uncle Sam, he bought a ticket for a 6:30 A.M. bus from Hyannis to Boston and prepared for his induction the next day with what he describes as "classic debauchery." But he overslept that morning, missed the bus, and so made a desperate call to a friend who had just gotten back from Vietnam and invested his discharge money in an MG. Eddie took him racing up Route 6 and the Southeast Expressway, averaging eighty to ninety miles per hour and weaving through weekday morning traffic.

Near Braintree, around a curve and on the back side of a rise, lay a steel I-beam that had just fallen off a truck and come to rest across the middle lane. Eddie swerved and sent the car skidding into the guardrail, where the impact shot both passengers through the MG's convertible canvas roof. Eddie hit the grassy slope on the other side of the rail and came away unhurt. Brian hit the highway and skidded on his shoulder into one of the driving lanes. There another careening car ran over his left leg.

The accident left Brian with a steel plate in his leg and a permanent limp; also with a 4F deferment from the draft, the impossibility of GI funding for college, and $5000 in medical bills. In a roundabout way, the accident also brought him a wife and child. He became friendly with a nurse at Cape Cod Hospital during his six-week stay there. Later that nurse's sister, Suzanne St. Amand, visited from San Francisco with her infant daughter. Brian and Suzanne met; they courted; they lived together off and on for several years. Six months after Brian and Suzanne were married, in 1973, adoption proceedings made Brian the child's father.

"By 1971 I was getting pretty serious about making money," he wrote. "I worked forty hours a week as a carpenter, two to three nights a week as a bartender, and tried to do odd jobs (painting, firewood, shingling) on weekends. This was, in part, prompted by what I thought to be enormous hospital bills, which I paid off with a few dollars here and a few dollars there every week. But the slow increments of financial gain I was realizing as a carpenter paled by comparison to the money fishermen could make shellfishing in the local estuaries.

"Though I had gotten somewhat burnt out by my summer indentures in the charter fleet, I always pursued the bass, the flounder, and the steamer clam whenever the time and opportunity allowed. In 1972 I quit my standard construction job and went to work building a charter boat with a Rock Harbor skipper. You can imagine that through all these years, and especially when boat building, every coffee break, every lunch, every beer, was steeped in conversations about fish, technique, and money. At that time I was strong, quick on the uptake, could drink prodigious quantities of beer or liquor without staggering, and had been brought up — while friends were waxing surfboards — to work twelve- to twenty-hour shifts, often under adverse conditions, frequently with a maniac screaming at me, and with rarely a day off. I felt that I could do anything."

Brian meant to go back to his carpentry work when the boat was launched in June that year. Instead he bought a small skiff and an outboard and went bullraking for littleneck clams, a.k.a. quahogs, in Pleasant Bay. He caught the three-bushel limit on his first day, made $40 or $50, and decided right there that he'd just made a career change — a fateful change, as he looks back at it now: "I've often thought about this early success from the perspective of a life spent luring a fish to a hook or a lobster into a trap."

Later that summer, while Brian was in the office of the Saquatucket harbormaster, who told him that good money was being made by the offshore lobster boats moored there, Harry Hunt came in. The old lobsterman groused about an engine problem, swore at the Russian factory trawlers circling the deep-water canyons he liked to fish off Georges Bank, and lamented his lack of a third hand to go out with him that afternoon. "Four hours later I was on the *Gertrude H* with Harry, Harry Jr. (still a high school lad), and an old drunk who had last shipped with Harry when Harry tried to bring a small quahog dragger around from Nauset to Cape Cod Bay by way of Race Point during the 1938 hurricane. For my services as an inexperienced lobster 'bull,' Harry promised me $200 per two-and-a-half-day trip. I was hooked," Brian explained.

I never met Harry Hunt, who died a couple of years ago, but I see him in Brian's talk, the stories of other Orleans fishermen, and their gleeful imitations of his manner and speech. Brian is of medium height and build, though if you look at his forearms, at their breadth and ropy sinew, you see the same sort of musculature caricatured in Popeye. Harry Hunt was no taller than Brian but thirty or forty pounds heavier, with much of that extra beef slabbed into his shoulders, the rest into his enormous hands, which he carried in front of him like the claws of the eighteen-pounders he hauled from the deep-water canyons a hundred miles offshore. Physically he suggested a troll out of the Brothers Grimm, Brian says, but his features were American Hero out of the Hollywood mold, so much so that his wife, Gertie, called him Duke: his jaw square and rocky, his eyes narrow and hooded and penetrating.

Hunt could be tough on his crew and tough on his family. Being a member of both qualified as double jeopardy. Carl Johnston, a friend and neighbor of Brian's, works on a dragger harvesting groundfish out of Chatham. Carl remembers once seeing Harry Jr., a grown man then, come sprinting up to his pickup truck in the parking lot at Wychmere Harbor, dive to the pavement, and wiggle underneath it like a spooked dog. Harry Jr. whispered to the amazed Carl that his old man was after him and please not to say anything about where he was. Down in the harbor, meanwhile, the elder Hunt raved from the foredeck of the *Gertrude H*.

As long as Brian knew father and son, the only words of affection he heard pass between them began with Harry's promise that if his son were ever in the hospital on a life-support system, he'd pull the plug. No doubt misty-eyed, Harry Jr. would vow that he'd do as much for his old man. But Harry suffered a stroke and passed his last years in a nursing home, living in a slack tide between life and life support, his famous misanthropy flat-lined into a blurred, drug-hazed cussedness.

Harry Hunt taught Brian how to make money at catching lobsters. As tough a master as Stu Finlay of the *Empress,* Hunt worked Brian to the last ounce of his wooden-headed zeal. When Brian finally quit, however, it wasn't the work that had gotten to him but the misanthropy. Yet he didn't blame Hunt entirely for the anger that always seemed to be going at a rolling boil inside him. "Hunter-gathering depends on continuous conceptualization, thousands of microanalyses going on in your head all day every day," Brian told me. "Fatigue, burnout, and the ever-present possibility that your analysis of something you can't see — lobsters on the bottom — can be thrown off by someone hauling your pot is the combination with which paranoia opens your door. When Harry fished inshore, where we fish, he would many summers haul pots a hundred days in a row without a break, without more than four to six hours of sleep per night. Getting bait, repairing engines, fixing stuff, were all done before or after the day's haul. When I first started fishing with Harry, he was sixty-two years old. At an age when many men are thinking of retirement, he was starting to run boatloads of pots a hundred miles offshore in a forty-three-foot boat. There are reasons why he became pretty crazed."

Brian himself, however, remains spooked by that craziness, and still feels pain over Hunt's turning on him when he finally quit. "By the time I came to know him, the great Harry Hunt had carefully ordered his universe into two vast realms: one, Harry Hunt, and two, everybody who was trying to fuck over Harry Hunt." Brian didn't quite know what to make of a man who was convinced that one empty pot in the middle of a fifteen-pot trawl being hauled from 120 fathoms had been pilfered by some cock-eatin' Portagee sons of bitches of bastards. "At first, being new to lobstering, and one hundred miles from land, I didn't care. I felt like I was muscle and barbed wire, I was making money, and I was learning new skills. It's tough to be near such a black hole. After a couple years of Harry's megalomaniacal, solipsistic, paranoid bullshit, I moved on, joining the legions of bastards who were and had always been marching against Harry. As he would say with lips drawn tightly against clenched teeth, 'The wind, the tide, the weather, and *every man* is against ya.'"

Part of the interest in hauling lobster traps lies in what else you might find in them. A common skate comes twisting out of a trap on Brian's second string. Known otherwise and variously as a little skate, bonnet skate, summer skate, hedgehog skate, old maid, and tobacco box, the raylike fish looks like home plate at Fenway but gritted over, grown eyes and a tail, possessed by a devil. Skates are on the move now, like the lobsters, swimming into shoal waters for spawning and summer feeding before retreating to deeper waters, thirty to fifty fathoms, in December.

This skate came for Brian's herring, part of a varied diet that includes crab, shrimp, worms, amphipods, mollusks, squid, and other small fish. Brian drops it on the pegging board while he duels with a lobster that has grabbed hold of one of the trap's parlor heads, the twine funnel inside the trap that keeps the lobster from getting out. The skate's left pectoral fin — more properly a wing — catches on a partition and the fish writhes onto its back. Its underside is albino white; its sole features, a grinning mouth and two eyelike ears, are impish and eerily human. Skates are predators of juvenile cod and competitors with mature cod. Like dogfish, the numbers of skates off the Cape have vastly increased this decade as the cod have declined. The fish have some slight commercial value, around sixty cents per pound right now (versus the five dollars that Brian might get per pound for a select lobster), and Brian sometimes uses skates for bait. But he doesn't want any now; I catch the fish at the joint of its fleshy tail and pitch it back into the water.

During that same string Brian pauses to admire a captive he pulls out of another trap. "That's a beautiful fish, isn't it?"

The black sea bass is the transvestite fan dancer of the bass family. Its kitelike dorsal fin runs the whole length of its spine, and its pectoral fins, broad and softly rounded, sweep all the way back to its anal fin. Its scales are limned in inky blue-black, the interior of each much lighter, closer to a milky gray. These line up like strings of dusky pearls stretched along the flanks of the two-pound fish. Hermaphroditic, the fish usually produces eggs at sexual maturity, but later its ovaries dry up and its testes begin to produce sperm.

The black sea bass may be found as far south as Florida and is one of many fish living at the northern limit of their range here near Cape Cod. This is the result of the Cape's position at the clashing juncture of those two flywheels of the west Atlantic, the Gulf Stream and the Labrador Current. Forty thousand years ago, when the Laurentian glacier was a mile thick and covered most of New England, this was precisely where the warm waters of the Gulf Stream halted the glacier's advance and eventually beat it back. One lobe of the retreating ice sheet left behind the morainal till — the boulders, rocks, and gravel — that became Cape Cod once the ocean level had risen again. But the till trails like a heap of tailings into the misaligned teeth of the Gulf and Labrador currents, which move in opposite directions, the Gulf Stream flowing clockwise up the East Coast and then out into the central Atlantic, the Labrador Current pouring counterclockwise down from the Canadian Maritimes. The meteorological sparks thrown off by the colliding currents confer upon Cape waters the fogs and gales that doomed the *Sparrowhawk* and too often thwarted Harry Hunt. They also make these waters unusually cosmopolitan, a place where the prettiness and the extravagance of the tropics swim side by side with the puritanism of the North Atlantic.

A trap on Brian's fourth string yields a specimen of the northern fish that most nearly belies that puritanism: the sacred and eponymous cod. "The 15th day we had again sight of the land," wrote an officer with the English explorer Bartholomew Gosnold's 1602 voyage to Virginia by way of Massachusetts, "which made ahead, being as we thought an island, by reason of a large sound that appeared westward between it and the main, for coming to the west end thereof, we called it Shoal Hope. Near this cape we came to anchor in fifteen fathoms, where we took great store of cod-fish, for which we altered the name and called it Cape Cod." Another officer reported, "We had pestered our ship so with codfish, that we threw numbers of them overboard again."

Throughout the Gulf of Maine and off Cape Cod, no fish was in those days so delicate and sweet, so congenial to salting and drying, no large fish so plentiful, nor any fish so reliably hungry and readily caught as the cod. And very few other creatures were so astonishingly fecund. Samuel Johnson's 1775 dictionary defines *cod* as "any case or husk in which seeds are lodged," and it was Henry David Thoreau's supposition that the word became attached to the fish because the females laid so many eggs. This eight-pounder from Brian's trap, if female, probably laid nearly a million eggs last winter. The fifty-pounders that fishermen of Harry Hunt's generation remember capturing could produce up to nine million eggs. Of those, of course, perhaps fewer than a hundred fry would reach their first year, but three hundred years ago this provided enough "great store of cod-fish" to lay the foundations of a nation. "Puritan Massachusetts derived her ideals from a sacred book," observes the historian Samuel Eliot Morison, "her wealth and power from the sacred cod."

No one now expresses any surprise concerning the recent disappearance of the cod from New England waters, given the ingenuity and work ethic of its fishermen, the remarkable improvements in their methods, and the snarled politics of natural resource management here. But a species of surprise still lingers like a long, drizzling rain in Orleans, in Provincetown, in Chatham, in Woods Hole, and off the Cape as well, in Gloucester, New Bedford, Cape Elizabeth, Point Judith, and elsewhere — surprise that after all these years, there is indeed a limit to the cod's wonderful, pestering resilience, as though God's mercy itself were found to be quantifiable and running low.

This particular cod is legal bycatch (lobstermen are allowed, at least as of this date, to keep and sell up to three hundred pounds per day of the groundfish that swim into their traps) and will be taken home for a friend's supper tonight. It lies stunned on the cabin floor against the transom, a fish nearly as lovely as the sea bass, though the cod's beauty is of a more idiomatic sort. Its color is somewhere between sepia and gold, mottled with dark spots and dashed with an undertint of olive. Its flank is bisected by a pale lateral stripe, like that left by a fingernail run through crayon, starting above the gill cover and curving down to the fleshy end of the tail. Its fins are not as extravagant as the bass's but are striking in their number: three on the spine, two on the belly, all set end to end like the sails on a schooner. The cod's scales are so fine that from a distance the fish seems scaleless, like an eel. A barbel dangles like a goatee from its thick lower lip, and the eyes of its flattened head have a goggled aspect. The fish looks something like Captain Nemo's submarine: sleek, functional, comfortable in black North Atlantic depths — down to 750 fathoms for the cod; I don't know about the *Nautilus* — but finished all the same with a Victorian flourish of pride, a rich and sensuous pleasure in fine ornament.

The same string that produces the cod also yields the first egger of the day: a big six-pound female lobster carrying a load of eggs stuck like spoonfuls of jam to the pleopods, or swimming limbs, under her tail. The lobster falls far short of the cod in the quantity of its egg production, with clusters ranging in number between three thousand and one hundred thousand, but the female tends them more carefully than any other inshore animal, carrying them about for nearly a year after mating, curling her tail protectively over them, fanning them with oxygen through the steady motion of her pleopods. The philosophy of the lobster industry is no less maternal, and the taking or sale of "berried" lobsters at any time has been forbidden in Massachusetts since 1880. Brian turns the lobster over and shows me her eggs, her berries, which look individually like pinpricks of purple ink encased in droplets of rainwater. He says these eggs are just about ready to hatch. Then he tosses the big lobster back into the water.

Other creatures come up from the bottom in Brian's traps: green, bristling sea urchins and beige-striped moon snails as big as my fist, then a primitive-looking cusk with its tail and fins run together, which gives the fish an unfinished, born-too-soon look, and its stomach blown like an inflated condom through its mouth. "This fish won't live," Brian says, and he lays it against the transom with the cod for supper. Hake also often fall victim to the change in pressure as a trap is winched to the surface, but on his fifth string Brian pulls out a gray, sleek-looking hake that looks to be healthy, and he throws it back in, saying, "I love playing God."

I tell Brian that mercy is divine, that the fish will be his friend for life, but Brian shakes his head. "He'll nibble my toes when I drown."

Aside from these hitchhikers, however, and a fair number of undersized lobsters, there is little in Brian's traps today. Many come up completely empty, though with their bait stripped.

He swings another empty trap over to the rail and says, "This is crazy." He takes off his gloves a moment to wipe his forehead, adjust his hat, and stare, mystified, over the stern along the line of the traps he's already hauled and, a mile back, those that he's just set in the water. "I've never seen it so bad."

After Brian quit on Harry Hunt, he went back to the shellfishing that had tempted him into the industry in the first place. But soon he was working offshore again, this time on the *Madonna,* a tub-trawler (long-liner) out of Chatham. "Tub-trawling is the business of setting hooked and baited long-lines on the bottom for cod, halibut, haddock, à la *Captains Courageous,*" he wrote. "This, too, was hard work for good money."

Brian found the captain of this tub-trawler to be better company than Harry Hunt. But he also found around this time that his incessant thirst, copious urination, and bouts of blurry vision signified the onset of diabetes. Nor did he find the separations from his family any easier to handle. Brian still grieves to this day that he was eighty miles east of Chatham, fishing with the *Madonna* on the Cultivator Shoals, when his daughter Bethany was born in April 1975.

With that, Brian returned to lobstering, working only inshore, in that three miles of water under state (as opposed to federal) jurisdiction, running a hundred pots off Nauset Beach and shellfishing on the days he didn't need to haul his traps. Eventually he caught on with a "highliner [that is, very skilled] inshore lobsterman who was (is) a recovering alcoholic, and as I was (am) a diabetic who doesn't imbibe too much anymore, this site worked out pretty well for two years. Unfortunately, the money on this site was insufficient for my family's growing needs, so I went back on my own."

The inshore fisherman can rarely afford the specialization of the blue-water or offshore fisherman. Not only is he more subject to the seasonal movement and availability of shoal-water species, but he is more vulnerable to the vagaries of state politics and shifting community demographics. "I would run two or three hundred lobster pots in the summer, dig clams or bullrake littlenecks on days between hauls, bay scallop or drag for flounders in the fall and winter," Brian wrote. "Also in the winter or early spring I would day-trip tub-trawl out of Chatham for cod and halibut. For a few years my most lucrative pursuit was dragging flounders. Pressure from the many retirees who built their dream homes close to the estuarine waters and didn't want to see commercial fishermen in them, and from those who proliferated in little tin boats with drop lines, made it politically incorrect for this fishery to continue. The state shut it off completely a few years ago. They never noticed the burgeoning populations of seals and cormorants that were targeting the flounder harder than the fishermen were. So it goes."

The loss of the winter flounder fishery was a bitter blow, robbing him of a reliable slack-season income of $100 per day, with an occasional $800 or $900 bonanza. In 1979 Brian had a poor lobster season made memorable by capsizing his skiff one day in the swells that often come in like boxcars over the sandbar outside Nauset Inlet. That same year, only a day or two after his son, Mike, was born, a site that Brian had lined up on a tub-trawler fell through. With that, he started looking offshore again, driving down to New Bedford in search of a big dragger or scalloper, finally signing on with a pre–World War II eastern-rigged scalloper named the *Cape Star.* When that boat subsequently went into the shipyard for major repairs, the captain helped Brian secure another site, this time on Olav Tveit's *Bell.*

"This was my introduction to the Squarehead (Norwegian) fleet, a society that once seemed to be as clandestine and exclusionary as the Masons or the Hanseatic League. I think the doors began to open for non-Squareheads when the children of the fifties immigrant Norwegians grew up to be fully assimilated in the American golden-arches, white-lines, and car-payment culture. But I believe the Squareheads I fished with came up through merchant seaman training in the old country, and they were seamen nonpareil. In heavy weather, nothing was left to chance. Hatches were lashed, companionways dogged, fish pens braced, loose gear secured. That sort of leadership resonates through a crew, and a respectful, cautious confidence develops (at least it did in me)."

Brian liked to joke that Cape Codders were known in New Bedford as at least the next best thing to Squareheads, and this Cape Codder, a child of the car-payment culture, did well among the Norsemen. When the engineer on the *Bell* had to be medevaced off the boat after taking a fish pick in the eye, Brian took over running the engine room, and Olav kept him on as engineer. Later, when Olav or his partner and nephew Arstein went on vacation, Brian would move up to the wheelhouse as mate. Eventually he worked it so that he could go inshore lobstering through the summer and fall, taking day trips out of Nauset Harbor, and then return to a site in New Bedford, on the *Bell* or the *Cape Star* or another big dragger, for the winter and spring.

Meanwhile, Harry Hunt wasn't the only one cursing the Russian factory trawlers in New England's offshore waters, particularly on Georges Bank, that strange drowned land at the edge of the continental shelf. Named after Sir Ferdinando Gorges, to whom the English Crown originally gave both the territory of Maine and its fishing rights, the great shoal has been as much a graveyard over the centuries as the Cape's Outer Beach. Henry David Thoreau made three walking tours of Cape Cod, in the course of which he heard many stories about Georges:

> Every Cape man has a theory about Georges Bank having been an island once, and in their accounts they gradually reduce the shallowness from six, five, four, two fathoms, to somebody's confident assertion that he has seen a mackerel-gull sitting on a piece of dry land there. It reminded me, when I thought of the shipwrecks which had taken place there, of the Isle of Demons, laid down off this coast in old charts of the New World. There must be something monstrous, methinks, in a vision of the sea bottom from over some bank a thousand miles from shore, more awful than its imagined bottomlessness; a drowned continent, all livid and frothing at the nostrils, like the body of a drowned man, which is better sunk deep than near the surface.

Georges Bank in fact was an island once, formed, like the Cape, as a terminal glacial moraine and then submerged, just barely, by rising sea levels. Massachusetts fishermen who attempted to fish on the bank found themselves swept off it by such powerful rip tides that they believed it was suicidal to anchor there — that a ship would be pulled under water like one of Brian's lobster buoys in a tideway. Nor were they reassured by the bank's tendency to froth at the nostrils and pile up in breakers in any small easterly blow, and finally they regarded the grounds with a superstitious terror.

So Massachusetts fishermen traditionally went north for their sacred cod, into the Gulf of Maine and Canadian waters. But in the years following the American Revolution, English warships blocked access to Canadian waters for Yankee fishermen, and an 1818 treaty formally shut the door to Labrador, the Bay de Chaleur, and adjacent waters. Harassed by the royal fish patrol, which occasionally confiscated their vessels and catch, Massachusetts men began to look more seriously at Georges Bank, only a day's sail from home. In 1821 Captain Samuel Wonson of Gloucester decided to take a chance. "Boldly he sailed to the Georges; boldly he anchored; safely he sailed home," intones the Cape historian Henry C. Kittredge. "A new era in New England fishing had begun; Georges Bank, where the churlish veto of British legislation could not be heard, was open for business."

What a business it was. For New England fishermen, the opening of Georges Bank was like the discovery of Siegfried's fabulous horde of gold in their own back yards. Cod are to be found wherever there is hard bottom, but the Georges' gravel pavement is laved by crosscurrents of oxygenated, nutrient-rich water, and its location on dividing lines not only between North and South Atlantic biota but also between shoal-water and deepwater forms makes the bank particularly delightful to a fish as catholic in its tastes as the cod (adults will eat almost anything that will fit into their mouths). Georges is also a wonderful spawning ground for cod, particulary the boulder-strewn sea floor of its northeastern section, where juveniles can find plenty to eat in relative safety.

Herring favor that northeastern section as well. Their eggs are not free-floating like the cod's but instead require firm substrate on which to anchor and well-oxygenated water in which to incubate. Sea scallops, which are plankton-feeders, also do well in the bank's brothy waters, and in the nineteenth century, enormous Atlantic halibut, which grew to nine feet and seven hundred pounds, did too. Present as well in boiling, swarming, pestering numbers were other sweet-tasting groundfish: haddock, halibut, hake, and yellowtail flounder.

In the winters throughout the 1980s it was to Georges Bank that Brian repaired on the boats out of New Bedford. "There were some mighty adventures on the *Cape Star* and *Bell* in those days — deckloads of scallops, massive sets of fish in winter gales, huge 'rimracks' of the fishing gear," he explained. "We repaired the gear, set it back into the heaving seas, basking in a momentary sense of pride, even heroism. And there was financial stability which the inshore man rarely sees."

By then Brian and other New England fishermen were reclaiming Georges as an ancestral ground after almost twenty years of sharing it with the industrial factory trawlers of not only the Soviet Union but also Poland, Spain, Iceland, Japan, Cuba, East and West Germany, and other countries. These great ships, which began appearing on Georges Bank when Brian was apprenticing at Rock Harbor, were indeed self-contained factories, built not only to harvest fish but also to process and package them. Big American boats such as the *Bell* bobbed like dinghies alongside distant-water trawlers that might be five hundred feet long and carry a crew of more than a hundred. They could deploy gear of awesome proportions: trawl nets wide enough to swallow the Statue of Liberty and capacious enough to hold a dozen jumbo jetliners. They worked twenty-four hours a day and sometimes lined up side by side, or even three abreast, trailing a two-mile plume of churned-up sea bottom behind them. In a single day a Soviet factory ship on Georges could produce 50 tons of fish fillets, 150 tons of fish meal, and 5 tons of fish oil.

The nineteenth-century naturalist Jean-Baptiste de Lamarck wrote, "Animals living in . . . the sea waters . . . are protected from the destruction of their species by man. Their multiplication is so rapid and their means of evading pursuit or traps are so great, that there is no likelihood of his being able to destroy the entire species of any of these animals." Lamarck is most remembered for his theory of the inheritance of acquired characteristics; he was wrong about that too. But he may be forgiven for failing to envision the advent on Georges Bank and elsewhere of ships so efficiently and murderously acquisitive as these; ships unprecedented not merely in the size and breadth and number of their traps, not merely in their capacity to absorb and hold fish — they were, after all, in the service of a world grown by nearly five billion people since the opening of Georges — but also in what might be described as the clairvoyance of their pursuit.

In this respect, though, the factory trawlers had little advantage over at least some of the tiny American and Canadian boats that pitched about in their midst like foxes among the lions. All benefited from World War II and Cold War military technologies adapted to commercial marine use: from the 1940s, the radios that allow fishermen to communicate with other boats and with buyers and processors onshore and to get swift help in times of distress; the autopilot and radar equipment that allow fishermen to steer straighter and more accurate courses and to see through fog to coastlines, buoys, and other ships; from the 1960s, the loran (long-range navigational) system, which turns the trackless sea into an electronic grid, allowing fishermen to pinpoint choice fishing areas to within fifty feet and return to them again and again, as Brian is doing effortlessly today with his lobster traps; navigational plotter systems that allow fishermen to find those fishing areas on autopilot; and marvelously percipient sonar systems that can see in any direction to any depth, allowing fishermen both to locate schools of fish and to identify what kind they are. Refinements from the 1990s include electronic sensors on trawl nets to indicate when fish are entering the net, and a global positioning system, tied to a fleet of satellites in synchronous orbits, which can take a fisherman within twenty feet of his desired location.

The wheelhouses of many larger fishing boats, running the gamut from the *Bell* to a factory trawler, are now equipped with enough screens and switches to suggest the cockpit of a space shuttle. Even the little *Cap'n Toby,* with its bare-bones radio, sonar, and loran, has something of a medium-range-bomber aspect to its wheelhouse. The military pedigrees of these technologies paint them all as doomsday weapons, arms in a war on fish turned now as much in favor of the hunter as Lamarck once saw it in favor of the hunted. But to the fisherman, they are tools that make his work easier, more efficient, and above all, safer. As more and more boats adopted them through the decades, they became necessary not only for staying competitive with other boats but simply in helping everybody to get back alive.

However, the presence in the 1960s and 1970s of ever increasing numbers of high-tech, stadium-sized distant-water trawlers on Georges Bank put souls less paranoid than Harry Hunt in mind of doomsday. For that matter, doomsday has not previously required the use of weapons as mighty as these. Colonial fishermen, for example, considered halibut a trash fish, unfit for consumption, and no doubt cursed like Hunt whenever an angry four-hundred-pounder took one of their hooks. But some anonymous Yankee gourmand changed all that when he boldly cooked a halibut steak, boldly took up his knife and fork, and pleasingly ate thereof. A Boston market for the fish developed in the 1830s, and soon individual schooners were returning from Georges Bank with twenty thousand pounds of halibut each day. Similarly, a century later, demand shot up for the tasty, slow-growing redfish found in the Gulf of Maine and off Cape Cod, with 130 million pounds of redfish coming out of the gulf in 1942. Both fisheries have yet to recover. And once-prosperous New Bedford, more than anywhere else in New England, is haunted even now by the free-fall decline and crash of the whaling industry.

The Atlantic halibut, the redfish, the sperm and right and blue whale: by no means have the entire species of these animals been destroyed, and Lamarck was right enough as far as that went, but they are all extinct from any commercial perspective, and have remained so for a sobering length of time. If that could be accomplished in small boats unaided by electronics, then what might be accomplished by the far-seeing factory trawlers working in the restricted waters of Georges Bank? In the 1970s the United States was only one of many nations concerned with the impact of distant-water fleets on the fisheries of the continental shelves. The harvests of these trawlers off New England were so obviously immense, however, and the complaints of American fishermen to their congressional delegations so persistently sharp, that in 1976 Congress preempted the United States' negotiations on International Law of the Seas treaties to extend territorial waters unilaterally to two hundred miles offshore.

This was the Magnuson Fishery Conservation and Management Act, and its effect seemingly was to turn back the clock. One morning in 1977 the factory trawlers were gone, steaming off to the Bering Sea, the southeast Pacific, the Patagonian Shelf, the Challenger Plateau between Australia and New Zealand, or the coasts of Namibia or Mauritania. The gray, choppy waters of Georges Bank looked much as they had when the men of Harry Hunt's and Olav Tveit's generation began their work on the shoal in the 1950s. And across their sonar screens moved — still in pestering, provident numbers — great shoals of cod.

By late afternoon the *Cap'n Toby* is working the Light, a stretch of bottom off Eastham opposite the Coast Guard station and lighthouse. Like the Can, this is an area already peppered with lobster buoys. It also contains a number of gill nets, which are anchored to the bottom at both ends and left to catch fish by their gills. A small boat from New Hampshire rocks in the sun several hundred yards off our port stern, slowly working one of the nets into its stern. Brian regards the gill-netter with unconcealed contempt.

"It's just a mess, another example of the yahoos taking over the fishing industry," he says. "You've got eight lobstermen working this area, and one gill-netter in here screwing everybody up. The tide pulls the lobster buoys under, and then he comes in and sets his nets right on top of our traps. He's probably got three hundred fifty-fathom nets set all over the place, up and down this coast, and he doesn't tend them but once a week, and the fish die, rot, and get sold just for scalers. It seems like gill-netting should be a clean fishery, but it's not when they're running that much gear and just leaving it."

Brian lifts another trap to the rail, this one containing a short and a small cull, a legal-sized lobster with one claw missing. He adds, "There always used to be a lot of skill involved in seamanship, but now with all the navigation and communication technologies on these boats, any yahoo can get out here and call himself a fisherman. Part of that skill, you realize, involved recognizing and respecting other people's territorial imperatives. Now it's just 'The hell with you, I'm here.'"

Part of Brian's anger stems from the increasing incidence of what the newspapers and industry regulators call gear conflicts: fishermen using different gear, and often targeting entirely different species, in the same stretch of water or bottom. Gear conflicts are older than Harry Hunt's curses hurled at the Russian behemoths, are as old as fishing itself, but nowhere have so many fishermen from so many different industry sectors been squeezed into such shrinking areas of legal and profitable activity as in New England. Only last December, the U.S. Commerce Department, taking emergency measures, closed 17 percent of New England waters to all kinds of fishing, including Jeffries Ledge and Stellwagen Bank off the coasts of New Hampshire and Massachusetts, the Nantucket Lightship area south of Cape Cod, and large portions of Georges Bank. Those fishermen who haven't lost gear in the crush already or been obstructed have gotten . . . well, paranoid. "That opinion was expressed by Brian Gibbons," says Brian, who goes on to provide me — for the convenience of gill-netters or yahoos wishing to retaliate against him — with the registration number of his traps and their loran coordinates here.

Another part of his anger stems from the poor return on his traps today, and a long spring of poor returns. Brian likes to get at least one pound of marketable lobster per trap in order to pay his expenses and provide a little something for the car-house-boat-payment culture. He's working now through his last string of the day, approaching nearly a hundred traps hauled, and he doesn't have to look in the holding tank in the stern to know that he's not even within hailing distance of breaking even.

In the afternoon light the sea around us has turned to platinum and chrome. The light itself has neither the pale translucence of early spring nor the aortal intensity of the Cape's high summer. It's something in between: warm, diffuse, effulgent, as absorbent as litmus paper of the day's late colors. The yellow of Brian's oilskin apron is radiant with the soft neon glow of a highway centerline. The ocher-orange of the red beard sponge that coats the warps and upper laths of the traps in this area — Brian calls the growths posies — glows with an almost chemical intensity. The traps themselves come up dripping and empty, the water ribboning off them like tinsel. Brian empties them of rock winkles and hermit crabs.

In the last trap of the day a small sculpin writhes, heaving itself about on its wide, bat-wing pectoral fins. The sculpin is scaleless, leathery, and has a chameleon's ability to change color. This one remains a stubborn, sullen brown as it gasps on the bottom of the trap. Brian's hand skirts the venomed spines that protrude from its fins. He throws the fish over the side.

Inside Nauset Inlet, as he ties the *Cap'n Toby* off at its mooring, another lobsterman comes in. Fair-haired and bespectacled, Bob Maraghy is the captain of the *Severance,* which draws alongside. Maraghy is wondering how Brian did today. "Terrible," Brian says.

"I thought it was just me. It sucked bad out there. You tried the Can?"

"The Can, the Light — I got maybe three quarters of a pound per trap."

"I hauled thirty-six traps today, and maybe seven had lobsters. I was gonna cut my wrists."

"Well, that makes me feel better."

Maraghy understands the layers of irony wrapped in that comment. Bob Maraghy is a colleague and a friend; he knows that Brian speaks only in jest. But he is also a competitor and an opponent, one who has gained no advantage over Brian in his work today; he knows that Brian feels better now as a matter of plain fact. Brian, after all, is a lobsterman, which means he redefines himself with each outing, much as an athlete does with each game, and he can't help being afraid that he's somehow lost his touch, that those thousands of microanalyses that go on inside a hunter-gatherer's head are suddenly being fed the wrong parameters inside his. Maraghy and all the other lobstermen in the inlet, however, are striking out as well. It must be the cock-eatin' Portagee sons of bitches of bastards, whispers Harry Hunt.

Brian lifts the plywood lid off his holding tank while Maraghy, feeling better himself, motors to his own mooring. When the *Severance* falls silent, the sweet, organlike note of the piping plover — a bird now rare, whose song the naturalist Henry Beston considered the loveliest musical note of any North Atlantic bird — is conveyed across the inlet from its nesting grounds on Coast Guard Beach.

A smile like a crease in a wallet stretches across Brian's face. His own voice is almost as sweet as that plover's, has a woodwind's understated warmth of tone. Its phrasings are crisp and unexpected, like a Charlie Parker solo. It sings even now with this morning's thirty seconds of just loving it, played out like a tonic chord through ten hours today and six weeks the day before of shorts and eggers and the wrong fucking fish and empty traps, and it asks of the evening now in its best pidgin French, "Ou sont les lobstairs de yesteryair?"

2

Ivy Day in the Committee Room

THE AMERICAN GOLDEN-ARCHES, white-lines, and car-payment culture runs riot along Route 1 just north of Boston: retail outlets, fast-food joints, thrift stores, rug emporiums, used-car lots, ballroom dance studios, pawnshops, health clubs, gas stations, and acres of white-lined parking lots. Inside the Marblehead Room of the Holiday Inn in Peabody, the walls are pink, the columns sheathed in mirrors, the carpet patterned in palm leaves. An ashen gray light, like the glow of a cathode ray monitor, sifts down from wells sunk into the ceiling or through honeycomb grates tacked beneath fluorescent tubes. People's eyes and mouths disappear in that light. Their brows beetle over their cheeks. Their noses drop like bulwarks to the points of their chins.

Brian's lap is heaped with the papers he picked up from a table at the entrance to the room: a listing of the membership of the New England Fisheries Management Council, a body established by the Magnuson Act to oversee New England waters; a listing of those members' committee assignments — Coastal Migratory Species Committee, Dogfish Committee, Enforcement Committee, Gear Conflict Committee, and so on; a draft public hearing document titled "Proposals for the Management of American Lobster"; reports from various scientists on various methods of reducing the lobster harvest; scientific analyses of the methods suggested so far by five Effort Management Teams, each with responsibility for reducing the harvest in a different portion of

New England and mid-Atlantic waters; and more, *ad chartam infinitum.*

Brian's beetling brow knots as he flips through the packet and looks at the first page of a scientist's report on the surmised effect of trap reductions:

> We used the model of Fogarty and Addison (1992; in press) which describes catch per trap haul as a function of entry and escapement:
>
> $$C_t = \left\{ \frac{a_t}{b_t} [1 - e^{-(1-m)b_t S}] \right\}^{1/(1-m)}$$
>
> where C_t is the catch at time t
> a_t is the parameter expressing entry rates
> b_t is the parameter expressing escapement rates
> m_t is a shape parameter describing density dependence in entry rates
> S is soak time
>
> For the case where m=0, the model reduces to Munro's (1974) model for unbaited fish traps:
>
> $$C_t = \frac{a_t}{b_t} (1 - e^{-btS})$$

"Whoever wrote that is a target for the Unabomber," he decides.

The council's Lobster Oversight Committee — eighteen men, most in sport coats and ties — sits behind individual microphones at a horseshoe-shaped table taking up the eastern end of the Marblehead Room. A heavy maroon tablecloth drops from the table to the floor, obscuring their legs. About thirty people, almost entirely men, are sitting on folding chairs in the audience. Attire in the audience ranges from coat and tie and wingtips to flannel shirts, blue jeans, sweatpants, running shoes. Brian occupies a politic midpoint in this continuum. His tieless yellow dress shirt is unbuttoned at the neck, his gray flannel pants pressed and creased, his Sperry Topsiders nicked and scuffed. He is part bureaucrat, part working man.

The chairman of the Lobster Oversight Committee is Eric Smith, a council member and assistant director of the Connecticut Department of Environmental Protection's Marine Fisheries Division. The committee's existence, and its business here today, are the result of a tense compromise between the National Marine Fisheries Service (NMFS, pronounced "nymphs") on one side and the lobster industry and the New England states on the other. In the last decade, while groundfish or catches have been plummeting, the landings of lobsters in the Northeast have been on a record-breaking climb. NMFS scientists have reached the same conclusion about this as Brian: that the climb is the result of ever-increasing numbers of traps in the water. While the total landings of lobster have gone up, the scientists say, a lobsterman's catch per unit of effort has dramatically dropped. This is the same ominous pattern that was seen in the groundfisheries in the early to mid-1980s.

NMFS responded as early as 1988 with the first in a series of gradual increases in minimum carapace size, or gauge length, for lobsters taken in federal waters, beyond the three-mile limit of state jurisdiction. These increases were intended to reach three and five-sixteenth inches by 1992. In 1990, however, both Maine and Massachusetts passed legislation against any gauge length beyond three and a quarter inches in their own waters. "Most of what the inshore lobstermen were catching in the Gulf of Maine were very small lobsters, right at that minimum gauge length, and they didn't want to lose all that stock," Brian explained. "Plus the dealers didn't want to lose out to Canada in the international marketplace. Canada has a three-and-three-sixteenth-inch gauge size, and if a hundred-pound tote of Canadian lobsters has ninety animals in it, while a hundred-pound tote of American lobsters has eighty-four, then international buyers are going to go for the Canadian every time."

That legislation, however, made it impossible for NMFS to enforce its own gauge length short of placing an observer on every boat in the industry to confirm that each small chick was indeed caught in state and not federal waters. So that same year a compromise was struck. The goals would be NMFS's: a large reduction in the lobster harvest in the Gulf of Maine and southern New England waters, and provision for contingency measures if and when other areas, such as Massachusetts's Outer Cape, were determined to be overfished. The means would be the industry's: an alternative lobster conservation and management plan forged by the industry itself (with the help of the council) and using methods other than a gauge increase to rein in the harvest.

In 1993 the council divided northeastern waters into five separate management areas and established an Effort Management Team for each area, made up of a mixture of scientists, public officials, and working lobstermen. At the same time NMFS established a deadline of July 20, 1995, for the industry to deliver its management plan, which is officially described as a "framework adjustment" to a larger amendment, Amendment Five, of the Magnuson Act. The council gave the EMTs only six months to arrive at conservation plans that would meet NMFS's goals, and reserved for itself another six months to pull common threads of the five plans together into a final regulatory package. This package, promised the council, would preserve the essence of each area's plan. The council would then present that package in public hearings throughout the Northeast and finally deliver the finished product to NMFS for implementation. If all this was not accomplished by July 20, warned the feds, NMFS would withdraw its own conservation plan, leaving federal waters entirely unregulated and the state fisheries vulnerable to the depredations sure to ensue beyond the three-mile limit.

The EMTs represented a radical new initiative in fisheries management, with biologists, bureaucrats, and fishermen — traditional enemies in that field of battle — working collaboratively in the same foxholes and trenches, and the five groups were all successful in meeting their deadline. No sooner had they done so, however, than NMFS announced that there would be no approval of a new federal management package without state regulations that were compatible with each other. In other words, the council had to do more than merely pull threads: it had to weave the contentious and grudging concessions of lobstermen from Maine to the mid-Atlantic states into one coherent fabric of inshore and offshore regulation — one that would not only satisfy NMFS but enjoy the general support of fishermen in all those different regions.

Time now is very short. The council had originally intended to sign off on a public hearing document by the end of March and then, after the public hearings, vote on the final package's major policy issues by the middle of May. It is already the middle of May, and still no public hearing document. Eric Smith wants to push today for a series of final decisions on that document before next week's Thursday meeting of the full council.

Smith is fair-haired, mustached, and slim, wearing glasses, an open-necked button-down shirt, and a blue blazer. He sits at the center of the committee table with his microphone raised like a charmed snake in front of him. At nine o'clock he checks his agenda, taps at the microphone, clears his throat. "Good morning, Peabody. Let's get started. I was walking down the hall with someone this morning who said to me that he wouldn't have missed the last meeting of this committee for anything. I mentioned I had a bridge in Brooklyn I could sell to him if he thought this was going to be our last meeting."

Smith reminds committee members of the urgency of their work today and welcomes a scientist from the Northeast Fisheries Science Center in Woods Hole. The scientist has come to present a report on the number of lobsters taken as bycatch by draggers. This information pertains to the question of how many lobsters fishermen in the mobile gear or "non-trap" sector should be allowed to harvest and sell under the new management plan.

Dick Allen, a veteran Rhode Island lobsterman and a member at large of the council, stands at a microphone set up in the audience and notes that proposals on this question are scheduled for two this afternoon, and since many of the interested parties on that issue aren't here yet, shouldn't this report be postponed until then? His amplified voice echoes hollowly about the room.

"Probably the only person from the non-trap sector who routinely comes, and I don't see him here now, is Jim O'Malley," says Smith. "Do you expect him?"

"I don't know," says Allen. "But you know how the non-trap people claim that things are done to them when they're not around."

"I would hesitate to wait until two o'clock to hear a technical report," Smith says, already sighing. "Nothing's going to happen during a report anyway. Let's just proceed with it. We may have to answer a few questions later, and that's okay."

"I'm very glad to be here," says the scientist.

"You may have cause to reconsider that remark after the meeting's over," Smith replies.

"Picture a pasture open to all," writes the ecologist Garrett Hardin. "It is to be expected that each herdsman will try to keep as many cattle as possible on the commons. Such an arrangement may work reasonably satisfactorily for centuries because tribal wars, poaching, and disease keep the numbers of both man and beast well below the carrying capacity of the land."

But finally comes a day, says Hardin, when those restraints are removed, and each herdsman faces the same question:

> As a rational being, each herdsman seeks to maximize his gain. Explicitly or implicitly, more or less consciously, he asks, "What is the utility *to me* of adding one more animal to my herd?" This utility has one negative and one positive component.
>
> 1. The positive component is a function of the increment of one animal. Since the herdsman receives all the proceeds from the sale of the additional animal, the positive utility is nearly +1.
>
> 2. The negative component is a function of the additional overgrazing created by one more animal. Since, however, the effects of overgrazing are shared by all the herdsmen, the negative utility for any particular decision-making herdsman is only a fraction of −1.

Math, intuition, logic — they all point to the short-term utility of a herdsman adding one more animal to the commons. As other herdsmen follow suit, the pasture is inevitably overgrazed and destroyed.

Taking this scenario from an 1833 pamplet by the English mathematician William Forster Lloyd, Hardin describes this chain of events as "the tragedy of the commons," using *tragedy,* he says, in the philosopher Alfred North Whitehead's sense of the word. "The essence of dramatic tragedy," asserts Whitehead, "is not unhappiness. It resides in the solemnity of the remorseless working of things."

Brian is familiar with Hardin's famous phrase and knows himself to be a herdsman of sorts, a wrangler of lobster traps, calculating a fresh cost-benefit analysis each season of how many traps to add to the heavily grazed commons that are the waters of the Outer Cape. With his interest in history and fascination with the arcane, he probably also knows that just as the eighteenth- and nineteenth-century wealth and power of Massachusetts had much to do with cod, so the sixteenth- and seventeenth-century wealth of the Netherlands had much to do with herring. The Dutch, in fact, may be said to have invented factory fishing with their discovery that if gutted fish are laid in barrels between layers of salt, they may be kept in great quantities for up to a year.

The English also had an interest in North Atlantic herring, however, and in 1609 James I forbade the passage of Dutch ships to the herring grounds through waters that King James was pleased to define as English. The Dutch won the war of words: the jurist Hugo Grotius of the Hague wrote that no one can own what neither he nor anyone else has previously owned; the sea, therefore, is the common property of all. In this he laid the foundation of international maritime law. But the Dutch lost the war at sea, and in the solemnity of the remorseless working of things, their power waned as that of the herring-fed English increased.

Grotius coined a phrase as well: *mare liberum,* or "freedom of the seas." It was out of respect for that phrase, and fear that

American merchant and military shipping might be denied transit of certain strategic straits, that American delegations to the United Nations' Law of the Sea Conferences in 1958 and 1960 resisted the urgings of many Third World nations for a two-hundred-mile exclusive economic zone, or EEZ, around national coasts. In 1972, however, the little North Atlantic nation of Iceland unilaterally proclaimed its own EEZ out to fifty miles in order to protect ancestral cod grounds from the depredations of British fishermen. The British responded, in a delicious irony, with angry citations of Grotius's *mare liberum.* In an international community beginning to question the implications of that phrase for the health of coastal fishing grounds, there was no clear victor in the war of words; the war at sea, however, was narrowly averted by a compromise that allowed small British draggers to fish within the disputed waters for two years but immediately banished British factory trawlers.

Despite the outcome of this abortive "cod war," and despite the remorseless workings of the international distant-water fleet on Georges Bank, the U.S. State Department still opposed the establishment of EEZs. By the time of the third Law of the Sea Conference, in 1974, however, New England fishermen had allies in Congress, held two seats in the American delegation to the conference, and were buoyed by an irresistible political groundswell from the Third World, which by then had grown more cynical than ever about fisheries management by international organizations. That year the United States finally went on record in favor of the two-hundred-mile EEZ, even as the State Department sought to restrain Congress from extending American jurisdiction that far until accord was reached on other issues at the conference. Georges Bank and the sacred cod, said the State Department, were safe in the hands of the International Commission for the Northwest Atlantic Fisheries, an eleven-nation organization founded in 1950 in response to drops even then in the North Atlantic harvest of cod, haddock, halibut, and redfish. Like hell they're safe, responded Yankee fishermen, who (like the Icelanders) had long since despaired of ICNAF's ability to control the factory trawlers. When the Law of the Sea Conference dragged inconclusively into 1975 and then into 1976, Congress finally acted, and with the Magnuson Act drew the two-hundred-mile boundary that transformed the gravel bottom of Georges Bank from an international commons into a North American one.

So began Brian's mighty adventures on the *Cape Star* and the *Bell* in the 1980s and his heady flirtation with the comfortably middle-class reaches of the car-payment culture. Soon he discovered, however, that once a line is drawn, whether in dirt or in water, another may be drawn just as easily. "The establishment of the Hague Line across Georges Bank in 1985 was the beginning of the end for the New Bedford offshore fleet," he wrote. "While winter fishing on the *Bell,* we would rarely even start looking for fish until we were well east of that which became the Canadian boundary. Now the 80-foot to 110-foot class of draggers were forced to work the western side of Georges, which formerly had been the province of the 70-foot to 80-foot class (Woods Hole boats and much of the 'Portagee' fleet)."

The establishment of EEZs by both the United States and Canada had prompted both nations to claim Georges Bank. The International Court of The Hague (an oak grown from the acorn that was Grotius) awarded most of the bank to the United States, but the richest portions, the Northern Edge and also the Northeast Peak, a great plateau friendly to scallop dredges and otter trawls, went to Canada. The decision was a terrible blow to Olav Tveit and other New Bedford captains, as it was to crewmen such as Brian. Brian, however, had already seen a telling increase in the number of boats working those areas and a corresponding drop in the enormous sets of fish Olav had once hauled in. To him the new boundary line represented a mere prompting of the inevitable.

"Make no mistake: the establishment of the Hague Line probably only served to hasten the progress of an existing trend in the fisheries," he explained. "In any event, I could see the offshore resource becoming a hardscrabble existence; so rather than spend my winters in a cigarette smoke–filled wheelhouse while my children were growing up (on my returning home, Nicole and Beth would run out of the house: 'Daddy, Daddy!' while little Mike would run out: 'The truck! The truck!'), I made up my mind to stay on hardscrabble Cape Cod."

With that, the large and steady paychecks Brian had earned with the Squareheads disappeared, though he had the consoling expectation of evenings at home with his family. But eventually he was disappointed in that as well. "During all these adventures," he wrote, "I became embroiled in local and regional politics. In 1983 many of us rejuvenated a somnolent organization, the Nauset Fishermen's Association, in response to some regulatory situations in Orleans. I, an offshore disciple of Ayn Rand, who believes that government functions best to the extent that its subjects can ignore it, was elected cochairman (at first I thought this was an honor). Town meetings, state legislation, hearings with guys in suits, ivy day in the committee room. In a way I traded those smoky wheelhouses and eighty-knot winter gales for that aforementioned menu."

Ivy day in the committee room: Brian sits in the audience today as a member of the Outer Cape Lobster Management Area's Effort Management Team and an alternate for his friend and fellow Orleans lobsterman Steve "Smitty" Smith, who is the Outer Cape's regular representative to the Oversight Committee. Steve is here today too but has asked for Brian's help and moral support in presenting the most recent work of their team to the Oversight Committee.

The Woods Hole scientist has dimmed the lights in the Marblehead Room and thrown a pair of bar graphs on the bed of an overhead projector. The graphs show the number of draggers reporting lobster landings in their trawl nets and the number of lobsters they land. The numbers are important to the share of the lobster harvest that the council may ultimately award to the dragger fleet, and are immediately called into question. Dick Allen, a trap fisherman like Brian, suspects they're too high. A draggerman from Gloucester counters that the numbers more likely are too low. Brian listens impassively to this early parrying, this probing at the eventual location of a new line in the harvest of lobsters. He himself was content with the original federal gauge increase plan. It was simple, enforceable, and seemed to be bringing more egg-bearing lobsters into the fishery. He admits, however, that since lobsters on the Outer Cape tend to be bigger than those in the Gulf of Maine or southern New England, the gauge increases were less painful for him than for lobstermen elsewhere — or would have been, he says, back in the days when he caught lobsters. When the scientist presents graphs on the distribution of pounds landed per trip by draggers, a committee member warns others not to be deceived by the occasional spikes on those graphs. They represent opportunistic fishing, he says, and one day of heavy landings may be followed by days of few or none at all. Brian whispers into my ear, "With me, you get just the days of none at all. There's no spike on that graph."

The Woods Hole scientist retires, yielding to a team of scientists, the committee's Technical Core Group, who present their analyses of the effects of proposed EMT conservation measures on population models for their areas. According to this group, there is trouble in the area with the largest amount of landings, the Gulf of Maine. In place of the federal gauge increases, the gulf EMT has offered a laundry list of alternative measures, one of which would be a modest cut in trap numbers, another the throwback of 3 percent of marketable female lobsters. NMFS, however, wants a 20 percent reduction in Gulf of Maine lobster fishing mortality within five years. Taken individually, each of the gulf's proposals falls far short of achieving that goal. Throwback, for example, says the biologist Joe Idoine of Woods Hole, should be much higher if it's going to work. Eric Smith listens with growing impatience. "That three percent thing," he says to the gulf representatives, "that just doesn't begin to get us there."

"That's just one thing, Eric, among all the other things we're doing," responds Patten White, the silver-haired director of the Maine Lobsterman's Association. "We need to see them all put in a package as to what effect they have all together on the population model. Until we know what fishermen are going to do to compensate, of course, we can't really know what'll happen. But to come out and say that one particular thing doesn't work doesn't give any guidance as to where we should be headed."

"My guidance is that there has to be serious consideration of the total number of traps being cut," Smith says, noting that the gulf's trap numbers start and finish much higher than those proposed by any other area. "That has to be addressed before any of the other things."

"There appears to be substantial increase toward the optimum if you reduce the number of traps," intones Idoine.

Hands come up in the audience: Dick Allen's and another's. "Yes sir, you haven't spoken yet, and Dick has," Smith says.

Smith knows what the gulf EMT is up against: recent years of record landings, high earnings, and a strong suspicion among lobstermen there that NMFS is flat-out wrong when it says that the gulf is overfished. The next speaker, a lobster dealer and processor from Portland, raises this issue, saying that over the last three years gulf processors have seen bigger lobsters coming into their pounds. "Has that been taken into account?"

"Most of our information comes from Massachusetts, of course," Idoine answers, "but in most of the landings we see in any one year, the lobsters are just one molt above the minimum size."

The processor claims that's not the case in Maine anymore; will the final management plan have the flexibility to reflect changes in the scientific data? Yes, says Idoine, but don't hold your breath — it would be a matter of due federal process and would take time. "You just can't say the science has changed and we don't have to obey the amendment anymore."

Howard Russell, a member of the Oversight Committee and the council's chief lobster biologist, points out that there could be other reasons than a growing population for this alleged shift toward bigger lobsters. "Water temperatures are higher now in the gulf, and the lobsters are molting more frequently. What you're seeing may be modified by how they're growing."

The processor and Russell warm to their arguments. When Russell says that there are many things still not known about lobsters and any number of ways to explain that size shift, if it exists, Eric Smith interrupts. "I'm going to have to crack the whip if we're going to get through the business before us today. After lunch we're going to have to make some decisions about what we've heard in order for me to have something to present to the council next Thursday. I think we need to move on to discussion of the EMT proposals themselves now, and before lunch we might be able to expedite the mid-Atlantic and Outer Cape's and — yes, Dick?"

Dick Allen has been to the microphone often this morning, and he takes his place there again. He is a round-shouldered man of medium build, and in his white button-down shirt and rumpled dark slacks, with his lined face and questioning eyes, he looks like a midlevel manager in the midst of a downsizing. There is a two-fisted roll to his step, however, that suggests he would not go quietly. He says, "I noticed there was no individual transferable quota analysis in any of the scientific reviews on each area, or any sheets or graphs that indicated the results of using an ITQ, and I only wanted to —"

"Only because the earlier assessment of the scientists was that an ITQ was one measure that clearly met the objectives of the management plan," Smith says.

"Oh, so the omission was not anything detrimental. Okay, I think that . . . thanks."

Smiling, amid general laughter, Smith suggests, "That concludes your remarks for the day, would you say?"

Brian so far has had no remarks. He sits calmly in an aluminum folding chair, out of the stale cigarette smoke and the winter gales of Georges Bank, while the fate of everything he has bet his life on is put on the table for negotiation. He cranes his neck, pricking his ears, now and then whispering to me both color commentary and play-by-play for the day's proceedings. He explains that the Magnuson Act defines overfishing as any instance when the assessed egg production of a particular species — lobsters, cod, scallops, whatever — is less than 10 percent of what it should be for an unfished population. By that standard, says NMFS, the Gulf of Maine and southern New England are overfished for lobster, while the Outer Cape, the mid-Atlantic, and the offshore canyons are almost overfished. So the former two areas are responsible for conservation plans to go into effect as soon as possible, the others for plans to take effect if egg production drops any further.

"So the Outer Cape drew up a plan that included some pretty drastic trap reductions," he says, "and with a lot of difficulty we built a consensus among the fishermen there based on what both the scientists and the Oversight Committee were telling us. The scientists were saying the same stuff to the other areas, but then people from the gulf, let's say, were calling up committee members, bitching about having to cut their traps, and the members were telling them, 'Okay, sure, we'll see what we can do. Don't worry.' That causes Steve Smith and me certain credibility problems. At the meeting last month Steve and I finally just said, hell, we weren't even going to bring a plan to the table if different areas are going to be held to different standards. That went over like a fart in church."

Brian approves of Eric Smith's impatience today with the proposal from the Gulf of Maine. He says the gulf's proposed trap reductions, from a cap of twelve hundred the first year down to eight hundred within five years, are nothing of the sort; that gulf fishermen as a rule are running fewer than eight hundred traps per permit even now; that the proposal simply defines a limit to which they can expand.

He listens to the Portland processor's and Howard Russell's

dispute about the size of lobsters in the Gulf of Maine and concludes, "It gets to be a circular argument after a while."

He sits without comment, his jaw working slightly, during Dick Allen's question about individual transferable quotas, which would assign each lobsterman the right to catch a fixed portion of the annual harvest in a region and allow him to transfer — that is, to sell — that right, that reserved portion of the annual harvest, that privatized corner of the New England commons, to another commercial interest. To Dick Allen and his allies on this issue, ITQs are the best solution to Hardin's tragedy: a fixed limit to the number of herdsmen on the commons, a fixed limit to the animals apportioned to each, and free-market upward mobility provided by the liberty to buy and sell these portions. Brian, however, is no admirer of ITQs.

Nor does he enjoy the literal circularity, from the committee room to Orleans and then back to the committee room, of his and Steve's arguments on lobster conservation. At the last Oversight Committee meeting, Brian and Steve were disappointed (to put it mildly) when they were told that the Outer Cape's trap reductions did not go far enough. But they have chosen to come to the table again, this time with an amended plan that Brian describes as a Bosnian peace accord, a plan arduously hammered out among thirty or forty suspicious and very uneasy full-time lobstermen, who finally pronounced themselves ready to live by the document if others did. The Outer Cape's plan starts off with further reductions in everybody's trap numbers, dropping them by as much as 20 percent by the end of five years, to a cap of 640. It also mandates tags on every trap for monitoring purposes, a maximum trap size, and a sizable one-and-fifteen-sixteenth-inch escape vent in each trap. The plan also requires a minimum soak time of five nights per trap and a seasonal closure between January and March.

Finally, the plan sets a limit on the number of lobstermen in the fishery, restricting permit-holders to those who were fishing traps off the Outer Cape during June of last year. Brian knows that a few Cape lobstermen are on the wrong side of that control date and that one of them, Chris Adams, is here today to complain about that. But the control date was difficult to negotiate as it was, and it succeeds in limiting the size of the fishery to its present number of active lobstermen. Adams was letting someone else fish on his permit in June of last year. Over the winter he told Brian that he was done with lobstering, wouldn't come to the EMT meetings. But now that those meetings are just about over, he has changed his mind. Brian advised him to appear at this Oversight Committee meeting and to speak up about the control date if he wants to, but not to get his hopes up.

The Outer Cape plan, if it were invoked, would be painful, but Brian accepts pain, evenly and fairly distributed, as an alternative to the death throes that have befallen New England's cod fisheries. The plan is also direct, enforceable, easy to understand, and fits within three pages. Dick Allen, to whom Brian has shown it as it has taken shape, has told him that compared to the plans of the four other areas, the Outer Cape plan, for all its brevity, is the most comprehensive.

And Brian takes a measure of reassurance in the fact that the plan has no recourse to the ITQs so admired by Allen, which disturb Brian even more than paranoia.

"You can call it a bit of a morale problem, but that's basically what it is," says lanky Steve Smith, standing at the microphone, one hand, muscled and gritted, clamped around its shaft. "When you abut an area that's using a trap cap that's out of sight compared to what our numbers are going to be, you can only hold the fort so long."

Steve describes the difficulty of returning to the fort last month with the task of demanding even lower trap numbers and then details the mathematics and logic of the new set of reductions. Eric Smith asks a member of the Technical Core Group if these new numbers will survive challenges in a public hearing. The scientist responds with cautious skepticism, saying he'd have to apply the formula and see what he comes up with. Smith says the formula needs immediate application. The scientist says he'll work on it over lunch.

Chris Adams, bearded, wrapped up in a heavy sweater as though for a dose of hard weather, takes Steve's place at the microphone. He says that he fished for lobster on the Outer Cape through the mid- to late eighties and has a problem with this plan's control date. "Right now I've got a boat again, I've got traps in my yard, I'm going to get back at it, and I'm told I'm not going to be able to do that. I've asked people to address it, and I don't think it's been addressed properly. It's also redundant to what the feds have already outlined with their regulations and deadlines. I've met all the criteria for a federal license. Now I'm told next year I won't be able to go. So please address it."

"This is the first meeting you've attended, isn't it?" Eric Smith asks after a pause. "We have discussed this kind of issue for the Outer Cape and other areas a number of times, probably for the last six months or so, as much time as we've spent developing new management plans. I don't know how to address all the problems of all the individuals who for one reason or another are on the wrong side of new lines being drawn.

"I have some problems myself with June '94, simply because it's a very narrowly constrained date. But I think you called me about a week ago and we talked about this. Weren't you working in a different fishery for five years or so? And now you want to come back and lobster. Well, as far as eligibility goes, that's very far outside the bounds of what any of the other areas are considering for control dates. The most liberal of them pick as much as a three-year period, or a two-month period, or a within-the-last-year sort of period, and if I recall, you stopped lobstering in '89. As I said up front, you've got a tough row to hoe here because you're so far outside the bounds of what any of these areas are developing." Smith pauses again, peering narrowly over his mike at Adams. "You can't solve every individual's problems all the time. I've just got to be straightforward with you."

"Okay," Adams replies. "I'm not gonna yell at anybody. I went forth in good faith, straightened out my finances, got on my feet, so to speak, thinking that I would be able to return. Now I'm told, 'Next year you can go jump off a bridge.' Maybe that's the answer."

Steve Smith steps in front of Adams at the microphone. He says that at the EMT meeting last week he made a motion to change "during" in the control date to "as of" June 1994. That would have accommodated Adams and others like him. "What I don't know is how much that increases the number of people who'll be fishing," he continues. "Every time there's an increase in that part of the equation, the number of traps everybody else is allowed goes down. The atmosphere I got at the meeting, when those questions came up, was kind of hostile at times."

Questions come from the committee concerning the exclusiveness of the control date. Eric Smith says again that all of the areas have control dates and are setting limits on who may participate; they're all just doing it in different ways. Someone on the committee asks if it's legal to have different eligibility standards within one comprehensive plan, which is what they're working toward here. Smith submits the question to Gene Martin, the Oversight Committee's legal counsel.

"You could justify totally different eligibility standards in different areas," Martin says, "as long as you can rationalize the differences based on the area and so forth. If it just appears arbitrary because of what parts the EMT was able to get through, or not get through, with their constituents, then that would be difficult or impossible to justify from a legal perspective."

Brian rises abruptly from his seat and replaces Steve at the microphone. He faces the committee and says, "As I mentioned at the last meeting here, that word *during* was originally *as of* in our plan, but the scientists recommended we change it to *during*. I think it's fair to mention that at our EMT meeting Tuesday night some ideas were brought forward on this, and it was voted by over two thirds of our membership to keep the plan the way it was without changing that language. One of the minority views was that you could let people like Chris come in at a lower trap level, not the level based on the historical number of their traps. But if NMFS or the council wanted to do something like that, I just hope it doesn't affect the rest of us, who are cutting back hard in terms of those trap numbers. I hope we can all have a little room to live that way."

Eric Smith asks, "Is there any way you can figure out how many people are in a situation —"

"I think there were five like Chris at our meeting the other night — just a handful. I'm sure there are other unused licenses that might crop up."

Smith has one elbow propped on the table. He rubs his temple with his hand, slides his fingers down the length of his face, props his chin on his fist. "Within the bounds of all you guys have to do, five people, one way or another . . ." he says finally. "Well, that's what I'm saying. Particularly with the Outer Cape, try and work these things out so that nobody's disenfranchised. You guys don't have to reduce fishing mortality just yet, so find a way somehow to smooth the edges of this without hurting everybody else."

He says that the committee will come back to the issue later that afternoon to hear what Steve and Brian might suggest, then adjourns the meeting for lunch. Chairs scrape and shoes scuffle under the microphones' pop and hum.

Chris Adams is already nearly out the door of the Marblehead Room. Brian and Steve catch up to him, and Brian taps him on the shoulder. "You're buying us lunch," he says.

Bill Adler of the Massachusetts Lobstermen's Association has an idea about what to do with the Gulf of Maine's problematic conservation plan: just toss it back to the Technical Core Group and let the scientists put in whatever numbers they like for throwback or trap reductions.

An exasperated Eric Smith pulls the microphone right to the point of his chin. "I'm trying to produce a public hearing document, not go back to technical review for another two weeks. We'll do that forever," he says. "Someone needs to tell me, whether by vote or consensus, either the hell with the July twentieth deadline or we do the best we can right now."

A momentary hush falls over the Marblehead Room, a hush blank and still enough to carry the whir and whisper of the traffic outside on Route 1. A chair is shifted, some paper shuffled, another microphone adjusted, a throat cleared. Finally someone speaks, and we apply ourselves once more to the July 20 deadline, to doing the best we can right now.

Throughout the early afternoon Smith has been the glowing iron between the hammer of the Technical Core Group, which is demanding greater and more specific concessions from the gulf EMT, and the anvil of the unyielding representatives of that group. Patten White has impressed him with the likelihood of wholesale revolt by his constituency if he goes back to Maine with tougher throwback or trap cap measures than the group's proposal currently contains. For his part, Joe Idoine, with more and more help from Gene Martin, the lawyer, has alerted the chairman to the equal likelihood of the full council's rejection of the document out of hand next Thursday — July 20.

Finally Smith proposes his own off-the-cuff revision of the gulf plan. He runs through a list of five conservation measures, one to follow the other over the five years of the plan if NMFS's goals are not met. Drastic trap reductions go into effect in year five. "So now to write that down?" he says hopefully. "Anybody got heartburn out there?"

Pat White's got heartburn. Over the next hour Smith's proposals are whittled back until finally they approach a teetering equipoise between the resistance of the gulf EMT and the demands of the scientists. Smith rattles off a watered-down version of his original list of measures. "Technically realistic?" he asks. "Socially realistic?"

"It's closer to where we were," White concedes. "But I don't want to go back to an EMT meeting to kick this around."

Smith suggests, "Maybe call some people, see how they feel." Then he laughs, adding, "Maybe you shouldn't call them. I don't want to raise any obstacles to Thursday. Joe, is this closer to credibility?"

Idoine grants that it's close. Martin, however, has served with the council through the groundfish portion of Amendment Five, 1994's torturous legislative attempt to reverse the downward spiral of the cod fishery. As White and Idoine have approached their peace accord, Martin's heartburn has gotten worse. "The problem is," he says, "this plan is still *much* looser than Amendment Five, which was uncertain anyway about its management objectives. This is loose, unpredictable, difficult to assess, and we're going to fight the battle every year of whether we really have good enough data to know if these measures are working. This is a very problematic way of getting this approved."

Smith pauses, looking as though the breath has been knocked from him. He bows his head, adjusts his glasses, looks at Martin. "Suggestion?"

For another thirty minutes suggestions are launched from all quarters, like gunfire in Bosnia, until finally Smith gives up. He returns wearily to his agenda and pounds his gavel. "All right, it's three forty-five. Let's revisit this Outer Cape thing and see if we've made any progress with that."

Chris Adams had stood for a fine lunch of pizza and beer at Bertucci's, a restaurant next door to the hotel. Conversation there tended to the day's last agenda item — how many lobsters to set aside for the non-trap sector, or draggers — and avoided Adams's quarrel with the Outer Cape's control date.

Now Steve Smith steps up to the standing mike. "Well, we wrote a couple things down." He proposes opening the fishery to anybody who put out traps between 1989 (the last year Adams did so) and 1994. In order to keep overall trap numbers the same, further cuts would be borne by part-time fishermen and those entering the Outer Cape from other management areas.

"Can you have this approved by Monday?" asks Eric Smith.

Steve's smile is ironic. "We love to meet on weekends. No problem — we'll fax it up."

"Anybody disagree?" Smith finds Adams in the back of the audience. "Does this solve your problem?"

Adams allows that it does.

Dick Allen's question before lunch did not conclude his comments for the day. Now that it is past four o'clock and the committee is only just getting to the agenda's thorniest item, "Discussion of Proposals for the Non-Trap Sector," Eric Smith finds that baiting Allen is one of the few pleasures left to him. "Tell me," he says, leaning toward the committee member seated next to him, "did Dick have a broken-off broomstick shoved up his shirt to keep his hand up for this long?"

"I've been training in the gym," Allen counters. "The other day I had to have another guy raise his hand for me to get recognized."

"Non-trap sector" draggermen such as Jim O'Malley, the East Coast Fisheries Federation director, who arrived in the Marblehead Room at lunchtime, and trap fishermen such as Dick Allen are the cats and dogs, the cowboys and Indians, the Yankees and Red Sox of the Northeast's fishing industry. In scouring their gear across the sea bed in pursuit of cod and haddock and other bottom-feeding fish, draggers can't help but catch some lobsters as well. Some draggers actually target lobsters, and have done so for years. Sometimes draggers catch lobster traps, and for such incidents the phrase "gear conflict" is only a dry, clinical shadow of the anger — and impotence — felt by trap fishermen. Brian remembers Harry Hunt once watching a dragger run over traps he had set off Hyannis. The enraged Hunt, who was prone to malapropisms, shouted to the dragger from his wheelhouse that first he was going to call the Coast Guard, then call his lawyer, and then "prostitute" the skipper of the dragger to the full extent of the law.

These days, however, with the collapse of Georges Bank

groundfish stocks, more draggers have taken to targeting lobsters, thus joining trap fishermen in an ever-growing directed fishery for that animal. Terms such as *quota* and *total allowable catch* (TAC) are now creeping into the vocabulary of lobster management, and the eventual division of any quota or TAC between the two sectors will probably be determined by what portion of the annual harvest each sector has claimed historically. So now draggermen who once protested that they caught very few lobsters are saying that the scientists' graphs on their catch reflect numbers that are too low; conversely, trap fishermen are suddenly conceding that, well, their brothers in the non-trap sector were right all along, and draggermen never did catch that many.

Mercifully, it is not Eric Smith's job to settle that debate today. Instead he simply hopes to define a cutoff point between draggers catching lobsters only as bycatch and those currently targeting them. Then only boats with documented landings higher than that cutoff point would be allowed to continue to drag for lobsters under the new plan. At its last meeting the committee was unable to decide on this cutoff and assigned the Technical Core Group to return with the data and graphs presented by the Woods Hole scientist this morning.

Dick Allen, finally recognized, has some advice for Smith. Gene Martin's earlier suggestion for the Gulf of Maine's plan was to define a total allowable catch either for that area or for the entire Northeast and mid-Atlantic regions, with lobster fishing to cease each year once the TAC has been reached. If that were the case, Allen advises, then keep the number of draggers in the directed fishery very small: "Otherwise boats with a history of large landings, twenty-five thousand pounds or more, are not going to have a chance to get their normal annual catch before the quota is reached. It's a question of equity."

Smith agrees, saying that when they finish, the fishery ought to look the same as it does now, with only a few draggers targeting lobsters. To that end, he proposes a cutoff of at least twenty thousand pounds in documented annual landings for membership in that fishery.

Comments swirl in reference to that and Smith's implicit endorsement of a quota for lobster landings. Jim O'Malley joins the discussion, his voice rumbling forth in a basso profundo familiar to all the members of this committee. O'Malley has just come from Washington, D.C., where he has been attending the Magnuson Act reauthorization hearings. "Quota?" he says. "It was my understanding we would get a hard-and-fast percentage of the overall catch, not a quota."

Smith says the draggers' quota would come from their historical percentage of the catch. "But if you're going to stop more draggers from moving into the lobster fishery," he adds, "nobody can figure out how to do that without a quota."

O'Malley looks over his shoulder at Allen. He stands at the microphone, smaller than Allen, senatorial with swept-back white hair and a profile like DiMaggio's. He wears an immaculate white shirt and a tie as red and narrow as a knife wound. "I'm touched by my brother Richard's concern for the mobile gear sector and their well-being," he says, "but I almost feel as if I should go into my preacher routine and warn ye, my brethren, there are satanic works afoot here. Because the fact of the matter is, when I hear words like *quota* and *TAC,* I become concerned that the mobile gear sector is being nudged into an ITQ system, and I warn you that you will find considerable resistance."

Smith laughs, enjoying O'Malley's preacher routine. He knows that for all their differences, Dick Allen and Jim O'Malley once fished together for lobsters in Narragansett Bay and remain good friends. Brian tells me that these are two very smart men. He admires them both, particularly the durability of their friendship in these bitter times, but in this question of ITQs he sides with O'Malley, no matter which sector O'Malley represents.

"It's the work of the Devil, no two ways about that," repeats O'Malley, his voice plumbing the recesses of the room. I suspect that O'Malley, who has been spending a lot of time in Washington lately, feels beset by devils. Down there the reauthorization of the Magnuson Act has presented Congress with an opportunity to tinker with the management structures imposed on American fisheries by the original law in 1976. Smart people besides Dick Allen have decided that the sort of privatization built into the individual transferable quota provides the best antidote to overfishing not only in the lobster sector but in any commercial fishery. Those people would now like to write that management tool into the Magnuson Act. Many also represent larger economic interests than those claimed by Brian Gibbons, Steve Smith, Chris Adams: large shore-based dragger/processor firms, or the big companies that own factory trawlers, or, above all, such global giants as Tyson Foods, Borden, and ConAgra, all of whom have seafood interests and all of whom covet the control over supply that an accumulation of quota shares in their own pockets would provide them.

The mid-Atlantic states' surf clam and ocean quahog fisheries have been managed by ITQs since 1990. Though NMFS is itself a leading advocate of ITQ management, in 1992 it mounted an undercover investigation of the surf clam industry in response to allegations of collusion, intimidation, price-fixing, and other activities geared toward hastening the movement of quotas from the hands of small owner-operators to those of larger interests. The results of the investigation were never made public. In 1994, O'Malley's East Coast Fisheries Federation requested copies of the investigators' notes under the Freedom of Information Act. O'Malley was not reassured to learn that the notes had been destroyed, by executive order.

Now, fresh from dueling with the devils in Washington, O'Malley fears he has been outflanked on his home turf. "This teaches me a lesson to pay too much attention to the Magnuson reauthorization," he fumes.

For his part, Smith thought it had already been settled that quotas and ITQs would be part of the discussion here. "Where were you on May second?" he asks.

"I don't want to tell you which Hill staffer I was strangling," O'Malley replies, adding that he also objects to assigning draggers strictly to either a directed or a bycatch fishery, since dragger captains target lobsters only on a seasonal basis and so need to move back and forth between fisheries.

Now Smith is relieved to see Dick Allen's hand. "Your brother Richard has his hand up in front. He might have an answer."

Allen says, "I'm wondering if we've had a report from the O'Malley-MacLeod-Avila group that was going to meet and come up with proposals acceptable to the non-trap sector."

Ed MacLeod, representing the Gloucester dragger fleet, explains why he's been quiet to this point. "I guess I've been paying penance for the four hundred meetings we've already had in the last six months. I did get in touch with Jim, and talked with people on the telephone, but trying to get people together for one more meeting is impossible. Everybody has the attitude now, what the hell are we going to meetings for? They've already got our destiny fixed. We're going out of business anyway, and we aren't going to lose a day's fishing while we've got the opportunity."

Smith sighs, checks his watch, harks back to the original question. "Can we just use twenty thousand pounds as a cutoff point now for the directed fishery, and then decide what we do with this seasonal thing? Because the disagreement on that cutoff throws this whole thing into the realm of strong opposition at the public hearings. And then deal with bycatch before we get done. It's five o'clock now. I'll go as long as anybody wants, but . . ."

Ivy day in the committee room: Brian has also done four hundred meetings, or at least a lot, in the last six months, and has lost another day's fishing besides. A mild southeasterly breeze was blowing off Nauset today, and the water was scalloped by spoonfuls of gladdening light. Brian would have liked to have hauled those lines of traps up by the Can today. As it is, with the Outer Cape's business buttoned up again — at least for the moment; at least until the other Cape lobstermen hear about that slight adjustment in the control date, from one month to six years — he decides that he might at least be able to get some bait today if he can get back to Orleans before dinner. He leaves in the middle of the afternoon.

The meeting lurches and veers and backtracks. Smith concedes that O'Malley has a point concerning fishermen who target lobster on merely a seasonal basis, says that the committee may have to sleep on that, recommends that they move on to defining a limit for the number of lobsters a draggerman can keep as bycatch. Smith suggests 160 lobsters per day, and the debate immediately turns acrid.

"Who's got the microphone here?" Jim O'Malley asks at one point as he turns to face some restive trap fishermen. "Glad to hear it. Back to the eighth grade." Another draggerman says that this is all smoke and mirrors, that between this and the tough restrictions built into the groundfish portion of Amendment Five, "you're out to cream the draggers. Just tell the truth. Thank you." He leaves the microphone and stalks out of the room, staring venomously all the while at the trap fisherman who spoke last.

Smith sits at his table with his eyes closed, as though facing into a stiff breeze. In the last few moments a pair of committee members have risen, pointed to their watches, apologized, excused themselves. Finally Smith says, "I'm going to stop public comment now and get us to make a decision. We're about to lose our quorum."

He puts a motion on the floor of one hundred lobsters per day as a bycatch limit for the non-trap sector. O'Malley says that's too low. Smith says okay, come say that at the public hearing. "Other comments on the motion?"

Someone on the committee has a hand-held calculator and has worked out the results of every dragger's catching a hundred lobsters every day. Isn't that a bit much? Smith's frustration is visible. "It's inevitable that when you set a limit, people do that multiplication. It's always wrong, because everybody catching their limit every day never happens. I don't know how to avoid that. I'm going to call a motion while we still have a quorum. All those in favor say aye. All those opposed? Thank you. We can keep talking, but we aren't going to take votes on anything."

The ayes have carried. Cheered to have salvaged at least something from the day, Smith tilts back in his chair, his arms hanging slackly at his sides. The light has precisely the same powdery quality that it had this morning, the faces of everyone in the room are squashed and foreshortened in precisely the same way, but the air is different: it has a stale and sweaty and faintly sulfurous fart-in-church sort of whiff to it. It's air that has run up and down too many windpipes, been beaded into too many syllables, been sliced into too many fearful men's equivocations, been hostage too long in too strange a place to the remorseless working of things.

Smith can smell the breeze in the parking lot on the other side of the Marblehead Room's hall door. Empty seats show in the audience and at the committee table. Like Brian, Steve Smith and Chris Adams are long gone. Jim O'Malley is packing up to go. Dick Allen is still here, with his hand up. Smith rights his chair and waits indulgently while the meeting breathes and talks its last.

A remaining committee member, who has been poring over the morning's technical reports, raises his hand. "Mr. Chairman, I know it's late in the day, but I'm concerned about what we just did." Smith fixes him with an open-mouthed stare as the member observes that in pounds per day, the median amount of lobster bycatch by the sampled draggers was eleven, and that 90 percent of the draggers caught seventy pounds or less.

"Where are you taking this?" Smith sputters. "And aren't we getting bogged down by something? Where are you going with this, so we can resolve it? Are you saying the motion we just took should be reconsidered?"

"Well, say you're talking about roughly seventy pounds a day.

If you divide that by an average of one and a half pounds per lobster, then you're talking about fifty lobsters a day, not a hundred. My recommendation from a scientific point of view is that you've doubled what the number ought to be."

"Okay, look," Smith says, leaning into his microphone. "We resolved this. We took a vote. Let's not belabor this, will you?"

"So —"

"What's your point?" Smith glares across the table, surveys the rest of the room. "Done? Any other orders of business?" In the silence he pulls at his collar, throws up his hands in a manner at once furious and conciliatory. "I'll stand corrected if people want to pursue it, but we don't have a quorum anymore. We debated it, we took an action that we can reverse at the council meeting if we want. But I don't want to get bogged down here in looking at minutiae of data and rethinking things we can't change anyway."

The agent of Smith's heartburn protests that members haven't had a chance to digest all their information on the bycatch issue.

"Okay, here's what I suggest. That's what you were supposed to do over the last month while you already had these reports. If you guys want to invest your labor to influence the decision we just took, feel free. But I'm getting to the end of my tether with going back and looking at the minutiae one more time. Let's just leave it. If you guys want to work on it, that's just going to confuse the issue, but like I said, feel free."

Smith bites off the corners of these last syllables and spits them into the microphone. His voice rattles off the empty seats. At that moment, I surmise, Brian is well along on the Southeast Expressway, the spring air whistling through the cab of his pickup as he steams toward the Sagamore Bridge. That low rise where Eddie's MG flipped over has long since faded into his rearview mirror.

"Other orders of business? Okay. Thanks for coming. See you Thursday."

3

Not as Bad as They're Crying It to Be

FOR ALL THE SWEET and airy lightness of its name, the *Honi-Do* is the musclebound pug of the Chatham dragger fleet. Only forty-five feet long, the boat is as wide as a boulevard, dense as an anvil. Eight years ago, when once the *Honi-Do* was hauled out at Outermost Harbor in Chatham, the big truck onto whose bed it was winched was later found with its tires flat, its wheel rims stapled into the pavement. Never again, said the contractor. Now the boat is hauled out onto a bigger rig at Ryder Cove in North Chatham.

The *Honi-Do*'s propeller has a four-foot-diameter nozzle around it, a great funnel that channels water straight from the stern as it explodes from the prop, and the prop's blades are flat, not flared as on most boats. The result is 25 percent more thrust from its 385-horsepower engine, enough oomph not only to move the *Honi-Do* at a serviceable nine-knot cruising speed but also to haul off boats stranded on the sands of Monomoy Island by conspiracies of fog and tide. What you gain in horsepower, says Carl Johnston, you lose in steerage, and whenever he's at the wheel of the boat as it negotiates its way between the shifting sandbars of Chatham Harbor, he feels like he is trying to steer a hippopotamus through a garden hose. But no complaints were heard from the captain of a forty-two-foot long-liner that grounded on Monomoy last year, not far from where the *May-*

flower's captain abandoned his intention of reaching the Hudson River and turned back to Cape Cod Bay. The good offices of the *Honi-Do* and its captain and owner, Mark Farnham, came much cheaper than the $20,000 it would have taken to hire a tug from Woods Hole, forty miles to the west of Chatham.

But even pugs give out occasionally, or at least their hearts do. Last week Mark was ashore, Carl at the wheel, and the *Honi-Do* just a few miles out of the harbor, off Monomoy Beach, geared for summer flounder. That was when the bolts came loose on a connecting rod and sent the rod flailing at 1600 rpms around the crankshaft. Picture an armor-piercing shell on a tether. When the rod snapped its tether and exploded through the engine block, Carl heard the report, felt the shudder run up and down the length of thirty tons of steel and fiberglass, and saw the needles on the water and oil gauges shiver and fall dead. He shut the engine down, pulled its cover, sucked in his breath, and sent out a distress call on the VHF radio. Then he called Mark on the cellular phone. That took fifteen minutes. The *Honi-Do* lay dead in the water.

It was towed back into the harbor by another Chatham dragger, the *Overdraft,* run by Chris Armstrong. The *Overdraft* happened to be the boat Carl had worked on as deckhand before jumping, or maybe being shanghaied, to the *Honi-Do* in 1993. By then he had served on the *Overdraft* for five years, sometimes under penurious circumstances. Once when the *Overdraft* blew its engine, Carl worked for six weeks without pay to help install another. "That's just a Chatham thing," said Carl, who as a crewman parks his pickup in the upper lot at Chatham Harbor, while captains and owners park in the lower lot closer to the dock. "It's just the expectation there that a crewman will work for free repairing a boat. I guess it has to do with historical supply and demand."

Also, since Armstrong took little time off from work, Carl had few opportunities to run the boat himself and got no extra pay when he did so. Then, two years ago, a wave came over the bow of the *Overdraft* as it was nosing its way out of the harbor, tore off the roof of the wheelhouse, blew out all the windshields, and laid the boat up in the yard again for several months. Armstrong arranged for Carl to work with the hired boat carpenter on a piecemeal basis, but, as Carl explained, "I couldn't get enough work that way. Just the same, I was still expected to show up at the boat every day, which meant that I couldn't go clamming either, and make a little money that way."

Mark Farnham happened to be looking for some good help on the *Honi-Do*. By that time in Chatham, the market had changed in regard to skilled and experienced crewmen: supply was lower, demand higher, though many of the old feudalisms remained in place. Mark offered to take Carl fishing with him until the *Overdraft* was ready to work again. Then he told Armstrong that Carl was quitting. "No one from the *Overdraft* would speak to me for a year, and I didn't have anything to do with it," Carl said, his eyes widening again with the wonder of it, as though that wave that poured through the wheelhouse of the *Overdraft* had literally carried him over the rail and left him sputtering in the cockpit of the *Honi-Do*.

But in truth, Carl liked the *Honi-Do* better. Within a few days Mark was letting him run the boat; he was the first crewman besides Dave Farnham, Mark's father, to be granted that privilege. And despite Mark's reputation for being close with a dollar, he paid Carl an extra share whenever he skippered, and gave him reasonable opportunity to do so by taking four to six weeks off each year for deer hunting in Maine, a vacation in Florida, some commercial shellfishing with the rest of his family. He paid Carl for repair work as well. "It wasn't much," Carl said. "Maybe it worked out to five dollars an hour, not as good as I could do clamming. But at least it was something."

Things have long since been smoothed over with Chris Armstrong, and Carl had hardly a thought for the irony of Armstrong's help when it took the good offices of the *Overdraft* to bring Mark's boat home. But he hated to make that phone call to

Mark, couldn't help feeling in some way responsible for the fact that those bolts had worked loose when he was at the wheel. Still, the engine had 13,000 hours on it, the equivalent of about 600,000 highway miles on a car, and Carl was grateful that Mark took the news of the breakdown well. "That's part of the ball-game," he told Carl. "It's the cost of doing business. Don't worry about it."

Carl estimates that this piece of business will run Mark more than $30,000. The *Overdraft* took the *Honi-Do* to Stage Harbor, just west of Chatham Harbor, behind Morris and Stage Islands, which drop like beads of sweat from the point of the Cape's elbow. At a dock there Mark and Carl cut a hole in the cockpit flooring and had a crane lift the ruined engine out of the boat. They pumped the bilges, transferred what usable parts they could to the new engine block, and dropped the new assembly — a six-cylinder diesel Caterpillar 3306T, just like the previous plant — into its housing, where a hired mechanic made the necessary connections. This week's task has been suturing up the wound: repairing the flooring, and today laying a new sheath of fiberglass over the seams where the flooring was cut.

Much of the flooring of the *Honi-Do* is now buried beneath clumps and layers of heavy tools, electric cable, plastic buckets, polypropylene rope, and sawdust. Carl shows me the guilty connecting rod, a gleaming inch-thick shaft of steel snipped clean at one end like a twig browsed by deer. Carl's beard is a warrior's beard, black and voluminous, a tangled palisade dropping halfway down his chest, and the man might look forbidding if not for the light in his brown eyes, something that in bearded men often suggests itself as a twinkle: partly an invitation to relax, partly the gleam of graces that the mouth can't convey so well, and partly, in Carl's case, an amused and ironic perplexity at the orneriness of men and circumstance, and at that combination of the two we call fate.

Harry Hunt became so angry at that orneriness that finally, Ahab-like, he became its mirror image. Carl witnessed some of

Hunt's shadowboxing with that image on the day that Harry Jr. wiggled under his truck. While Brian Gibbons tries to gather all that he can still muster of his boyhood faith and optimism in standing fast against what Herman Melville describes as "all truth with malice in it," Carl Johnston raises his eyebrows, scratches his head, shakes his great beard, tries to keep moving in order to stay ahead of it all.

Brian calls Carl a survivor and admires him for that. Carl hands me the sheared rod and smiles at my transparent awe of the mechanical forces that could accomplish that. He gestures to the port rail, against which lie some jagged pieces of the old engine block, each an inch thick, each bent and torn like a broken bit of seashell. Then he hands me one of the steel shards with the air of a monk instructing a novice in the long, bitter roots of truth.

Mark Farnham stands where the *Honi-Do*'s bulkhead once stood, before it was torn out to clear a way for the engine transplant, and rigs an electric drill with a screwdriver bit. Mark is in his mid-forties, some ten years older than Carl. He wears a Jacksonville Jaguars sweatshirt underneath red suspenders, and fills the shirt the way it might be filled by one of the slabbed iron otterboards hanging by the gallows frame of the net reel. If some men look like their dogs, Mark looks like his boat: broad and powerful and careful around the corners.

He surveys the area at his feet, where the flooring was cut with a power saw in a rounded rectangle big enough to uncover the engine housing and where that excised rectangle has now been puzzled back into the floor again. He hands the drill to Carl so he can start screwing the piece down and takes up another drill himself. He tells Carl that he's going to have to send him out to buy some fiberglass matting to plaster over this seam. He smiles wickedly and adds, "There's nothing you can't fix by plastering over it."

Both Mark and Carl worked in construction before they came to commercial fishing. Mark's specialty was plasterwork, while

Carl framed houses, and the relative merits of the two disciplines are a running joke between them. Mark smiles again while Carl kneels to begin driving screws, telling Carl how his nine-year-old son hooked a thirty-six-inch striped bass the other day while fishing off a boat ramp and was nearly pulled into the water before he landed it.

Carl has told me that the Farnhams are a true-blue fishing family. He says that Mark's father, Dave, in his seventies now, can still work at the backbreaking job of harvesting steamer clams and easily collect two bushels between tides. Mark's wife, Sue, hangs nets for several boats in Chatham's gill-netter fleet. This involves skillfully knotting gill nets onto their float and lead lines in such a way that the nets hang straight and don't twist in the water. She rises at three each morning to work so that she'll have time on the beach later with their kids.

I ask Mark if his two sons are going to be fishermen. "No, they're going to be shysters," he says proudly, then tells Carl how his kids recover golf balls from a swampy area at a local golf course and sell them back to the players. "The club owner insisted they sell the balls to him at a lower price," Mark says, "so they sell him just the grungy ones and keep the good balls for the premium market out on the course."

Fishing? Mark says there's no future in it now. "I've got a quarter million dollars invested here, and I'm not getting the return I should." And yet, despite it all — the crash in the groundfish stocks, the poor spring for lobstermen, this recent misadventure for the *Honi-Do* — he doesn't think that things are as bad on Cape Cod as they're widely claimed to be. "I've got one thing to say about this so-called fisheries crisis," he says. "You see all the new pickups in the parking lot up there? You see all the new fishing boats out in the harbor here? It's not as bad as they're crying it to be."

I know that Carl has his eye out for where the immersion suits are stowed. He kneels on the flooring, bent over his drill. One screw pops through the plywood and spins free on the other side, finding no purchase, the drill whining. He yanks it back and tries again, wearing out his jeans at the knees, smiling once more at the orneriness of things.

In the 1990s, several sectors of the American economy have been shocked by the cold exigencies of what has been termed the global marketplace — a world more fluid than before, more oceanic, where capital, goods, and information flow with increasing ease across national borders; a world where large corporations have assumed the aspects of sovereign states; a world, therefore, where domestic consumers, wage-earners, and small businessmen find themselves increasingly at the mercy of vast and distant currents.

But American fishermen have been swimming, and often floundering, in these bracing waters since the end of World War II. Cries for help — particularly from the capital-intensive offshore groundfish fleet — have for even longer been part of the routine background noise in the lives of congressmen, the prelude to occasional red welts around the necks of intervening Capitol Hill staffers. During much of this century the federal government has actually been a great friend to the American fisherman — albeit the sort of friend, sometimes, who obviates the need for enemies.

The first overtures were made during the Depression, after the national banking system collapsed beneath its burden of risky loans. On hardscrabble Cape Cod, fishing had been largely a poor man's occupation anyway since the end of the Civil War. But in Boston and Gloucester and other deepwater ports, it was a business in which muscle-and-barbed-wire youngsters such as Brian once was, or savvy crewmen such as Carl still is, could expect to make some money — at least until Black Friday and the bottoming-out of fish prices that followed. In Boston the big trawlers of the Atlantic Coast Fisheries Company stayed tied to the dock throughout the summers of 1931 and 1932. If a crewman could find work, his annual earnings averaged $600 in

1933, paltry even for those days; only domestic workers and Dust Bowl farmers did worse. The following year the Gloucester dragger *Gertrude L. Thebaud* sailed up the Potomac to Washington, there to present boat records to Franklin Delano Roosevelt and Congress to prove that, even excluding wages and fixed costs, income from fishing could not possibly cover operating expenses.

Certain new technologies began to help in the years leading up to World War II. Clarence Birdseye, an eccentric New Yorker who went to Labrador to work as a fur trapper after dropping out of Amherst College, discovered there (by accident) that if he put cabbage into a basin filled with salted water and exposed it to Labrador's arctic wind, he could freeze the cabbage without destroying its tissue and flavor. In 1925, after refining this "flash-freeze" process and adapting it to fish, Birdseye moved to Gloucester and founded the General Seafoods Company. He continued to experiment with meats, fruits, and vegetables as well, and his fortune was made when the daughter of the founder of the Postum Company, a food-processing business, tied her yacht up in Gloucester and was served a goose frozen at Birdseye's plant. This delectable bird led to her father's buyout of Birdseye for $22 million and to the merger of Postum and General Seafoods to become General Foods. New inland markets for seafood opened up with the advent of frozen fish, and at the same time new filleting machines made such tasty but hard-to-process fish as yellowtail flounder a valuable commodity. When New Bedford opened a Birdseye-style freezer in 1937, and when its processors introduced flounder filleting in 1938, that city's fishing industry expanded for the first time since the decline of commercial whaling.

World War II brought shortages in engine parts and crewmen and danger from German U-Boats, which sank three large New England draggers in 1942. But it also brought boom times for those Yankee fishermen who could get out on the water and back again. The military and the lend-lease program snapped up canned fish as soon as it came on the shelf, and consumers pinched by meat rations turned frequently to fresh and frozen fish. Demand and prices skyrocketed. Seafood was consumed so widely and so zealously during the war years that when the war ended, there seemed no reason why the jackpot should not keep pouring forth. The fishing industry "is destined to enjoy a steadily increasing demand for its products, as a result of the public becoming better acquainted with the fine qualities of fish and shellfish," cheered the *Atlantic Fisherman,* an industry journal. The demand for flash-frozen fish would not only "take up present productive capacity . . . but will provide an outlet for still larger catches in the future. In fact, it is possible that with the great expansion of the frozen food business, fish consumption could be increased to two or three times its present level."

But demand immediately fell after the war, as the military cut its purchases of fish and as consumers did likewise, returning to meat and poultry once they became available again. Most distressing to New England fishermen, however, was the tide of flash-frozen groundfish fillets that washed in from abroad, chiefly from Canada and Iceland, at prices that consistently undersold their own fillets — at prices, in fact, that would compel American fishermen to tie up at the docks again. Why were these fillets so cheap? Because of lower wage structures in those countries, explained industry spokesmen, along with lower production costs, non-interest-bearing government boat loans, and the government-subsidized construction of processing plants and freezers. Birdseye's frozen fish, it turned out, could not only cross mountains and plains; they could also cross oceans. The trade in frozen food had expanded in a way not anticipated by the *Atlantic Fisherman.* And the global marketplace had opened for business.

Just before the war, the federal government had set a protective duty of 1.875 cents per pound on imported groundfish. This was just a small cost of doing business to foreign processors, especially now that fish were coming in filleted, rather than "in the round," and therefore at a much higher value per pound. Boat owners and seafood processors tried to stay in the game by keeping wages low. In 1946, Boston fishermen went on strike against General Seafoods, which still existed as a division of General Foods and at the time was the largest boat owner in New England, running eighteen offshore trawlers out of Boston Harbor, all of them many times the size of the *Honi-Do*.

The strike dragged on for five months, and finally General Seafoods simply left town. The company sold off some of its trawlers and moved the rest to Maine and Nova Scotia. It also took its Birdseye processing operations across the border to Canada, as did other wholesalers and processors. Boston withered as a fishing port, and never recovered. Through the late 1940s and early 1950s, while strong labor unions in other industries helped the incomes of many American workers to rise, elevating them into an expanding and prosperous middle class, those of fishermen declined.

During these years fishermen sought relief in what historically had been the most common and effective form of government aid to domestic industry: tariffs, or at least firm government quotas on the volume of imported groundfish. But they found no relief, because Washington had . . . well, bigger fish to fry. In 1948 the United States signed the General Agreement on Tariffs and Trade, opening the world through the doctrine of free trade to the might of American manufacturing and the entrepreneurial craft of American corporations. Eventually the Eisenhower administration reduced tariffs further even than Truman had, boosting free trade not only as opportunity writ large for business but as a bulwark throughout the West against the spread of communism.

In truth, however, Washington was not dogmatic about free trade. It all depended on who needed relief, as demonstrated by the curious career of the imported fish stick. In 1954, General Foods was stunned when the federal Bureau of Customs ruled that the new frozen fish sticks pouring in from abroad fell into the legislative category of "preserved fish," as opposed to "processed fish," and were therefore not subject to any sort of tariff. Massachusetts Senator Leverett Saltonstall promptly proposed an amendment to a tariff bill that specifically attached a fee of 20 percent on uncooked fish sticks "if breaded, coated with batter, or similarly prepared," and 30 percent on cooked fish sticks. The amendment passed, the bill was signed, and cheap imported fish sticks swiftly disappeared from American freezers.

This was an impressive turn of events for New England fishermen. Margaret E. Dewar, the author of *An Industry in Trouble: The Federal Government and the New England Fisheries,* zeroes in on what was wrong with that picture: "The fresh groundfish industry had pressured publicly for years for higher tariffs without success despite its financial problems, yet the fish stick interests obtained very high tariffs on fish stick imports without ever appearing in a public hearing, testifying before a congressional committee, or demonstrating that their industry suffered harm."

In lieu of the tariffs so swiftly provided to the fish stick interests, the federal government throughout the Eisenhower, Kennedy, and Johnson administrations offered the man on the pier programs (in research, marketing, education, vessel safety, and so on) and insured loans (for mortgages, operating expenses, refitting, and so on) designed to make him individually more competitive with his counterparts abroad. This man was generally conceived of in terms of the offshore groundfish industry, and of the government's offerings, the most successful was the series of loans authorized under the Fish and Wildlife Act of 1956, though merely because its goals were subverted.

Only about half of the Fish and Wildlife loans went to offshore boats, a good number of them New Bedford vessels similar to the *Cape Star* and the *Bell* — boats that were retooling for lucrative new markets in yellowtail flounder and sea scallops. The other loans were all for less than $10,000 and were disbursed up and down the New England coast to boats similar to the *Cap'n Toby* and the *Honi-Do* — small craft with a single owner-operator, able to work the inshore waters for their eclectic yield of cod, haddock, hake, fluke, winter flounder, herring, whiting, mackerel, bay scallops, quahogs, shrimp, and above all, lobsters. The lobster sector enjoyed a modest boom in the two decades after the war, with trap numbers more than doubling, and inshore fishermen in general found that their small-scale operations were more cost-competitive with the Canadian industry.

In Washington this was all well and good, but when federal legislators thought about fishing, it was still in terms of the big offshore draggers, who were, after all, in the greatest distress from imported fish. By the mid-1960s, moreover, it was precisely those vessels that were suddenly coming face to face with, and being muscled aside by, the gargantuan factory trawlers of the Soviet Union and East Germany and everywhere the hell else. If one were to witness the assembled multitude of the ships of these and other nations on Georges Bank, reported a Gloucester fisherman, "it would look just the same as a large city with thousands and thousands of lights as far as one can see over the horizon." Those lights belonging to Eastern Bloc ships, moreover, were strongly suspected of burning the midnight oil of Cold War spy operations.

In 1964 the Fishing Fleet Improvement Act provided direct subsidies (up to a third of the total cost of the vessel) to cover the difference between the costs of boat construction in the United States and abroad, so long as the new vessel worked in a sector harmed by imports (such as Georges Bank groundfish) and was of advanced design. With that, the goals of federal fisheries legislation underwent a sea change: no longer so directly concerned with helping individual American fishermen realize a profit on their work, lawmakers and bureaucrats more often spoke in terms of national honor and national competitiveness. "We were formerly in second place among the fish producers in the world," said the Bureau of Commercial Fisheries director, Donald McKernan, in 1964. "We are now in fifth place. Our share of the world catch has dropped from thirteen to seven percent in the years since 1956. By way of comparison, the Soviet Republic has more than doubled its fish catch."

By 1969 this winning-is-the-only-thing philosophy was even

more explicit. Edward Garmatz, the chairman of the House Committee on Merchant Marine and Fisheries, opened hearings on the extension of the Fishing Fleet Improvement Act with this observation: "Since the 1940s the United States has slipped from first to sixth place among the leading fishing nations of the world. We are now outranked by Peru, Japan, Red China, Russia, and Norway, respectively. In view of this deteriorating situation, it is imperative that the Fishing Fleet Improvement Act be extended."

It was extended, just a year after a company named American Stern Trawlers, a New York–based subsidiary of American Export Industries, had weighed in with its own American gargantua, courtesy of the Fishing Fleet Improvement Act: a subsidized seagoing equivalent of the Saturn moon rocket, a 296-foot, all-bells-and-whistles, state-of-the-art factory trawler christened the *Seafreeze Atlantic.*

It was a bad idea from the beginning.

The excised section of flooring has been nailed and screwed back into place, and now both Carl and Mark are on their hands and knees, working trowelfuls of Guppie, a puttylike filler, into the open seam circling the *Honi-Do*'s new engine. Carl and I are just back from the industrial park, a nondescript loop in south Chatham colonized by the fishing industry: boat repair shops, bait shanties, fish-packing operations. Carl has a friend there who lets Carl use his repair shop when needed because Carl once helped him work on a skiff. The owner has told Carl where the key is hidden so that Carl can go in any time, take what he needs, pay when he can.

An old ChrisCraft cabin cruiser, stripped down to bare wood, was up on blocks inside the shop. The air was sweet with sawdust, pungent with varnish and paint thinner. We took a pint of Guppie, a gallon of polystyrene resin, and nine feet of fiberglass matting. Carl made a note of this on a pad on the owner's desk, and we went back to Stage Harbor.

Mark is less casual in his own transactions and gets to thinking when others aren't. While we were gone, he found that he didn't have any acetone for thinning the resin, but no matter — Mark Simonitsch, the owner of Chatham Fish Weirs, a fish-trapping operation at Stage Harbor, told him to go into his dockside warehouse and take as much as he needed. Mark relates this to Carl with the sort of wonder that implicitly asks what the hell Simonitsch is up to. "My buddy Mark," this other Mark says twice, as if the words have a funny taste in his mouth.

I remember a conversation Brian Gibbons had with Mark's buddy Mark after the Lobster Oversight Committee meeting in Peabody. Brian had come down to Simonitsch that day looking for mackerel to use as bait. Employing a method of fishing as old as history, Simonitsch harvests mackerel, scup, butterfish, squid, and herring from a system of wooden weirs he has set up a short distance down the coast. The previous night, however, forty-knot winds had scoured the Cape, and Simonitsch was visibly distressed by all the expensive hickory posts that the seas had sheared from his weirs, which now lay in splintered piles on the dock. For his part, Brian was still wearing the disappointment of the inconclusive outcome of the Oversight Committee's meeting.

"No action, huh?" Simonitsch said, his horn-rimmed glasses and his curbstone chin jutting out from the hood of his orange raingear. "That's okay. You know what I think? I think this is a democracy. When you're going to put people out of business, you move slowly."

"Okay. In the meantime the resource goes to hell," Brian countered. "In the meantime, Hillary Clinton and Tyson Foods stake out everything that's left. It's like Neville Chamberlain coming back from Munich waving his piece of paper while Hitler mobilizes his tanks, except this time it's Dick Allen mobilizing his ITQs."

Simonitsch wasn't worried about that. "That's a worst-case scenario. Tyson's lost $250 million in the Pacific Northwest. They'll be getting out pretty soon. My father told me that best-case and worst-case scenarios both never happen. Hillary? Hell, Reagan, Bush, Clinton — there's no difference. Reagan tripled the deficit. All I know is that squid's down eighty percent, butterfish are down, there's no scup at all — I'm signing off. I'm going to New Zealand. The hell with all this. Of course, they got their own problems down there."

Carl himself has no plans to sign off, nor any dreams of New Zealand. He likes it fine here on the Cape, likes what he's doing. He grew up in Attleboro, Massachusetts, a town within shouting distance of Rhode Island's northeast corner, and like Brian came with his parents to the Cape for simple living: summer vacations, weekends, sometimes just for a fugitive evening. His father, Tom Johnston, was a toolmaker in a Texas Instrument factory, but his heart lay in the ponds of Brewster and Sandwich, the tidal flats of the southern edge of Cape Cod Bay. Occasionally, if the tides were right, Tom would load Carl, his only child, into the station wagon after work and race down the highway to the flats of the bay to fish for striped bass all evening, the two of them walking the tide out and then walking it back in. Carl would sleep in the back of the wagon during the two-hour ride home.

In the summers they rented a camp on a pond in Brewster for four to six weeks at a time. Carl stayed there with his mother, Eleanor, and an aunt and a cousin. Tom would stay the first week and then return for weekends afterward, fishing with Carl for trout in the pond and bass and flounder in the bay. On the bay they rowed out with an old-timer to spots opposite the Dennis bluffs, the Rock Harbor bell, the bell buoy, and the submerged shoals of Billingsgate. Then once a week they rowed to the end of the pond and walked a few miles down the railroad tracks to a store, where they laid in groceries for the week.

Carl graduated from high school in 1976, went to work for a local construction company, and got back to the Cape whenever he could, particularly in the winters, when he was laid off. There he cut scallops for spending money and hunted deer, rabbits, ducks, and geese in his spare time. By 1978 he could see that the housing market in Attleboro was falling off, and that was enough for him to make the sort of move that Tom Johnston had always dreamed of making but had instead deferred for a lifetime. He bought a sixteen-foot skiff, rented an apartment in Orleans, and supported himself with anything he could do with that skiff: bullraking for quahogs, digging for soft-shell clams, dredging for bay scallops, jigging for cod. When jigging — fishing with hooked lures — he borrowed a friend's twenty-seven-foot Nauset boat because that took him farther out, all the way to the well-frequented places along the western escarpment of the Great Southeast Channel known to fishermen as the Ridges, the Lemons, and the Figs.

He also met and began dating Bindy Thompson, a certified special education teacher who had worked for a year as an elementary school tutor and an aide to the multihandicapped in a center in Caribou, Maine. Bindy grew up in Washburn, a hamlet fourteen miles southwest of Caribou, but she quit her tutoring job to join some college friends who had moved down to the Cape. In Orleans she was working variously as a checkout girl at Stop 'n' Shop, a chambermaid at a hotel, and a waitress at a Chinese restaurant, carrying out a shore-based version of what Carl was doing: scaring up a paycheck here and there, scrounging to pay the bills.

It was during his courtship of Bindy that Carl began finding work on offshore boats, usually sea scallopers such as the seventy-eight-foot *Gertariva* out of Provincetown, or on one occasion the *Cape Star* out of New Bedford, which Brian had served on. The winter of 1981–82, however, was hard on those boats. Too often they had to steam into Gloucester or ports in Maine ahead of bad weather, and the crew would then spend three or four hours chipping ice off the rigging and the otterboards. Midway through that winter Carl jumped to the *Sea Dog,* a fifty-five-foot long-liner that went out to work the Franklin Swell and the southeast edge of Georges Bank. Carl knew that the *Sea Dog*'s skipper had a reputation as a crazy man, but the boat made good money, and he decided to risk it.

The *Sea Dog* was built to an experimental West Coast design. The vessel carried a flume tank mounted on top of its pilothouse. Water ran from this in a system of baffles over enclosed work areas to provide additional stability. But it also gave the boat a queer, double-roll motion in deep swells that made Carl sick at the start of each trip. The *Sea Dog* also had an intricate mechanized system for baiting hooks automatically on board (rather than by hand on shore) and paying out the baited long lines by rollers over the transom. In this way the vessel could set as many as thirteen thousand hooks in two hours, though hauling back and cleaning that many hooks could take sixteen to eighteen hours.

Carl worked the rail on the *Sea Dog* or else cut fish as they came aboard. Others tied knots or straightened hooks. Carl knew his way around the boat, was paid a full share, and says he was treated well by the *Sea Dog*'s mad skipper. But many of the other hands, often refugee Vietnamese, were pretty green, and while the boat was in port Carl saw some of them chased off the deck with a baseball bat if they happened to complain about their half-shares. On one occasion that winter the captain sent a Vietnamese crewman up to the foredeck in rough weather to chip ice off the forward stay, knowing full well what was likely to happen. With one blow, enough ice shivered off the stay to knock the man off his feet and leave him lying stunned on the foredeck. He was dragged back into the cabin by a line tied around his waist. "The guy thought that was funny," Carl said.

That winter the *Sea Dog* averaged ten to sixteen thousand pounds per day of cod, cusk, and hake. Carl put a lot of money away, but the antics of the skipper wore on him, and the boat's automated bait-and-set system kept breaking down. When a friend bought a little thirty-foot combination dragger and longliner the following spring, he was glad to be hired to run it for a while, and he was not surprised a few years later to hear that the *Sea Dog* had finally gone broke.

He and Bindy married at the end of October 1982, and Carl, who enjoys teasing Bindy, rarely misses an opportunity to remind her that their wedding and honeymoon in Jamaica caused him to miss the first week of the bay scallop season that year. That made the honeymoon more expensive, since that year (and the next) brought an extraordinary boom in bay scallops in Pleasant Bay and Orleans's Town Cove. After the honeymoon, Bindy knew she wasn't in Jamaica anymore when she found herself working long hours deep into the winter at Carl's side in his skiff on the bay and then going nights to work in a residential home for the retarded. Both she and Carl had scallop licenses, each with a five-bushel daily limit, and they had no trouble harvesting their combined limit of ten bushels each day. Carl remembers how fastidiously Bindy cleaned their scallops for the buyers, dressing them so well that the buyers took their scallops all the way into January that first year. He also remembers how hard and cheerfully Bindy worked beside him and what good company she was during the bitter days out on the bay.

After leaving the *Sea Dog,* Carl found a site on the *Lonely Hunter,* a fifty-three-foot dragger out of Stage Harbor. In the spring he quit to devote six months to building a house for himself and his wife on a lot near Baker Pond in Orleans. Then he went to work on the *Miss Molly,* a forty-two-foot gill-netter. That was his site for the next four years, until in 1988 he moved to the *Overdraft,* and from there finally to the *Honi-Do.*

On the way over to the industrial park I asked Carl if he wished that he had his own boat to run, like Mark did. He said that he was satisfied with where he was, with what he had, and that he was glad he didn't have a lot of money tied up in as dicey a proposition as even such a modest boat as the *Honi-Do* had become. "I wouldn't buy a dragger now, with all the rules they have on groundfish harvesting and all the rules still to come," he said. "What am I going to put up to finance it? My home? No frigging way."

Nevertheless, he was disappointed that nearly all the state and federal programs for distressed fishermen being bandied about, such as a recently proposed pilot program involving a $2 million federal buyout of fishing vessels, concerned and benefited only boat owners and little of the infrastructure beneath them: mates and deckhands like Carl, the pier workers at Chatham and Stage Harbors, the various buyers and packers and shanty workers at the industrial park, and net hangers such as Sue Farnham. "The owner says, 'I've got a $28,000 mortgage on this boat.' And the crewman says, 'Well, I've got a $35,000 mortgage on my house, and car payments, and insurance, et cetera.'" Those debts are no less pressing, Carl said, and are tied to things even more fundamental than a business.

But only the owners were being heard now. "I was at a meeting about fishing quotas a while ago, and a guy raised his hand to make a suggestion. 'Are you a boat owner?' they asked him. 'No, I'm a crewman,' he said. 'Well, let's leave this discussion to the boat owners.' The crewman couldn't say a word."

Carl has played it close to the vest, minimizing his financial risk, sharing modestly in the profits of those such as Mark, who are out on a longer limb than he is but who are not inclined as a group to share the life raft with him if they all go under. And he has hedged his bets with a small venture into what is now being trumpeted as the future of fishing on Cape Cod: aquaculture.

Somewhere out in Pleasant Bay is half an acre of muddy bottom that belongs, at least provisionally, to Carl. The property is stable, sheltered from storms and ice, and washed by waters rich in the nutrients required by the 172,000 seed quahogs Carl has planted there so far.

Carl was awarded that half-acre last year as a grant from the town of Orleans, which divided thirty acres of such bottom among a waiting list of people eager to go homesteading: to plant their lots with seed quahogs, make them productive, and later enjoy an option of expanding to two acres. Right now Carl looks forward to the arrival of another 56,000 seeds he has ordered from a supplier on Long Island, New York. If all goes well for the three years that it will take those seeds to mature to market size, he expects that his grant will provide him with as much as 20 percent of his annual income in good years, and in bad years with at least some unemployment insurance.

Carl also has his eye on a sliver of the money soon to be distributed by the National Marine Fisheries Service in the form of fishing industry grants, or FIG grants. NMFS has $4.5 million to fund new entrepreneurial ideas in the industry, and Carl shares the pleased surprise of others on the Cape that this time a federal agency is actually soliciting fishermen — even crewmen — for these ideas rather than imposing solutions from the top. NMFS is soliciting scientists as well, but Carl isn't worried about that. He has arranged to team up with a University of Rhode Island marine biologist on a project involving the use of submerged fine-mesh tents to collect and protect drifting clam larvae in either Cape Cod Bay or Pleasant Bay. The biologist would supervise some graduate students in researching the factors governing the success of such a project, while Carl and two "displaced groundfishermen" — Chris Adams, who is already lobstering again, and Carl's friend Dan Howes — would maintain the tents. Carl calculates that he can set the project up and hire the fishermen and students on a part-time basis, along with a bookkeeper, for a little more than $48,000. He also says in his proposal that he could eventually derive 80 percent of his income from the project and so be able "to discontinue fishing for regulated species," which is to say groundfish.

Not everyone on Cape Cod is applauding this new kind of farming. Some raise the same concerns about quahogs that have been raised elsewhere about farmed salmon, predicting disease and the genetic degradation of wild stocks. Others, including Brian Gibbons, see this as the next step down a slippery slope toward a social transformation of American fisheries. Brian points out that Massachusetts law calls for common access to the commonwealth's fish and wildfowl. In 1981, however, the state's Division of Marine Fisheries implemented a limited-entry system in its lobster fishery, setting a ceiling on the number of lobster licenses it would issue. To Brian's mind, this has accomplished nothing for lobster conservation but has allowed profiteering on the sale of lobster licenses and also made it hard for the youngsters of Orleans and other towns to enter the fishery. His own right to harvest lobster has been transformed into something approaching private property, and he sees the same end being served now in towns' readiness to assign tracts of bay bottom — formerly a public resource — to at least temporary private ownership. "But in any crisis in public resources," he says, "it's been the history of government to give money — or the resource itself — to private individuals."

Regarding quahogs in particular, Brian argues that aquaculture's current glamour has seduced many towns on the Cape into neglecting the health of their wild fisheries. Orleans at least has taken the trouble to seed its wild shellfish beds over the years, and in recent years the Nauset Fishermen's Association has experimented with various methods of protecting seed quahogs. The association also funds its own propagation program, buying and planting seed (as they are commonly called) and then supplying volunteers to build and maintain "bottom boxes," screened trays sunk into the mud containing beds of seed quahogs. Cheap and simple, the screening protects the seed quahogs until they are large enough to be safe from crabs, drills, and sea stars. Last month the association planted a hundred thousand seed in the public bottomlands of Town Cove. At Brian's urging, it has also applied for a FIG grant, requesting $87,000 to plant a million more in the wild fishery over the next two years and to hire someone to monitor the bottom boxes regularly.

How much is a million seed? Carl told me that Mark Simonitsch has some three million seed in his warehouse at Stage Harbor right now. Simonitsch has a deepwater grant in Nantucket Sound obtained from the town of Chatham, and he has also leased part of his warehouse as a hatchery to the leading supplier of seed quahogs to fishermen here, the Aquaculture Research Corporation, of Dennis. The seed rest in long refrigerated vats of seawater, and Carl showed them to me just before we rejoined Mark on the *Honi-Do,* lifting the cover of one vat like the lid of a treasure chest. The seed were hardly bigger than bits of gravel, each an exquisite Matchbox version of an adult quahog, and in the semidarkness they lay heaped to the brim of the tank in unimaginable, unquantifiable profusion.

Small-arms fire breaks out on another front when a lanky, bearded man appears on the dock next to the *Honi-Do.* I'm cutting the fiberglass matting into narrow strips; Carl and Mark are laying the strips in a triple ply over the cut in the flooring and painting them down with resin. The man on the dock wears a Seattle Mariners baseball cap and carries a sheaf of papers under his arm. He stands over the starboard rail and asks in what sounds like a Cockney accent if the owner is on board. Mark looks up and scowls.

The man in the Mariners caps says he's a New Zealander and was a fisherman until an ITQ management system was instituted there in 1986. Now he's one of many small owner-operators squeezed out by the three corporations that currently control 50 percent of New Zealand's catch.

"You want to just tell me what you want?" Mark says.

"Well, I'm working for Greenpeace now, and I've got a petition here for your Senator John Kerry." He says the petition outlines Greenpeace's objections to the pro-ITQ language currently contained in the Senate version of the Magnuson reauthorization bill.

"I gotta wonder what a New Zealander cares about the Magnuson Act."

"Well, sooner or later it affects us all, doesn't it? If I could just have a few minutes —"

"No, I don't have a few minutes today. Sorry."

The New Zealander lifts his hand as though blessing the boat and moves on. Mark goes back to his glass work, murmuring to Carl that even if he did have a few minutes, he sure as hell wouldn't be donating them to Greenpeace.

The distrust and animosity that lie between urban-based environmental organizations such as Greenpeace and those who make their living directly from the lands, the forests, and the seas run long and deep. In New England it was an aggressive, Greenpeace-minded legal think tank, the Conservation Law Foundation, that filed suit against the federal government in 1991 for mismanagement of American fisheries under the Magnuson Act. A settlement between the CLF and the Department of Commerce required the New England Fisheries Management Council to put an end to the overfishing of the region's groundfish stocks and then develop a credible plan for rebuilding them.

That led to the constraints on groundfishing contained in Amendment Five of the Magnuson Act, and now the looming shadow of Amendment Seven: additional and more draconian restrictions, which will further depress the return Mark has been getting on the *Honi-Do,* and accelerate a winnowing-out process in the New England fishing fleet that will be livid in ruin and human costs. Mark's anger was not nearly so piqued by having his engine blow apart last week as it was by the appearance of this (to his mind) sanctimonious foreign activist with his tree-hugger petition.

Greenpeace, however, has come to pay court to Mark Farnham and skippers like him, hoping to build an alliance with at least a portion of the commercial fishing industry. The nub of it has to do with precisely that giveaway of public resources feared by Brian. Greenpeace can see it coming: if Amendment Seven is accompanied by an ITQ-style management system in New England waters, then inevitably cash-strapped fishermen, saddled with boats they can hardly give away in today's market, will sell their fishing quotas in order to get out of the industry. Who will buy those quotas? Tyson Foods, a company with close ties to the Clinton White House, is interested, as are several other corporate food producers. These companies currently support a too-numerous fleet of factory trawlers, northern Pacific descendants of the *Seafreeze Atlantic,* in the Bering Sea. There were eleven such vessels harvesting Bering Sea pollack in 1986; by 1992 there were more than seventy. Tyson is desperate for another fishery — such as Georges Bank — in which its great trawlers can work.

Between a deep-pocketed devil like Tyson and the deep blue sea, Greenpeace prefers a more traditional version of the deep blue sea: boats such as the *Honi-Do,* which fish cleaner, with less bycatch, than the factory trawlers; men like Mark Farnham, whose small, low-cost operations would be easier to rein in as fish stocks rebuild, and whose roots in the community would guarantee a long-term interest in the viability of the industry and the health of the stocks. Jim O'Malley is delighted by this stance and appreciates all the help he can get in arguing against his friend Dick Allen and against Tyson Foods' lobbyists in Washington. But O'Malley isn't sure that the alliance is going to work. He told Brian that Greenpeace and commercial fishermen are like two dogs circling around and sniffing each other's asses with the hackles up on both their backs.

The New Zealander has strolled down the dock to a gill-netter that has just come in to unload. Bill Amaru, who owns the *Joanne A,* another small dragger moored in Chatham Harbor, is down there with his own sheaf of papers, a bundle of pamphlets regarding an upcoming public hearing on Amendment Seven. Amaru is dressed neatly in a flannel shirt and pressed chinos. A member of the New England Fisheries Management Council's Groundfish Advisory Committee, he has a bluff and engaging face notched with tight, mournful eyes. He listens earnestly to the New Zealander, his brow knitted, his head nodding at intervals, while behind him pods of golden cod begin to pulse down a chute.

4

The Ghost of Harry Hunt

Maybe it's only paranoia, but Brian has begun to suspect that he's enduring more than just the general run of bad luck in his lobstering this season. He and Greg Wade stand beside their pickups on the beach at Snow Shore. The inlet stands at a dead calm on a morning in late July. A sticky, vaporous fog hangs over the water, clings like frog spit to the cordgrass and beach pea and salt-spray rose, rubs in a floating mass against the bluffs of Nauset Heights. The fog has halted the three of us at the water's edge, shut out the vehicle traffic behind the heights and the boat traffic beyond the beach, left us alone and talking in a soundproof vault.

"I hauled one whole line of traps that were empty, but they were all clean on top," Brian tells Greg. "The doors weren't tight on some of them, and one had its top torn off, just one lath left."

Greg is a fellow lobsterman and the current president of the Nauset Fishermen's Association. He wears an association T-shirt and a look of uneasy deliberation. "Maybe a dragger went over it?"

"Maybe, but . . ." Brian is unpersuaded. Both he and Greg know that a dragger's net might indeed rip the top off a trap, but what about the other traps in the line? Only caged lobsters have the leisure and the inclination to clean wooden laths of their usual growth of red beard sponge and barnacles, what lobstermen call mung; only someone who removed those lobsters would leave doors unsecured, someone working hastily, perhaps, and that wasn't Brian, who works carefully and is attentive to doors. The ruined trap might be another casualty of gear conflict, or it might be vandalism, an example of meanness not satisfied with merely stealing from a man's traps.

Brian tells Greg that he's the only other fisherman he's spoken to about this and asks him not to say anything to anybody else just yet. He adds that he has an idea who might be robbing him, if in fact that's what's happening. He says he'll look for scrapes of paint on his buoys that might match the bottom of the suspect's boat when he hauls up there again, and on the next clear day on which that boat is working, he'll take his spotting scope up to the sand dunes in Wellfleet and keep an eye on his lines from there.

Among lobstermen, *pariah* is too kind a word for a trap robber. The act arouses a loathing in direct proportion to a lobsterman's vulnerability to that sort of theft, and also to the difficulty of doing anything about it inside the law — particularly, in Brian's mind, within the bounds of what fisheries law has become in the hands of the courts and the enforcement officers. "They won't just take my word for it, and even if they catch him in the act, they don't do anything about it," he told me last night. "The regulations are a joke. Game wardens don't want to enforce them, and to the judges there's no such thing as a bad boy. There's always some technicality, some extenuating circumstance. It makes me wonder what groundfish and the other fisheries would be like if the regulations had been enforced from the start. There's no way of proving it, of course, but do you really think fishing would be as scratchy as it is now?"

Greg is no more optimistic, citing the virtual impossibility of arranging for a game warden (now called an environmental protection officer) to catch a trap robber in the act. "I tell them exactly when I'm going, that I can't screw around with the tides in here, and they can't even get themselves organized to get to the boat on time. How are they gonna catch a trap robber?"

"Instead you get lynched yourself if you make a report," Brian says.

"Not if you're straight, but it's hard to be straight all the time."

"They'll find something. Aren't I supposed to have a life preserver in that dinghy?"

Brian falls silent and joins Greg in staring out at the inlet, their eyes probing the fog, which has consumed the *Cap'n Toby* and every other boat beyond the shoal water's line of skiffs and dinghies. "I'm not going out today anyway, but if I was, I wouldn't in this stuff," Greg volunteers.

"Well, I've got a lot of bait that's going to go over the side if it gets any riper. I might as well haul."

Brian believes the fog will thin out over the water. Greg leaves, and Brian, without the benefit of a life preserver, shuttles out to his skiff in the dinghy. The splash of his oars rings like a bell in the shrouded hush. Then the hush falls away in the rattle of the skiff's outboard, only to return again, now with a throbbing undertone, when Brian kills the outboard to let the skiff drift into the beach. "Hear the sirens?" he asks. "Want any wax for your ears?"

My first thought is that the environmental protection officers are already on their way, maybe about that life preserver. But then I recognize the allusion. The undertone resolves itself into the battery-powered boom box of a shellfisherman, the rock-and-roll clamor of a clammer working the exposed flats on the far side of the inlet. I remember how three bushels of quahogs tempted Brian into a life on the water without granting him the clairvoyance that the sirens promised Odysseus. He motors the skiff out to the *Cap'n Toby,* fires up the boat, and then steers past the musical clam digger and out into the fog.

Earlier Brian picked up copies of the *Cape Cod Times* and the *Cape Codder* at the convenience store where he gassed up his pickup. I looked through them on the way to the inlet and saw that it was a busy good news/bad news sort of day in the local newspapers.

Bad news: The federal government rejected Governor William

Weld's request for disaster aid for Massachusetts's commercial fishing industry. Officials from the Federal Emergency Management Agency said that Weld needed to prove that natural causes, not overfishing, had led to the decline of groundfish on Georges Bank. This Weld failed to do, they said.

Good news: Cape fishermen say that the cod fishing inshore has been as good this summer as any in recent years.

But what's bad about that: Good cod fishing on the Cape will neither deflect nor delay the approach of Amendment Seven. The working lives of draggermen such as Mark Farnham will change under any of the alternative measures being discussed for Amendment Seven, but the most attention-getting alternative so far is a total allowable catch of only 4.4 million pounds for Georges Bank cod. This would amount to less than 14 percent of the 32 million pounds fishermen landed there in 1993.

"All the alternatives are just variations of bankruptcy for the groundfish fleet," Maggie Raymond, a spokesperson for Maine draggermen, complained to the New England Fisheries Management Council. "I know you all think you're doing the right thing and that the human fallout is an unfortunate consequence of all of this. But now you should just do what's right. You should say, 'We're not going to do this until we do something for the people.'"

On the other hand: A good cod season on the Cape won't stop Amendment Seven, but the decision-making process itself might. The council meant to agree on a public hearing document for groundfish at its May 17 meeting, precisely when it was also supposed to sign off on a similar document for lobsters. When no consensus was reached at that meeting, the council established June 9 as a drop-dead date for approving a package. That meeting was derailed, however, by debate on small points in the recreational fishery. Frustrated council members vowed not to leave the next meeting, on June 28, without a document in hand. That was followed by a vote on June 28 to postpone action until August.

It could be worse, and it is: Brian isn't surprised that the same sort of runaround has caused the lobster management process to become not only invisible but irrelevant. Another problem arose at the May 17 meeting in addition to the expected one, the Gulf of Maine's unsatisfactory conservation plan. This problem lay in offshore lobstermen's inclusion of an ITQ system in their own conservation measures. The language of Amendment Five states that ITQs can only be implemented through another full-blown amendment to the Magnuson Act, not through a framework adjustment such as this.

Faced with an industry revolt in the Gulf of Maine and handed a hot potato by offshore lobstermen, the council postponed approval of the public hearing document to June 28 and then postponed action again. The National Marine Fisheries Service's July 20 deadline for having a comprehensive regional plan in place has come and gone. But currently it's groundfish, not lobsters, that dominate the headlines. The weekend work of Brian and Steve Smith and other lobstermen in the EMT process has been put on a political back burner and turned down a notch below lukewarm.

Good news if you think so: Two seats have come open on the council. One has been vacated by Dick Allen, whose term has expired, and one is about to be filled by Bill Amaru. Responding to criticisms that the council has been paralyzed by too much representation from recreational and big dragger interests, Massachusetts Congressman Gerry Studds suggested that some representation from the small owner-operators of Cape Cod might be expedient. Amaru has already done committee work for the council, and has been openly critical of a top-heavy fleet structure throughout New England in which 15 percent of the boats harvest 65 percent of the fish. Last March he applied for one of the available seats, and now Governor Weld has submitted his name for approval to U.S. Secretary of Commerce Ron Brown.

Brian greeted this news with a lift of the eyebrow, saying that a lot of guys in this part of the Cape are actually going to be unhappy about it. The reasons are partly cultural, partly political. Amaru came to the Cape twenty years ago, I learned from the *Cape Cod Times,* with college degrees from the State University of New York in Albany and the New England Conservatory, where he studied clarinet. He also brought an enchantment with the history and romance of fishing. But in Chatham, where he was the only college-educated Italian in a town of swamp Yankees, Brian said, Amaru was immediately regarded with suspicion, and certain breaches of fishermen's etiquette that he committed in ignorance in his early years have never been forgotten. "He also published an article in, I think it was *Saltwater Sportsman* a while back in which he said that commercial fishermen have got to cut back. It made a lot of sense, actually. Of course the sport fishermen all loved it. Most of the local guys were pissed," Brian added.

When Amaru joins the council, the more alienated elements of the Chatham fleet will view him as someone who has already sold out to big-boat interests. But Brian admires Amaru's experience in a variety of fisheries, his conservationist ethos, and his willingness to remind all fishermen that by their typical resistance to regulation through the 1980s and the readiness of too many to break the rules, they at least helped to make the threadbare bed they now occupy. "He's always worked as hard as anybody," Brian concluded. "He'll be all right on the council."

Brian is right about the fog, which thins out just outside the inlet, where we stop to fill bait bags. To the north a flock of black-capped terns are hovering and shearing and then knifing into the water as they feed on a school of sand eels. Brian says that probably some striped bass are also feeding there. In the event that we wander into a school of bass — in case of in case of — Brian carries a couple of fishing rods on board, along with some lures still in their packaging, Mambo Minnows with "that wide, slow wiggle ideal for stripped [*sic*] bass."

I know a lot about stripped bass, having gone face-to-face with a number of them just this morning at Rock Harbor. Brian and I got there at six o'clock so we could get out again, he said, "with the blood and the smell before the hundred-dollar-per-day charter-boat patrons arrive." Brian has a deal going with some of the charter-boat captains: he gives them choice culls from his lobster traps, and they leave their filleted racks of bass or bluefish in blue plastic bait barrels that Brian leaves on the docks. In the wispy early stillness the charter boats sat in their berths like items in a department store window: chromed, cushioned, immaculate, their outriggers bright and spidery against the sky. Brian gave me the choice of scooping out the carrion from the barrels on the east or on the west line of docks: "Do you want Jurassic Park or Wild Kingdom?"

During late July in Wild Kingdom, Brian's covered bait barrels become solar ovens, packed with residual heat and fetid with ripening fish. Each of the three barrels was half full of a malodorous mixture of the remains of bluefish running eight to ten pounds and big twenty-pound striped bass: unblemished heads and tails wired weirdly together by the white lacerated combs of the fishes' bare spines and ribs. To get the fish into a plastic tote, I had to tip each barrel on its edge and then lean in headfirst to get my boat hook into hard flesh. Once secured, the fish would slide on the gurry of slime and blood into the tote, the tails of the stripers whip-cracking a dollop of the gurry into my face each time. The tote filled rapidly. The barrels also contained gratuities from the sportsmen: a half-eaten sandwich, an empty can of mixed nuts, some crumpled beer and pop cans, and the like.

I pulled the tote behind me the length of the dock and then up the gangway to the parking lot. Brian was there already, smiling (wearing no gurry), in conversation with Dick Woodland, the tanned and angular skipper of the charter boat *Madame B.* Woodland apologized for not having more bluefish, which Brian prefers to bass, in his barrels. "We caught forty or fifty yesterday, but these guys tossed 'em all back," Woodland explained. "They only kept four or five."

Brian laughed, a chortle like the sound of his winch when hauling. "Well, keeping me supplied is the whole point, isn't it?"

Brian's thin supply of bluefish heads was mingled judiciously into the general run of bass that went into the bait bags. Then we headed south of Nauset, to an area off the beach where Brian keeps only a few traps and where knots of competing currents confuse the tides and make it hard for lobstermen to know which direction to haul their strings in. Sometimes the tide will be running north at one end of the string and south at the other.

By ten o'clock the tide is nearly at flood and Brian is working easily on a southerly tack in light fog, no wind, gelatinous seas. The terns have disappeared over the horizon. A herring gull shadows the *Cap'n Toby* through much of the morning, regarding our proceedings with the calculating eye of a canny investor, dipping down to the water to snatch up the occasional rock crab tossed out of Brian's traps. A northern moon snail comes up in one trap, a fist-sized mollusk whose rubbery foot can spread to bathroom-mat dimensions as it moves across the sea floor in search of clams. Brian pitches it into a bucket, saying if he catches a few more he'll try them as bait. In another trap a skate is hooked on a sportsman's fishing line that has gotten wrapped around a medium-sized cull. Brian unwraps the cull, saves it for Dick Woodland, and throws the unhooked skate back.

He guns the *Cap'n Toby* down to the next trap in the string and says that lobsters can play mind games with you. "You go out one day to move your gear and you find it all full of lobsters. So you leave it, and then you go out the next day and your traps are empty. It's a business designed to drive you around the bend."

Brian's mind can't leave what might be happening again to his traps up at Wellfleet today. He says it's been a slow season, that rookies in this fishery always think their traps are being robbed when they come up empty, and that the lobsterman he now suspects of preying on his traps has never gotten beyond that stage. "He just goes out and retaliates against some poor bastard instead of doing his homework — trying different baits, different locations, different soaks."

Sometimes trap robbing begins with peeking — pulling other men's pots just to see where the lobsters are, what baits people are using, what design innovations somebody might be trying out in his traps. "That infuriates me too," Brian says. "Lobsters that have been gathering around a pot scatter once it's pulled up. It's not going to fish as well during its soak time as an undisturbed one. And anyway, how do you call yourself a fisherman if you have to resort to spying like that to make your catch?"

The ghost of Harry Hunt whispers in Brian's ear that all men are spies, trap robbers, rule-breakers, backstabbers, hypocrites, and false fishermen. Just as peeking blends almost imperceptibly into theft, so ordinary prudence darkens into paranoia and then vindictiveness. A few years ago the lobstermen of Orleans were riven by a run of trap robbing that was handled badly, says Brian, by law enforcement, by the lobstermen themselves, and by the newspapers, once they got wind of it. That season everyone came away dirtied, Brian believes, including himself. "What good is wisdom if you don't use it?" he wrote in the event's aftermath. "I knew of St. Paul's letter to the Romans. I knew of Father Ferapont's fable (in *The Brothers Karamazov*) of the souls that couldn't escape Hell because they wouldn't help each other. I've spent hours thinking of these things."

My own thoughts run to the French existentialists: André Malraux describing "a fatality discoloring all forms" in *Man's Fate*; Jean-Paul Sartre's fictional Antoine Roquentin reporting, "Things are bad! Things are very bad: I have it, the filth, the Nausea."

Things got a little discolored this morning inside those blue barrels at Rock Harbor. They were bad outside the bar at Nauset Inlet as we filled the bait bags with the *Cap'n Toby*'s motor running, in air that had no breath or movement to it, that was thick and acrid and queasily sweet with its blend of diesel fumes and decomposing bluefish. We occupied an inert atmospheric system, one that had inflated over the boat and was then toggle-switched up and down by a slow and easy swell beneath the keel.

Now things are very bad. Brian opens a trap, removes the old bait bag, and hands it to me for emptying. The contents of the bag — mackerel, or what's left of it after a five-day soak — are swarming with amphipods, which Brian calls sand fleas: carrion eaters, minute and bone-colored, segmented, appendaged. They curl themselves into chitinous sprockets and then unsnap, popping into the air from the midst of the gore like bubbles from a carbonated drink.

The bait bag rests on the pegging board in a brown slurry of seawater, mud, crushed rock crab, and the curdled juices of previous bait bags. We've already used up the bags we filled this morning. Now we're filling fresh bait bags as we go along. I watch from some distance as my hands reach for the bag on the pegging board, open it up, dump the fishbones and remaining sand fleas into a plastic tote at my feet, and then go into the bag to scour bits of old flesh, a snagged spine or jawbone, from its mesh. My hands reach into another tote behind the engine cover to pluck out the head of a bass. This goes into the empty bait bag. The bass stares at me from behind the pink mesh in gaping, open-mouthed amazement. The boat rolls with a heavier swell than this morning. Diesel fumes curl up my nose and coat my sinus membranes.

I've filled lots of bait bags for Brian before this, usually in heavier seas than this, but for some reason, today — because of the heat, the lack of wind, the alignment of the planets, the sneeze of a butterfly in Mongolia — things are very bad. My breakfast (fried eggs, toast, and coffee) has been running at a rolling boil throughout the morning. Now and then, it bubbles to the top of my throat, causing my spine to suck in toward my belt buckle. Then the whole stew drops down the length of my gullet, there to roll and simmer some more.

I try not to look into the tote at my feet, where all the old bait bags have been emptied, but I do anyway. I'm thankful that the bones are generally as clean as they are. The tote's jumble of ribs and teeth and sockets and tails seems at once unearthly and discount-store tacky: the disassembled polystyrene components for glue-together models of alien life forms. The lobsters whose claws I have to band — many of them today two to three pounds or bigger, what Brian calls deuces, as opposed to chicks, which weigh in at a pound or so — are amphipods swollen to monstrous proportions. On the pegging board they spread their claws and stand with the shingled plates of their mouthparts slowly working. Carrion eaters themselves, they study me through their stalked compound eyes and realize they only have to wait. The diesel fumes have penetrated my sinuses and roil about in the spaces between my brain casing and my skull. I hope the lobsters don't have to wait long.

Brian glances at me sympathetically during our runs from trap to trap, when I put my head outside the wheelhouse and try to suck in as much fresh air as I can. I have no inkling — mercifully — of the stories he will tell me tomorrow. Once he was fishing in December on the *Bell* with Olav Tveit, going directly from the flatlands of Cape Cod to the mountains of Georges Bank. Olav sent him below to check a fuel saver that Olav had bought out of the back of someone's station wagon. One line was pinholed, and Brian got a faceful of fogged diesel fuel. It was worse than it might have been because he hadn't gotten his sea legs yet. "I ended up puking up my asshole," he says.

On another occasion he was long-lining with Bill Amaru and using squid for bait. Amaru asked him how the bait was, and Brian bit into one of the raw squid, turned to him with the mollusk hanging out of his mouth, and said, "Pretty good." Brian ended up puking up his asshole.

Brian has seen men so sick they vomited blood. He remembers that Harry Hunt believed the way to cook an egg was to spoon grease all over it as it fried. That was all right with Brian, but not with many of the men who shipped with them. "I think I saw more eggs come up than went down."

My own eggs come up at the end of the day. By 4:30 the fog has burned off entirely, and beneath freshening skies we halt as we approach Nauset to clean the boat. The tote of old bait is lifted to the rail and dumped over the side. I fall to my knees at that same rail, puking up my asshole, my belly boiling over into the bait's viscous, sinking stain. A flock of gulls, their breasts heading into the wind, hovers over my shoulders and above the oil and bones and spreading bile like a chorus of angels. A pair of bass fishermen are attracted by the gulls; they approach from the south in a speedboat, make out the *Cap'n Toby,* and turn and disappear again to the south.

In 1607, five years after Bartholomew Gosnold found himself so pestered with cod near the eventual site of Provincetown, the English ship *Mary and John* anchored off an island near Nova Scotia. There, in three feet of water at low tide, using only a single long gaff, sailors were able to capture in less than an hour fifty "great lopsters" of a "great bigness I have not Seen the Lyke in Ingland," wrote a ship's officer.

Throughout colonial times four-foot, forty-five-pound lobsters, probably at least a century old, were common in the Northeast. Reports of five- and six-foot leviathans were considered apocryphal until the 1950 discovery of similar giants in the offshore canyons that Harry Hunt used to prowl. Along the beaches, in the aftermath of storms, thousands of stranded lobsters could be found piled in windrows. Though their flesh was acknowledged to be "sweet, restorative, and very innocent," they would be pitchforked into baskets for fertilizer. At any other time, wrote a minister of colonial Salem, "the least boy in the plantation may catch and eat what he will of them." There was even an eighteenth-century petition to the General Court of Massachusetts that indentured apprentices not be served lobster more often than twice a week. "The animals were so common that it is not surprising that their value was not appreciated," wrote Francis Hobart Herrick, the author of a classic 1911 natural history of the lobster.

This was an example of the first stage in a landmark four-stage population model advanced by Herrick: "Period of plenty: lobsters large, abundant, cheap; traps and fishermen few." Northeast Indian tribes had learned to preserve lobster meat through a delicate process of smoking and drying; immigrant Europeans, however, contented themselves with the much more easily preserved cod. The lobster was too easily caught, too arduously saved, and too little esteemed, good only as fertilizer or starvation food — at least until after the American Revolution, when markets for fresh lobsters as novelty items began to grow in such places as Boston and New York, where the animals had finally become scarce.

At this time lobsters were caught from dories by fishermen using simple hoop traps, a net bag hung from an iron ring, two to three feet in diameter, with bait hung from the intersection of two half-hoops arcing vertically over the ring, and hauled every fifteen or thirty minutes. The lobsters were shipped live to the cities in sailing smacks, which were small vessels whose hulls contained smacks, or wells, through which seawater could circulate. Most of the lobsters for this trade came from Cape Cod, while most smackmen, at least originally, came from Long Island Sound. The trade was soon lively enough, and volumes were high enough, for Cape Codders to notice a decline in their native lobster stocks. In 1812 the residents of Provincetown persuaded the Massachusetts legislature to make it illegal for anyone not a citizen of the commonwealth to take lobsters from their waters without a permit.

Herrick observes that this first conservation measure was a law to protect fishermen, actually, not lobsters, adding that this and subsequent nineteenth-century measures to hold the trade in check — closed seasons, length limits, prohibitions on the sale of egg-bearing females — were of little avail in Massachusetts: "By 1880 the period of prosperity had long passed, and few lobsters were then taken from the Cape. Only eight decrepit men were then engaged in the business, and were earning about $60 apiece. This great local fishery was thus rapidly exhausted by overfishing, and it has never recuperated."

The period of prosperity, or "rapid extension" — "greater

supplies each year to meet a growing demand; lobsters in fair size and of moderate price" — is the second stage of Herrick's model, and by the middle of the nineteenth century that period had passed up the coast from Cape Cod to the Gulf of Maine. At the same time certain innovations in food preservation brought about a quantum leap in demand. In the 1820s, the Boston merchant William Underwood began to sell jams, jellies, sauces, ketchup, and mustard, all in the sort of hermetically sealed glass bottles and jars used by the London firm of Mackey and Company, where the observant Underwood had recently apprenticed. He also packed lobsters in quart jars, selling them for $4.50 per dozen and shipping most to overseas luxury markets. Because it was difficult to achieve a reliably airtight seal in the jars, however, Underwood began experimenting with cylindrical metal canisters, their seams soldered by hand.

Underwood's tin cans soon revolutionized the food packing industry, as the technology spread via corporate defections and the same sort of espionage Underwood himself had practiced. In 1842 a Maine entrepreneur felicitously named U.S. Treat learned how to pack lobsters into the new cans, though Treat was soon forced to sell out to Underwood. By 1844 Underwood was operating a lobster cannery himself in Harpswell, Maine, and with that the lobster — thanks to its reputation among gourmets — became one of the first items canned commercially in the United States.

Many other canneries followed, operated by a small number of companies privy to the new industry's trade secrets. In 1880 twenty-three Maine canneries packed 2 million pounds of lobster meat (along with quantities of mackerel, clam, salmon, and chowder) from 9.5 million pounds of lobsters, more than doubling the volume of live lobsters sold in Maine that year, with most of that meat still going to Europe. By then the hoop trap had been replaced by the lath trap, which first appeared in Casco Bay around 1830, while dories had been partly supplanted by sloops or Hampton boats — double-ended, lap strake, keeled vessels equipped with two spritsail-rigged masts.

Lobstermen worked longer seasons in deeper waters with more capital-intensive equipment, but prices were high — two cents per pound and rising — and competition was keen between canneries for the services of good fishermen. An 1870 photograph of a cannery wharf on Isle au Haut shows big lobsters heaped nearly to the cannery's eaves by the fishermen of that island. As early as 1870, however, there were signs that the Gulf of Maine lobster fishery was in trouble. The packer George Burnham, who owned twenty-one canneries, said to the U.S. Fish Commission, "In the fall of 1854 I went to South Saint George, on the coast of Maine, to pack lobsters, and sent a smack to Deer Isle, where the fishermen used hand nets, and 1,200 lobsters then caught filled the smack's well. It would take of the lobsters we now catch from 7,000 to 8,000 to fill the same well."

Trap numbers were still climbing — there were more than a hundred thousand in the gulf by 1880 — and lobster sizes were going down. Traps also needed to be spaced farther apart and shifted more often. This was the third stage of Herrick's model: "Period of real decline, though often interpreted as one of increase: fluctuating yield, with tendency to decline, to prevent which we find a rapid extension of areas fished, multiplication of fishermen and traps and fishing gear or apparatus of all kinds; decrease in size of all lobsters caught, and consequently of those bearing eggs; steadily increasing prices."

In the 1870s Maine instituted a series of conservation laws similar to those that had failed to save the Cape fishery and that were to have little appeal for Herrick. Regarding the protection of berried lobsters: "It is an easy matter to brush or comb off the eggs . . . and thus evade the law." Regarding a minimum gauge length: "Most people are aware that the gauge law has not been rigidly carried out, and that the illegal sale of short lobsters has become a trade of big proportions." Regarding closed seasons to promote breeding success, Herrick observed that the lobster spawns only once every two years over a wide portion of the calendar and then carries its eggs for nearly an entire year. "Closed seasons . . . serve merely to restrict the total amount of fishing done in the year, and do not touch the root of the difficulty."

But a seasonal fishing closure from August 1 to October 15, enacted in 1874, was in fact a bitter blow to an industry that relied on a steady supply of lobsters. Then an 1879 law specifically prohibiting lobster canning from August 1 to April 1 spelled the end of that industry along the Maine coast. The canneries quickly moved north of the border, into the Canadian Maritimes, where conservation laws were looser and where there was less competition from the still-expanding live market for lobsters. By 1885, sixteen of George Burnham's twenty-one canneries were in Nova Scotia. By 1895 lobster canning had ceased altogether in Maine.

The canneries left behind them, however, an excellent representation of Herrick's unhappy fourth stage: "General decrease all along the line, except in price to the consumer, and possibly in that paid the fisherman." The canneries also left behind a live-market sector that by then had learned the tricks that would eventually make canned lobster meat once again a novelty item: the shipment of fresh ice-packed lobsters by steam and rail, and the storage of large numbers of live lobsters in saltwater lobster pounds until market conditions were most favorable. In Canada, meanwhile, the efforts of the canneries to meet European demand quickly brought that fishery to the point of collapse, and Herrick, writing in 1911, could conclude, "The history of Cape Cod has been repeated on one and another section of the coast, from Delaware to Maine, and is already well advanced in the greatest lobster fishing grounds of the world, the ocean and gulf coasts of the British Maritime provinces of Canada, especially of New Brunswick and Nova Scotia, and in Newfoundland."

Brian's son, Mike, fifteen now, likes the irony of David Letterman and the hell-for-leather pace of a good John Woo–directed action movie. Last night's pre-Letterman offering on cable TV, *Hard Helmet,* was merely a thin imitation of a John Woo film, but it sent me to bed with the idea for a script that I describe to Brian as we head for the inlet the next morning.

You get a firm ID on the trap robber, I say, and you're just about to slide a bomb up the tailpipe of his pickup — Saint Paul's letter to the Romans will not be a necessary reference here — when FBI agents step in and bust you in the act. But the G-men say they're not really after you, Mr. Gibbons, or even after that trap robber. That robber's just a little chick, they tell you, and they're after the big deuce: the Mafia don who's running trap-robbing operations all up and down the East Coast out of an office in the Big Apple's Fulton Fish Market. They've got him nibbling some bait, they say, but they're not ready to haul their trap just yet, and they may never be ready without a good man on the inside. They also mention that you can get twenty years to life for doing what they just saw you doing.

After lengthy ethical deliberation, you realize where the greater good lies, and pretty soon you're robbing traps yourself, then fencing the hot lobsters to the don. You rob so many traps so well that you climb up the organizational ladder, and pretty soon you're the don's right-hand man, and the don's beautiful daughter, Gina, who is devoted to her father, has fallen hard for you. That's when you get word from the FBI that it's time to haul in. You think about what this is going to do to Gina, and it breaks your heart. You've even started to like that old bastard the don. But then you ask yourself what Harry Hunt would do in this situation, and you understand that if trap robbers are ever to be effectively prostituted, this is your last best chance.

But then, nearly at the last second, one of the G-men happens to mention that he wishes there were tariffs on imported fish, and in a flash it hits you that these guys aren't the feds after all. Instead you suddenly realize that you've been set up by a rival Mafia family that wants to move in on the don's operation. This supposed arrest is actually going to be a gangland execution. But not to worry, because the next thing you figure out — in moments of reflection during the car chase and the gun battle and your against-all-odds rescue of Gina — is that you haven't really infiltrated a Mafia family after all. This whole thing has been an FBI sting operation whose sole purpose was to trick the target family into showing its hand. Operation Crusher Claw is a huge success, permanently putting the target family out of business, devastating most of the New York City waterfront, and restoring the entire lobster industry to stage two of Herrick's population model.

In the end Gina wants you to stay with her in the Big Apple, I tell Brian, where she's finishing her study of marine science and ecology at Columbia. But all you can think about is Suzanne, the kids, the *Cap'n Toby,* your friendly buyer Ronnie at the Nauset Fish & Lobster Pool, and of course all those bait barrels at Rock Harbor that would otherwise never get emptied. Fade to credits as you walk away. Music by Miles Davis, something easy and wistful, maybe "Gone, Gone, Gone," from the *Porgy and Bess* album.

Brian nods, proposes several revisions that I silently reject. He swings the pickup off Tonset Road and down to Snow Shore. A white Lincoln Continental, early 1980s vintage, is on the beach, parked at the water's edge, its long trunk thrust like a diving platform over the incoming tide. Two men are transferring clam rakes and baskets and other gear from the trunk to an outboard skiff. They look like goodfellas going straight. "There must be better money in clamming than in lobstering," Brian suggests.

The writer and ornithologist Wyman Richardson devotes a chapter of *The House on Nauset Marsh* to an occasional event he calls a "do-nothing day" — a day for little more than watching marsh hawks and butterflies from the porch; a day where the closest thing to hard labor is a walk to the boathouse to watch the streams of shorebirds pass. Brian finds the concept enchanting, and refers to it often. The thought of enjoying a do-nothing day himself shines above the mountainous topography of his event horizon like a star, bright and beautiful and impossible to reach.

"In this business there's always something you've got to do," he says.

Today is as close as he gets to a day like that. He had a relatively good outing yesterday hauling his lines south of Nauset, even if I wasn't much help: a little more than two hundred pounds, most of that big deuces that he sold for $5 per pound to Ronnie Harrison at the Nauset Fish & Lobster Pool. All his other traps now are in the middle of their soaks. With nothing to haul, he can attend to some of the details: run to MacSquid's, a fishing tackle and gear store, for a clam rake to replace the one he can't find; drop a load of culls off for Dick Woodland, the charter-boat skipper; scrounge up some bait ("In this business, too much bait is never enough"); and perhaps find out something about who's been tampering, maybe, with his traps.

Last night, just before the start of *Hard Helmet,* Brian packed a spotting scope and tripod into the cab of his pickup. "Understand that this is not a typical activity in a lobsterman's life," he told me. "And ninety-nine percent of the time it's fruitless anyway."

The day began with a reconnaissance of the suspect's boats at one of the other coves on the inlet. His lobster boat was at its mooring, but his skiff was gone. Then to Snow Shore, where Brian stood next to the Lincoln, gazing across the inlet and the beach and out to sea. The weather report gave today's visibility at three miles, but the morning heat and a mild westerly breeze have raised a puffy haze across the length of the horizon. Brian estimated the visibility out to sea at no better than a mile.

Other lobstermen were gearing up to go out. Steve Smith was one of them, and Steve told Brian that another Lobster Oversight Committee meeting is scheduled for July 28, the Friday of this week. Steve said the target date now for a sign-off on a public hearing document is the full council's August 10 meeting in Danvers. He and Brian discussed the prospects of an ITQ-style management scheme's being included in that document: on the one hand, diminished by yesterday's appointment of Jim O'Malley to fill the at-large seat on the council vacated by Dick Allen; on the other hand, augmented by the recently declared willingness of other council members to change the format of the lobster plan from a framework adjustment to a full-blown amendment in order to accommodate those sorts of measures. Andy Rosenberg, the newly appointed Northeast regional director of NMFS, gave strong support to such a process at the last council meeting. Jim O'Malley rose from the audience to complain that the maneuver would be "an absolute betrayal of the message the council gave to this industry a year ago."

Brian knows that he'll have to be hauling traps on Friday. He's not inclined to spend another payday on one more inconclusive meeting, nor to allow whoever might be plundering his traps to feel secure about his whereabouts. He asked Steve if he intended to go to the meeting. Yep, said Steve. "Good — you're our representative," Brian said. "Anything they do up there, you'll get the blame."

Steve pushed off in his skiff, and then Kurt Martin showed up. Baby-faced and blue-eyed and soap-opera handsome, Kurt stopped to talk about tuna fishing, scuba diving, and all the mussels that have been eaten this year by bottom-blanketing hordes of starfish. I know from Brian's stories that Kurt used to go fishing on occasion with Brian and Harry Hunt, and that one such occasion was nearly his last. Early in his career, Brian says, Hunt was as flexible and imaginative and innovative as any lobsterman on the Cape. As he aged, however, he became set in his ways, often to the point of resisting such easy-to-like advances as synthetic warps and wire-mesh traps.

Nowadays Brian's warps, and those of any other lobsterman, are made of two different kinds of rope: polypropylene, which is water-resistant and buoyant, on the bottom part ensuring that the warp floats free of the trap, and on the upper part a polypropylene-Dacron blend that sinks and keeps the warp free of the buoy and passing boat propellers. Harry Hunt, however, wasn't sufficiently impressed with the new composite lines to spend money on them and made do with old warps that were buoyant but hung with lead weights. The weights would be strung halfway between each trap in the long trawls, or connected strings, of traps that he laid on the bottom.

He used these also on the pots in which he caught sea bass, though the weights — which tended to catch in bights of rope, or snag on protrusions in the cockpit as the pots went over the side and the warps shot out after them — were always difficult and dangerous to handle. Once, while setting a bass pot trawl near Nantucket, Kurt found himself in the thick of a writhing snarl of lead weights as several pots went overboard at once. The weights would have pulled him straight down if he had gone over the side with them, and all Kurt could do to save himself was howl at the top of his lungs. But Hunt just gunned the motor harder until Brian got to the wheel. "You should've told me," Hunt said as he kicked the boat into neutral and his shaken crewman disentangled himself from the rope and lead weights.

"I was screaming!" said Kurt.

"You got to scream louder, boy," growled Hunt.

Wire-mesh traps appeared in the early 1970s, first in the offshore lobster fishery, then in inshore waters, where they proved more durable and productive than wood-lath traps. Brian and Kurt knew, however, that Hunt would never adopt the traps without some cognitive management. After prevailing on the skipper at least to try a couple of them, they made sure to complain aloud every time they hauled one, saying that the fucking things were no good, that they never caught lobsters, that they should just cut 'em all loose. "No, no, now give 'em a chance. They'll be all right," counseled Hunt, who finally decided to buy more of them.

Kurt didn't seem to be going out today, was down there perhaps simply to see who was. Brian had seen enough, and concluded that his spotting scope wasn't going to do him any good. "Visibility's not good enough," he said as we rode back toward the center of town. "And even on the clearest days it's hard to pick out the color of a buoy in the glare of the sun and with the mirage effect."

Then it was time to share some of yesterday's wealth. We stopped at Dick Woodland's house, a neat clapboard affair with a velvety lawn, a small swimming pool, and a sign that said, "One fisherman and one nice person live here." Brian said that Woodland is a former lineman who runs the *Madame B* to supplement his retirement income. I remembered from the *Cape Codder* that Woodland and his wife, Barbara, were the chief volunteer organizers the previous week of the Rock Harbor Special Olympics Fishing Derby, an annual event in which many of the Rock Harbor charter boats now take part.

Barbara came to the door, looked at the twelve culls Brian had for her, and said, "Oh, Brian, that's overkill. We can't take that many."

"You might as well — they're bursting the laths on my holding car," Brian replied. "Sorry I haven't had too many to bring over till now."

At MacSquid's Firearms and Sporting Goods, where Brian picked out a clam rake for $40, discussion surrounded cod prices, which started out the summer at a dollar per pound and have been falling ever since. "Fifty cents now," said the clerk at the register incredulously. "I never thought I'd see that happen. I don't know why they're so low."

Brian suggested that maybe the draggers were hitting the cod hard. "No, they're not coming in with much," said another customer.

"I hear New Bedford's bringing in a lot of frozen pollack from Alaska," Brian says. "Maybe that's having an impact."

Shrugs all around, and then it was south through the summer's mad traffic to a packing plant at the industrial park in Chatham. "I don't really need bait now, but there's always so much demand for it, you've got to be proactive," Brian explained.

His friend Mike Russo needs a lot of bait, but thanks to such dismal cod prices and a lot of political uncertainty, he doesn't know whether to be proactive or not. Brian came out of the packing plant just a moment ago with a tote full of premium fish scraps: swordfish, bluefish, cod. Now he stands in the parking lot and listens sympathetically while Mike leans against his pickup and describes his dilemma.

"I don't know what the hell to do," Mike says, visibly distressed. "If I spend eight thousand dollars on bait tomorrow and then find out that the fishery's shut down, I'm fucked. And if I don't and the fishery stays open and I've got no bait, I'm fucked that way too. I can't live like this."

Mike is a young long-liner who lives in Orleans and moors his boat, the *Susan Lee,* in Chatham Harbor near the *Honi-Do.* Brian just paid five bucks for this tote of fish scraps, but Mike can't afford to scrounge for his bait like a lobsterman does; a long-liner needs too much of it too often. Instead he contracts for it in large quantities, baiting three to four thousand hooks per day with pieces of squid, paying a crew of baiters to accomplish that for him before each trip, also paying the rent on a shanty here in the industrial park where he can keep a freezer full of frozen squid and his crew can do their baiting.

Right now Mike can't decide whether to go into the packing plant or climb back into his truck and drive out of the lot. Fifty cents per pound for good groundfish may be high enough for factory trawlers out in the Bering Sea, but it doesn't pay his expenses on Cape Cod. That's bad enough; worse may be in store at the New England Fishery Management Council's August 2 meeting in Peabody, where the council will review a public hearing document on four separate proposals for dealing with the groundfish crisis, four different possible versions of Amendment Seven. Alternative One, the harshest, has stolen the thunder from the low TAC previously discussed. This measure would involve a complete and immediate shutdown of all dragging, gill-netting, and hook fishing for groundfish in New England waters, similar to the measure Canada imposed last year on its economically devastated Maritime provinces. A vote for Alternative One next week would transform $8000 worth of bait into toxic waste.

Mike throws up his hands in anger and confusion while the

sun beats down hard on the parking lot. Brian has no advice to give beyond the observation that "immediate" conservation measures have historically proven less so; for better or for worse, the task of putting people out of business around here has, as Mark Simonitsch would have it, proceeded slowly and with great deliberation.

"Sure," Mike says. "So I'll buy my bait, and this one time, this one day in history, the council will just snap their fingers, and poof! I'm a goner."

Maybe this is a do-nothing day for Mike, and then again maybe it's not. Brian bows his head, rubs his neck. He knows that the only thing sustaining the lobster fishery right now is price, not volume, which, after all, is the way prices are supposed to work, both in Herrick's population model and in Economics 101: as supply goes down, price goes up. Brian finds himself puzzled — and disturbed — to see both of those variables bottoming out at once for groundfishermen like Mike.

The glare of the sun and the mirage effect: these account for the third figure I see, a heavy-shouldered man with his face blotted out, who leans out from behind Brian's left shoulder and whispers almost audibly to Mike: *You got to scream louder, boy.*

5

Wampum

IT IS ONLY JUST getting light at Rock Harbor, and Dick Woodland, normally a placid sort, is beginning to lose his temper. He stands at the top of a gangplank beside two young men in their twenties. The gangplank leads down to the dock where Woodland's *Madame B* is berthed, and the plank's angle of declension is growing steeper every moment as the tide runs out of the harbor. "Where the hell are they?" Woodland wants to know.

"Well, they said their alarm didn't go off," one of the young men replies. "They were just getting out of bed when we stopped at their place at four this morning."

Maybe their alarm goes off, maybe it doesn't, Woodland notes; in either event the tide goes out. "In ten minutes we aren't getting out of this harbor," he adds. "There won't be enough water."

The others laugh and moralize to the effect that when you party hard on Saturday night, you're late to church on Sunday morning. "Man, are we going to roast them," they promise.

The August sun climbs bright and orange and roasting hot behind the harbor. The channel into Cape Cod Bay through the mudflats outside the harbor is marked by a winding line of scrub pines, the trees driven like stakes into the bottom, their foliage intact. Their reflected crowns shimmer in the still and glassy water; their shadows trail thinly to the west. Brian told me they were quahog trees: lucky shellfishermen stop and simply pluck clams off their branches.

Woodland grows even more exasperated as the minutes tick by, and his clients try to draw him into their banter. Finally he explodes: "Jesus Christ! Well, you know what? We aren't going fishing."

At that instant a blue Mazda Miata races through the stop sign at the end of Rock Harbor Road and turns, squealing, into the parking lot. Its riders bring laughter, apologies, grocery bags of food, six-packs of beer, and there is some preliminary roasting. Woodland, still grumbling, leads a hasty caravan down the gangplank to his boat.

The *Madame B*'s painstakingly careful passage out of the harbor confirms my suspicion that I've missed my own boat. I thought I was supposed to meet Dan Howes, a shellfisherman, here in the parking lot twenty minutes ago. I walk past the charter boats and follow the estuary as it doglegs to the north, where the working boats are slipped: lobster boats, quahoggers, a gillnetter or two. Two fishermen in a small, flat-bottomed quahogger call out to me: "Are you Dan's friend?" A minute later, only a moment ahead of the tide, I'm speeding past the quahog trees and up the eastern side of the bay to the point off Eastham where Dan's boat is already working the bottom.

Built in 1980, the *Last Resort* is a twenty-five-foot converted lobster boat that Dan bought only a month ago in Gloucester. This is his first season of quahogging. He started in the spring with the *Sea Biscuit,* a twenty-five-foot open skiff with a pilothouse and a new 220-horse outboard motor powerful enough, he believed, to haul a 300-pound iron dredge through hard-packed bay bottom. When that engine blew after only a hundred hours of operation, he considered looking around for a boat with an inboard diesel engine, only to decide on another 220-horse outboard for the *Sea Biscuit.* That one lasted four hundred hours before it threw a rod. Now the *Sea Biscuit* is for sale, and Dan is at the bottom of a hole, making payments on two ruined motors, his new boat, and the house he built for himself in Orleans in 1991, all with Amendment Seven looming over the horizon, at a time when small fishermen are looking just to survive, let alone catch up.

I can't help but think that the *Last Resort* is aptly named, though it isn't of Dan's choosing; the boat came with that name, and Dan's choice was to keep it. The vessel is white with slate-gray trim, and with its straight wheelhouse, its stubby mast amidships, and the short boom from which the dredge is suspended over the side, it has a jaunty and capable look. Its skipper welcomes me aboard, explaining that he thought I was going to come look for him in his slip at the northern end of the harbor, but it's okay, it all worked out. I thank Chris West and Dave Reed, who watched for me in the harbor and then brought me out here, and I'm hardly out of their quahogger and into this one before Dan has the sledge dropped into the water and his vessel in motion again.

This eastern edge of Cape Cod Bay washes over a sandy shelf that extends at one point, from North Eastham to the tip of Billingsgate Shoal, as far as seven miles from shore. Depths over the shelf vary from six to sixteen feet at low tide, and right now, according to the screen of the Si-Tex electronic depth sounder in the wheelhouse, the *Last Resort* is running in about ten feet of water. Dan says he's working the guzzles, or depressions, in a series of parallel ridges that run through this section of the shelf, and he stares all the while at the screen of the Si-Tex for the red and green spikes that indicate patches of eelgrass, near which quahogs are often found.

Unlike Brian and Carl, Dan Howes is a native Cape Codder, born and raised in neighboring Brewster. Precisely how native he is, how deep his roots extend into the soil here, into what cellar holes of local history, he can't say, though Howes is one of those names, like Nickerson and Eldridge, that you hear all over this part of the Cape. I like to think that his blood links him to the venturesome Thomas Howes, who in 1639 became one of the three original settlers of Yarmouth; or to the redoubtable Alvin Howes, whose house in Barnstable was a stop on the Underground Railroad for runaway slaves smuggled on board ship in Norfolk or Savannah and landed at night on Cape Cod; or to the unfortunate Captain Benjamin P. Howes, a Dennis shipmaster who commanded the clipper ship *Southern Cross* when it was captured and burned by the Confederate raider *Florida* in 1863. After the Civil War, Captain Howes took the clipper brig *Lubra* on a trading errand to the South Seas, but that ship was captured by Chinese pirates, and its captain was murdered in his cabin.

The names suggest a world much larger than Cape Cod Bay or the Nantucket Shoals. Benjamin P. Howes was an actor in the last scenes of a decades-long drama in which Cape Cod, through the prowess of the American merchant marine, had commerce with all the nations of the earth. That story began with the noisy prelude of Alexander Hamilton's tonnage duties, heavy fees first levied in 1790 against foreign vessels entering American ports. These duties drove English ships away from the ports, as their critics promised they would, but also kindled an explosion of shipbuilding in Cape Cod and elsewhere in Massachusetts. Soon the English ships were no longer necessary: sturdy, well-crafted Massachusetts vessels were performing their offices in carrying salt cod, whale oil, tobacco, leather, and Northwest fur to the West Indies, the Mediterranean, China, and the South Seas, and then returning home laden with tea, silk, indigo, wine, sugar, molasses, rum, and spices.

The captains of those ships won wide admiration not only for their seamanship but also for their force of character and entrepreneurial zeal. Most were part owners of their vessels, with full responsibility for trading, and according to Samuel Eliot Morison, this degree of personal investment helped to make "a Yankee shipmaster, in 1840, the world's standard in ability and conduct." Because of the risks they ran and the rewards they could reap, such merchant shipmasters were the matinee idols of their day, and it was said of any pretty and accomplished Cape Cod girl, "She's good enough to marry an East India cap'n." Certainly there were many to choose from: in 1837, Thoreau counted in the single town of Dennis, where Benjamin P. Howes was growing up, the names of more than 150 local shipmasters scattered about the world.

The discovery of gold in California in 1848 opened up a new and impatient market to New England traders, one ruled by whoever could make the fastest passage around Cape Horn. The clipper ship, with its sleek, tapered hull and towering masts, revolutionized marine design and established a whole new echelon of achievement in seamanship. "The clippers were the most beautifully finished and expensive vessels that had ever been built," says Henry C. Kittredge, "and were the pride of their owners' hearts — a pride nicely blended with anxiety when they considered the investment that each ship represented. Handling these wild beauties was a ticklish business, requiring far more dexterity, nerve, and experience than was needful for the old-style barrel-bottomed craft that bobbed along comfortably under stumpy masts and seldom came to grief."

The men who handled these beauties were the idols of the matinee idols, the aristocrats of an aristocracy, almost entirely of New England stock and a great many from Cape Cod. After dropping their cargoes in San Francisco, they usually brought their ships home by way of the Orient. There, says Morison, the new clippers "came into competition with British vessels, and the result gave John Bull a shock worse than the yacht *America*'s victory":

> Crack British East-Indiamen humbly awaited a cargo in the treaty-ports for weeks on end, while one American clipper after another sailed proudly in, and secured a return freight almost before her topsails were furled. When the Yankee beauties arrived in the Thames, their decks were thronged with sight-seers, their [speed] records were written up in the leading papers, and naval draughtsmen took off their lines while in dry-dock.

The clippers, however, while built by Yankees and commanded by Yankees, were not often manned by Yankees, or even Americans. Following the American Revolution, an able seaman was paid $18 a month on Pacific voyages, enough to attract ambitious American youngsters to sea and help cultivate the subsequent wealth of great merchant captains. But in the slack economic times that immediately followed the War of 1812, this wage dropped to $12 a month, and it remained there even as trade was restored and grew. "An increase of tonnage in the Thirties required more seamen," says Morison. "Instead of raising wages, to compete with the machine-shops and railroads and Western pioneering that were attracting young Yankees, the shipowners maintained or even depressed them, until ordinary and able seamen on California clippers received from eight to twelve dollars a month. . . . The shipowners could have obtained American crews had they been willing to pay for them; but they were not. Like the factory owners, they preferred cheap foreign labor."

It was the free market at work. An 1817 federal law sought to quash this trend by requiring that two thirds of the crew on an American vessel be American citizens, but the law was disregarded and never enforced. By the time of the christening of the first clippers in 1850, the crew of an American merchant vessel had become no better than that of a whaler: an international human hash of the criminal, the depraved, and the habitually drunk, men fit only to keep bread from molding, performing work so hazardous and for so little pay that even a number of these unemployables had to be illegally shanghaied aboard against their will.

But the ships themselves were glorious. In the clippers, Morison says, "the long-suppressed artistic impulse of a practical, hard-worked race burst into flower. The *Flying Cloud* was our Rheims, the *Sovereign of the Seas* our Parthenon, the *Lightning* our Amiens; but they were monuments carved from snow." The clippers indeed were so glorious that the shipbuilders of Massachusetts could not see past them to the slower, unprepossessing, but much more economical steamship. "Far better," sniffs Morison, "that the brains and energy that produced the clipper ships had been put into the iron screw steamer (in the same sense that Phidias had been better employed in sanitation, and Euripides in discovering the printing press)."

When the gold rush faded in California, when the steamship emerged from its trial by fire in the Civil War, when shipowners found that the profits they had skimmed from seamen's wages were required instead for the mere maintenance and repair of a clipper's intricate devices, the great ships became anachronisms. Boston and Salem ceased to be rivals in international trade. Donald McKay, New England's Phidias of the shipyard, the designer and builder of our Rheims, our Parthenon, our Amiens, died in poverty and neglect. The proletariat that had once scaled the heavenly masts of the clippers found themselves shanghaied instead to tramp steamers out of New York, Philadelphia, and San Francisco.

"We ceased to be a maritime nation," laments Kittredge, "and one by one the Cape commanders furled their sails and squared their yards for the last time, left their vessels to the mercy of the ship-breaker, and came home to live." After strutting the world's stage, they returned without fanfare to towns and villages similarly in eclipse. Most of the cod caught on Georges Bank now went to the domestic market rather than to Europe. For lack of rushing rivers, the industrial revolution largely bypassed the Cape, and it wasn't until 1873 that a rail line from Boston reached Provincetown. Meanwhile, Gloucester had a branch line in 1846, by which it could ship its fish, and by the end of the Civil War that city — with an expanding immigrant population, a fleet of 342 cod and mackerel schooners, and an annual catch of $3 million — had far surpassed any town on the Cape in importance as a fishing community.

One Captain Zebina H. Small of Harwich is remembered for his decision in 1845 to sell his fishing vessel and set out a cranberry bog instead. Where once wise men on the Cape knew the way to China by sea better than they knew the way to Boston by land, by the end of the Civil War the region was on its way to becoming the sort of backwater where wise men knew the way to neither, and from which smart young men went west. Kittredge also notes, however, that even in retirement the Cape's sea captains were impressive, bringing to civic affairs, those nineteenth-century versions of ivy day in the committee room, a beneficent breadth of spirit:

> One such man could make village affairs hum, and there used to be not one but scores in most of the Cape towns. Town meetings in those days were handled in quarterdeck style. Narrow-mindedness found barren soil in a district where two houses out of every three belonged to men who knew half the seaports of the world and had lived ashore for months at a time in foreign countries. They had gained a perspective that measured matters against far horizons, and they kept the Cape mentally alert for as long as they lived.

The horizons that Dan Howes surveys today are more circumscribed. The *Last Resort* turns from one point of the compass to another as it runs in clockwise, overlapping circles over the ridge guzzles. The inner beaches of Brewster, Orleans, and Eastham range across the south and east. To the north lies Wellfleet Harbor, scooped like several ounces of flesh from the inside length of the Cape's raised forearm. Ten miles distant, on the northwest side of Great Island and Jeremy Point, lies the curled fist of Provincetown. Only a mile to the northwest, looking from this distance like another small fishing boat, the shell-pocked superstructure of a World War II Liberty Ship, the S.S. *Longstreet* — a freighter, scuttled here and once used by the navy for target practice — rises above the waves. Then, to the west, the bay spreads oceanically to the horizon, rounding up to clouds being shoved in purple broken shelves toward Plymouth by a mounting northeast breeze.

Dan says that one year there was a bumper crop of scallops in this part of the bay. "One guy loaded up his sixteen-foot skiff so high that he sank on his way back to the harbor. Later another guy dredged up some of his bags of scallops, and then somebody dredged up the skiff itself, and finally somebody else got the rest of the bags."

I find that a perversely beguiling image: scallops dredged from the bottom and already sorted and bagged, suggesting lobsters hauled from Brian's traps already banded and boiled, or ready-made flounder fillets spilling from the nets of the *Honi-Do*. This is the sea as a Shop 'n' Save, a dispenser of commodities. The quahogs that Dan roots out of this bay are not growing on trees exactly, but seem only a step removed from the sort of convenience-food allure those sunken scallops had: sedentary, sweet, bite-size, in packaging that might have been made a little less stout for ease of consumer use. But the package is certainly sturdy and has a handsome look in its half-shell format.

The boat heels slightly to starboard as it circles about, because of the purchase of the dredge on the bottom. The towline, made of one-inch nylon and adjusted to three times the depth of the water, runs from a cleat on the starboard rail down to the dredge, while another double-braided half-inch line runs from the dredge to the boom and then to a hydraulic hauler similar to Brian's Hydro-Slave. When Dan cuts the engine and throws a handle on the hauler, the boat heels more sharply as the dredge is tugged up through the five or six inches of bottom it has sheared into, and then shivers with a motion that is just perceptible in the sole of the boot as the dredge breaks free.

The water today is the color of Spanish olives. A billowing plume of ashy mud gathers in its depths, builds toward the surface. From the midst of that plume the dredge explodes, looking from the rail like a coffin being shot out of a missile silo. It pops through the surface and hangs swinging from the boom like a lobster trap on sled runners — narrower than a trap, longer, but roughly the same volume, steel-ribbed instead of wire-framed, a sort of maximum-security facility for lobsters too incorrigible, perhaps, for wire or oak. But the dredge has no lobsters inside, nor any of a lobster trap's spiderweb interior architecture. It's just a cage open at one end, with a ten-inch downward-sloping knife edge for its lower lip. The steel of that edge is burnished to a damascene brightness; the belly of the cage, dripping above the water, is clogged with a sticky black plug of bottom mud and eelgrass.

Dan hits another handle and swings the end of the boom over a table-sized plywood culling board built into the starboard gunwale. The dredge rocks over the board, and then another adjustment drops it far enough so its foot rests on the board, its mouth hanging upright from the boom. The plug packed into its ribs is pebbled with the protruding rims and broken ends of shells, squirming with the motions of crabs. Dan reaches over the board and hits a lever on the dredge, releasing a door hinged into its belly. The quahog-laden plug loosens and spills out.

Even in the shapeless bulk of his gray sweatshirt and orange rain pants, Dan is tall, straight, fair, and fence-post lean, as spare as the trunk of a quahog tree. His gloved hands sweep like a card sharp's over the culling board, sorting fist-sized quahogs from the smaller and higher-priced littlenecks and cherrystones, dropping the sorted clams into separate wire baskets.

These are all the same hard-shell clam, *Mercenaria mercenaria*. The two-year-old 'necks and three-year-old cherries are destined for the gourmet raw-bar trade; the tougher 'hogs are used in chowder and such specialty items as clam pies and clam fritters. (Fried clams are usually made from soft-shell clams, or steamers, while clam strips are prepared from surf clams, which are bigger than quahogs.) Once Northeast Indian tribes made beads from the shells of these quahogs. The beads, called wampum, were then strung into necklaces and belts used only for ceremonial purposes, at least until white traders introduced the notion of using wampum for money. In that new economy, purple beads from the edges of quahog shells were worth twice as much as white beads, and *Mercenaria mercenaria* became an early agent of globalization.

There are no more mature clams in this haul. Dan pushes the seed clams, less than the one-inch minimum hinge size, over the rail, along with rocks, gravel, clumps of mud and eelgrass, scuttling orange rock crabs spotted with drops of India ink, broken bay scallop shells arced with splashes of Toledo silver.

The only other vessel in sight today is another quahogger to the south, one twice the size of the *Last Resort* and running dual dredges, both with big twenty-inch knives, off its stern. Too big to work the ridges as precisely as Dan's boat can, the *Comanche* simply sweeps back and forth across this side of the bay like a harvester across a field of grain, concentrating on volume of large quahogs rather than the supplementary mix of littlenecks and cherries that the *Last Resort* can get. "He's a good guy," Dan says about the captain of that boat, "but he worries too much about what other guys are doing."

Knowledge of what the competition is up to is one of the virtues of entrepreneurship. In the fishing business, however, exactly where one builds a fence between fraternity and camaraderie, on the one hand, and prudent self-interest, on the other, is a matter for the kind of internal microanalysis that Brian sees underlying every gesture of hunting-and-gathering. Harry Hunt built that fence high and topped it with razor wire. Lobstermen in general, with their traps laid out unsecured, are prone to do so. But to Dan Howes, the signal beauty of this quahog fishery is that he can nearly tear that fence down entirely. He's pleased these days to worry hardly at all about what other guys are doing.

Like Carl, Dan fishes because his father loved to fish. Everett Howes worked briefly on a long-liner out of Chatham, but fishing as a business didn't hold him. For paying his bills, he preferred the steady, guaranteed money of running a crane at the Nickerson Lumber Company's yard in Orleans; for feeding the soul, he liked angling for striped bass on the same Brewster flats where Tom Johnston used to fish with Carl. Dan's grandfather

Otis Howes was an electrician, and his advice to Everett was that he should do whatever he wanted, live where he wished, so long as he didn't go across the bridge at Bourne, where the Cape Cod Canal crisply divides this land from Boston and China and everywhere else. I don't know if those same words were ever offered to Dan. Maybe they just came dissolved in his blood.

On his mother's side, Dan has a view of the wide horizons of college-preparatory schooling. His maternal grandfather, Samuel Bartlett, was the founder and headmaster for thirty-five years of the South Kent School in Connecticut. His uncle George went to South Kent to teach in 1961 and later served twenty years as headmaster himself, retiring in 1989. His mother's brother-in-law, Rip Richards, was formerly the director of plant operations for the Holderness School in New Hampshire. Dan remembers playing hockey on the school's ice rink during vacations there in winter.

Those schools, I think, were too far from the striped bass, or else simply on the wrong side of the bridge. Dan went to Nauset Regional High School. In the summers he worked on the charter boats at Rock Harbor, indenturing there ten years after Brian under the notorious Stu Finlay. "Stu had his own way of doing things, and he could be pretty hard on you, especially if you were a kid," Dan told me. "But I worked for him four summers, and I guess I caught on. The last two summers he never raised his voice to me." And if Everett Howes could do what he pleased this side of Bourne, then so could his son. "As far as fishing goes, my parents actually didn't encourage me or discourage me. I don't think they really approved, to tell the truth, but I guess they figured I was going to do what I was going to do."

Dan tried the University of Rhode Island, enrolling there in 1978 for a two-year associate's degree in commercial fisheries. He stayed a month and then came home. "I might have stuck it out if I'd had a little more seasoning," he explained, "but I was into the skiff fisheries — I liked to be home at night — and I didn't see why I even needed to know about dragging, for example, with all the other stuff I could already do." He bought a twenty-foot skiff, went back to Rock Harbor in the summers, and used the skiff through the winters to jig for cod, dredge for bay scallops, bullrake for quahogs.

In 1984 he quit the charter boats for a site aboard the *Sea Lust,* a forty-foot fiberglass gill-netter out of Chatham. In the summer the *Sea Lust* worked the Great South Channel, the shipping lane in and out of Boston that angles down into the mid-Atlantic as a broad valley between the Nantucket Shoals and Georges Bank. There, where the *Southern Cross* once sailed, they caught cod. Occasionally in the winter they ran forty miles east to such drumlins in the channel as Jim Dwyer's Ridge, where they caught pollack. "We used to pound away inshore, catching two thousand or twenty-five hundred pounds of codfish each trip, and make good money there all winter," Dan said. "But then we'd go out for pollack, get boatloads of that, and just get it out of our system. We'd realize we could make the same money within ten miles of the beach, and so we'd go back to that for a while."

But by then the Hague Line had been drawn across Georges Bank and groundfish harvests were already in decline. When offshore areas such as Jim Dwyer's Ridge were identified as haddock spawning grounds and closed to all winter fishing in the mid-1980s, the *Sea Lust* followed the example of other Chatham gill-netters and converted to long-lining for the winter. But Dan didn't like long-lining any better than his father had. "I don't know," he said. "It always seemed like we were catching a shitload of fish hooking, but we never made any money. We got a lot of small fish, and we had a lot of overhead with all the bait you had to buy. There wasn't any overhead with gill-netting — the expense of the nets was taken out automatically every trip, and your gear was always as good as baited. I hated the hooking, hated everything about it."

In 1986 he quit the *Sea Lust* and caught on eventually with another Chatham gill-netter, the *Destiny.* Its skipper, Doug Matteson, shared Dan's distaste for hook fishing and was skilled enough at gill-netting to be able to pay his bills with that even during the lean winter months inshore. At the same time, Dan bought the *Sea Biscuit* and took time off from gill-netting each September and October to hunt bluefin tuna from the new skiff. In 1987 he caught eight of the great fish, in a fishery where few fishermen catch any. One of Dan's bluefins that year topped 900 pounds, dressed out at 720, and sold at the dock for better than $9 per pound.

But the cod continued to dwindle. The 1988 season was the first in a series of years in which not even Doug Matteson could find fish along the beach. Reluctantly, in 1989, Dan went back to the *Sea Lust,* where he remained until just last year, taking one year off in 1991 to build his house in Orleans. In the winter, when the *Sea Lust* went long-lining, Dan fished by himself in the *Sea Biscuit.* It seemed to him by then that the summer run of cod in the Great South Channel was making a comeback. But he didn't at all like the way things were going politically in the fishery. The New England Fisheries Management Council was clamping down hard on groundfishermen with Amendment Five, which was finally implemented last May. This began a stepped series of cuts in the number of days fishermen were allowed to go to sea to fish each year, imposed certain gear restrictions, required captains to keep extensive logbooks and call federal agents at the National Marine Fisheries Service when they were leaving or returning to dock, and made provisions for a satellite-based vessel-tracking system whose on-board devices boat captains would be required to buy.

It could have been worse. By and large, Amendment Five concerned only boats bigger than forty-five feet. Therefore the *Sea Lust* was spared the chaos that attached to delays in NMFS's mailing of new photo ID operator permits and issuing of logbooks. It was also spared the near impossibility of penetrating NMFS's overloaded phone lines, the vague status of fishermen's appeals of individual days-at-sea allocations, confusion regarding the effective dates of many of the amendment's separate provisions, and so forth. But now Grendel is to be followed by Grendel's Dam: only twelve months into this five-year regimen, fishermen are learning that Amendment Five is not harsh enough, that an even fiercer Amendment Seven must be drafted immediately and put in place, and this will be a regulatory package that catches even small boats like the *Sea Lust* up in its maw.

Whether there are cod in the channel or not, gill-netting these days does not look like a growth industry to Dan Howes. Nor does any fishery that works in federal waters. More jaundiced in these matters than Brian, Dan speaks for many fishermen in his contempt for scientists and their estimates of the unseen, a contempt that answers the scorn fishermen themselves feel from the academy. "All we hear from them is 'This is what the science says. We're right, you're wrong, so shut up and go home.' This is the same science that told us there weren't any tuna left in the world," Dan says, referring to NMFS's gloomy stock assessments of giant bluefin populations in the early 1990s — assessments that were loudly disputed by fishermen. Now, working collaboratively with full-time tuna fishermen, scientists are concluding that bluefin travel farther and spend more time feeding on the bottom than was previously thought, that estimating populations on data derived from surface sightings may skew the numbers downward, and that quite possibly the fishermen were right a few years ago and tuna stocks were larger than they were credited to be.

The fact that over the last twenty years scientists have been proven right in their gloomy stock assessments of every other commercial species is incidental to Dan's fears for the fate of offshore fisheries. He says that science and politics are oil and water and should not be mixed, and that just such a polluted mixture is routinely dispensed now by NMFS. Sprinkle that over enough regulation to strip a fisherman of his treasured independence, and enough bureaucracy and paperwork to make compliance with those regulations difficult, and you have an indigestible recipe calculated to rid the waters of vessels such as the *Sea Lust,* the *Cap'n Toby,* the *Honi-Do.* "The federal government doesn't want to deal with small owner-operators," Dan complains. "All they want here are six giant factory trawlers with observers on each one. Amendment Seven is just a way to get us out of the picture."

Dan says he doesn't have to call NMFS ("as far as I know") whenever he leaves Rock Harbor in the *Last Resort,* and that he'll mail in his permit on the day he does. In abandoning the Chatham gill-netters for this little-subscribed quahog fishery, which takes place only in state waters and at the moment includes only a half-dozen or so other boats working full-time, he believes that he has not only moved out of the gunsights of the federal government but found a small and profitable niche away from the paranoia of the groundfish and lobster sectors — a place where he doesn't have to worry much about what other guys are doing.

Dan checks his watch and starts winching his dredge in again. He says that the tide is just turning; this is the best time of day for him, since the clams move closer to the surface for feeding. He looks back over the transom at his towline. His blue eyes peer hawkishly from beneath his watch cap, glinting like stones above the hawkish thrust of his nose. The dredge comes up gravelly and seamed with shells.

He says some days are better than others out here, but there is always something. You can always make at least a day's pay. I ask if he isn't worried that a lot of other people will move into this fishery as Amendment Seven settles in.

"Not enough money in it," Dan says. "You're not going to gross more than four or five hundred dollars a day out here, and for most of those guys, that's not going to be enough to service their debts." He throws the boat into neutral, swings the dredge over the culling board. The *Last Resort* pauses in its interlocking circles, each one an infinitesimal version of a clipper ship's circumnavigations. "And you've got to have the right sort of temperament to work in this fishery. It's not as exciting as some."

Speaking of excitement, I think of Mike Russo, the Chatham long-liner, who decided after all to gamble on that bait. That turned out to be a smart move, since the council did not invoke any peremptory shutdown of groundfishing at the August 10 meeting and is instead working through a public hearing process on Amendment Seven's alternative forms. But I also hear that Mike's young deckhand has quit in order to enroll at Massachusetts Maritime Academy in the fall, and that Mike hasn't been able to find any experienced help in his place.

I wonder if Mike is fishing alone today somewhere in the Great South Channel. The stagnant heat of the morning has dissolved beneath the force of this northeast breeze, which has lately kicked up to fifteen knots. Now the windward flanks of the waves are scrimshawed with the probing weight of its gusts. "It's kicking up worse than this on the Atlantic side," Dan says. "It always blows ten knots harder out there."

He positions the dredge on the culling board and releases its door. Its contents clatter and ching onto the board like the jackpot out of a slot machine. Dan's face betrays the inward smile of a man who has learned to run money off in his basement while others chase it to the far corners of the earth, or else into the bureaucratic mire of federal waters. He bends over the culling board and sets about paying off his boat and motors, twelve to eighteen cents at a time.

6

Piss and Vinegar

When Brian thinks of the way he hopes the Cape fisheries will look in ten or twenty years, he thinks of Mike Russo and hopes Mike will be as full of piss and vinegar then as he is now.

The fishermen of his own generation, says Brian, are world-weary and cynical; they complain all the time that they can't catch a fish. "That's why a young guy like Mike and some of those other long-liners out of Chatham now are so refreshing," he explains. "When the cod's not there, they go out for dogfish, and they do all right. They don't bitch about it. It's just 'There's the fish — let's go get 'em.' It's a whole different attitude."

In Mike the piss and the vinegar exist in roughly equal proportions. Add to that a dash of a young gun's kick-ass brashness, a helping of learning-curve humility, a teaspoon of a Zen monk's absorption in detail. Then flavor with a garnish of that native buoyancy that so impressed Alexis de Tocqueville, who characterized young America as a land of wonders, a place where every change is an improvement, every novelty an amelioration: "No natural boundary seems to be set to the efforts of man; and in his eyes what is not yet done is only what he has not yet attempted to do."

These days several boundaries, both natural and regulatory, attend the efforts of a groundfisherman, but Mike Russo has generally declined to bitch about it. As a long-liner, he makes his living setting lines that may run for a quarter mile or more, punctuated with baited hooks at three-foot intervals, across good stretches of hard bottom. In doing so, he is proud to follow a calling that is not only the most ancient of all the Cape's major fisheries but also widely regarded as the most technically demanding. From the cockpit of the *Susan Lee,* he studies the habits of older and more experienced fishermen, the fleet's acknowledged superstars, what fishermen call highliners — studies them with rolls of light-sensitive paper shuttling through his eyes, later cross-referencing these high-resolution images to the written notes he keeps on his own habits: where he set his gear, when he set it, how much he set, and each conceivably relevant circumstance of tide, depth, current, temperature, orientation, and character of the bottom that might influence his yield. These are then cross-referenced in turn to whatever the scientists say, to anything he can ferret out from books — field guides, reference works, natural histories — that he didn't know already, that maybe even the highliners don't know, about the cod's movements, habitats, feeding preferences, spawning habits.

This is part of doing his homework and climbing the curve — the hunting-gathering equivalent of John Coltrane alone in a hotel room practicing his scales. Mike enjoys all that for its own sake, as well as for the good of his wife, Susan, and their daughter, Abigail. He says he doesn't care so much about becoming a highliner himself; he is wary of the sin of pride. "If you start out with the aim to be the best, you end up eventually believing your own bullshit, and sooner or later that blows up in your face," he says. "The important thing is to do well for your family."

But because he loves so much what he does for his family, each turn of the ignition key on the *Susan Lee* is a gesture of moment and portent. Everything else, from the fall of the Roman empire to his first cup of coffee that morning, has led inevitably to that point. Beyond that, the music takes over: the seismic bass of the engine, the percussive slap of the chop against the bow, the overlapping treble notes of the pursuing gulls. Mix in the scat-singing of the hauler as it pulls in his lines, as the cod mount up one by one to the rail and then go thumping into his bins, and you have, to Mike Russo's sensibilities, the sort of gig that a fisherman can ride up to heaven.

He also loves the cod's beauty, its delicacy, its history, its reliable market value (at least until this summer's strange nosedive), loves both its substance and its ethos. The present scarcity of cod is simply another sort of boundary, a degree-of-difficulty gradient challenging his skills and the capabilities of his boat. Each two-thousand-pound load these days feels like a standing ovation, an unlooked-for pleasure above and beyond the calling of his art. Each cod that he lands is another crisp and triumphant note blown out into space, animating the void, defying the naysayers, damning the skeptics and anyone else who might suggest that the fishing is all done, that there's no more money to be made around here.

Catching dogs is another thing. If the cod aren't biting, as they usually aren't off the Cape in the early part of the summer, and if the price on dogfish is better than thirteen cents a pound, which is Mike's break-even point for those fish (the price was as high as thirty-two cents in June; now it's down to around eleven cents), then catching dogs is a living, at least. The dogs travel in packs and chew up his gangions (the short lines that connect his hooks to his trawl line). They clog up the nets of draggers, purse seiners, and gill-netters and tear right through lobster traps. They also prey voraciously on sand lance, mackerel, herring, haddock, lobster, and — unforgivably — cod. Mike doesn't like dirtying his cockpit with them, doesn't so much like selling them, but at least he approaches killing them with a certain dark relish. He approves as well of the Yankee pragmatism that leads to ventures into alternative fisheries, underutilized species; appreciates their faint tang of wonder and novelty and amelioration. He hasn't bitched about it.

But his prosperity has been built on cod, which is why the prospect of a boundary as terminal as Amendment Seven's Alternative One — the complete shutdown of all groundfishing, just unplugging the lights and walking away — is enough to shrivel even a spirit like his. He sits this afternoon on a couch opposite a clean brick hearth and a mantle spangled with photographs of the young family. The picture window to his right reveals a flagstone path, a neat lawn, and a neighborhood of snug new houses shaded from the August sun by second-growth oak and locust. One-year-old Abigail coos at her father's feet on the pile rug. Susan is in the kitchen, making coffee. Her spinning wheel stands like an emblem of fate to the left of the hearth in the rented house. Its spool is wrapped with a silvery yarn she carded, combed, and spun herself, a yarn the color of the spray-flecked aluminum on the boat Mike named after her.

Abigail has her father's sea-blue eyes. Mike's hair is flecked prematurely with gray. His thighs are big, his shoulders wide, his features American hero out of the Paul Newman mold. He glances down at Abigail, looks up. His eyes at this moment are Cool Hand Luke's, full of a wintry despair, the piss and vinegar sucked out of them. "If they do that, I'm done," he says, motioning to the newspaper and its stories about Alternative One. "I don't even like to think about it. I can't think about it."

Mike says he did well last year, fishing with a good crewman on board, an eighteen-year-old named Ty Vecchione. He adds that he's still learning the grounds out there at the channel, that the whole trick, in fact, is to be learning new grounds throughout your career, always to be on the curve. At the channel he's going against guys with ten to fifteen years more experience there, and he's just starting to compete with them, and yet even with that, and even with all the eulogies that have already been offered over the demise of the cod, last year he had the best year he has ever had.

Now he can't get away from the eulogies. The newspapers and radio stations are full of Amendment Seven and the public hearing due to take place in Hyannis next month. This sort of stuff had to come sooner or later, Mike supposes, but it's too bad from his point of view that it had to come right now, on the heels of a good year that didn't come cheap. After buying the *Susan Lee* in 1992, for a mere $27,000, he has spent three years and an additional $30,000 converting her piece by piece into the kind of vessel he needed her to be. That meant gutting her and replacing everything but the hull, forward bulkhead, and wheelhouse: new deck, new engine, new structural timbers, new fuel tanks, new electronics, and more.

Then, as soon as everything was finally in place on the boat and he felt in a position to take a big chunk out of his debt with another good year, things fell out of whack elsewhere. The three baiters whom Mike employs to coil his lines and dress his hooks with frozen squid had been working out of the freezers at Chatham Fish & Lobster, down in the industrial park. But this has been a hot summer, and at Chatham Fish & Lobster the sun shines directly onto the floor space where his crew worked in the afternoon. Often they weren't able to keep the gear and bait cool enough as they worked, and later Mike found himself fishing with contaminated bait. Finally, at some additional expense, he had to find a shanty with a freezer set-up in the park and rent that for his crew to work in.

Then, right around the time Mike was dealing with that, Ty told him he was quitting to go to college. At first Ty said he could work until the end of the summer, but then he had to leave earlier than that to get his affairs tidied up before he reported to the maritime academy. Mike hasn't been able to find any other experienced help ("It's hard to find good crew when there are a lot of clams around the shore"), and so he's doing now what he did for five months last year before Ty came aboard: abandoning the wheel and letting the *Susan Lee* run whenever he has to move aft to set gear. He's organized his bins so that afterward he can clean his catch right at the wheel. When heading home in weather too rough for that, he cuts the engine a few miles from the beach in order to dress off his catch there. "In a way, actually, I kind of enjoy all that," he says. "But it's tough on your back. Well, any kind of fishing — it's not an old man's game."

Susan comes in with the coffee, then plucks Abigail up into her arms. I think one part of Mike enjoys fishing alone precisely because of the extra work — work that right now he's young enough and strong enough to do; work that he relishes so much for its own sake that he is jealous, maybe, of the deckhand he's paying to do it. But that kind of fishing also militates against young men's living long enough to become old. Abigail sits half hidden behind the curtain of Susan's straight brown hair. Mike looks at her and says, "Forty miles out there, all alone. Kind of pushing the envelope of safety, I guess."

Susan nods.

Mike knots his hands and stares down into his coffee like a diviner, as though all the future — the next twenty years, the next twenty days — were concealed in its blackness.

7

The Solace of Outward Objects

THAT LIGHT is on again in Carl's eyes, the one that deems it only a matter of time until the foot and the banana peel find each other. He wants to know if I saw the picture in the newspaper of the fisherman from Alaska at the meeting Greenpeace organized in Chatham last night. He says Greenpeace brought that fisherman and a couple of others down here to talk about the impact of ITQ-based management schemes on fishermen in the northwest Pacific. "Did you notice the bandage on his finger?" he asks.

Carl says this guy visited aboard the *Honi-Do* on the afternoon before the meeting. "Mark was talking to him in a sort of half-assed way — well, you know what Mark thinks of Greenpeace," he tells me. "Anyway, we'd dragged up some lobsters, and this guy was watching me take them out of a barrel, and he took this big four-pounder out of my hand and asked Mark, 'Why do you band these things, anyway? We've got crabs in Alaska this big, and they don't do nothing to you.'

"And Mark just kind of looks at him and says, 'Oh, no need, really. Just so they don't crunch anything in the barrel, I guess.'

"'Are they fast?'

"Mark says, 'No, they're not that fast.'

"'Mind if I take the band off?'

"Mark says, 'Go right ahead.'

"So the guy takes the band off the sawtooth claw, and he's kind of tickling it with his finger, but this lobster is still pretty fresh. It grabs his finger and jeez, he lets out a yell and throws the lobster down on the deck, but not before his finger got pretty well shredded. Then everybody at the meeting that night in Chatham was raising their hands to ask how his finger was."

Carl's smile explains that it couldn't have been helped, knowing *Homarus americanus* and knowing Mark; it was probably mapped out in the zodiac — Scorpio was in the House of Farnham, and somebody had to pay, preferably the innocent. I missed the picture in the newspaper, but I can see the white slab of the Alaskan's finger and hear Mark reminding Carl, "There's nothing you can't fix by plastering over it."

Carl stands as straight as Aquarius at the stern of his sixteen-foot wooden skiff, which is floating in one of the lower corners of the Town Cove, a two-mile finger of water dropping southwest from Nauset Harbor into nearly the heart of downtown Orleans. At the end of the cove lie the docks and white clapboard buildings of the Orleans Yacht Club. Opposite us, between the calm of the cove and the furious traffic of Route 6A, is the Goose Hummock Outdoor Center, which sells clothing and supplies for fishing and boating from its low-slung, window-fronted blue warehouse; the Orleans Inn, with its tall red chimneys, Victorian eaves, and rooftop cupola; and the *Blue Heron,* a sixty-passenger tour boat tied to a dock in front of the inn. Carl's hands are wrapped around the aluminum handle of his bullrake, which is now dropped straight into the water. He works the handle back and forth, his back arching, as though he were pumping the handle of a railroad handcar.

Maybe the zodiac has something to with Carl's not being out on the *Honi-Do* today. The boat is fully operational again, and recently he and Mark have been out in federal waters, fishing for cod, hake, gray sole, monkfish, and plaice. By now the early morning rain has dwindled away, though the sky remains gauzed over in gray, filmy clouds. The summer air is heavy and moist, stirred into grudging motion by a slow southwest breeze, a fair-weather breeze, a good fishing breeze. For whatever reason,

Mark never called, is not going out today, and Carl is filling the day in with some clamming instead.

He brings the rake up from the bottom, working hand over hand on its long aluminum handle, letting the recovered length of the handle extend out from the bow and over the water like an outrigger. The dredge, when it finally appears, is a skeletonized dustpan, sixteen inches wide, with eight sharp tines and gracefully curving steel ribs. It comes up clean, with hardly any mud or weeds: a few rocks, a half-grown horseshoe crab, and a little more than two dozen quahogs, in all that species' market gradations in size, including the barely legal littlenecks, which are two to two and a half inches in diameter; the slightly larger cherrystones, around three inches in diameter; and one big chowder, this one nearly as thick and round as a baseball.

Roughly half of what Dan Howes brings up on the *Last Resort* in Cape Cod Bay are the big quahogs, which processors buy by the pound. But Carl's haul here is almost entirely 'necks and cherries, purchased by the piece at prices that may range from fourteen to twenty-two cents each. The clams come up dripping and already rinsed white, their striated shells burring against each other in the dredge, and are dropped, clattering like crockery, onto the skiff's culling board. Carl transfers them in random handfuls into one of the wire baskets, saying, "Don't sort them. The buyers have a machine that will do that."

It's getting toward the end of August, and any day now NMFS will be announcing the names of the recipients of its FIG grants. Carl's thoughts are never far from the fate of his application for growing quahogs in the bays, but his talk is mostly about run-ins he's had with property owners while fishing for quahogs here around the cove and elsewhere. In Massachusetts, the rights of shorefront property owners extend to the low-tide mark, though shellfishermen are allowed to work the intertidal zone. Carl says he stays away from that zone all the same, is careful to work beyond the tide mark, but he gets lots of complaints anyway.

"One time the tide was going out, and no matter how hard I tried, I couldn't keep the skiff over this patch of bottom I needed to work," he says. "So finally I just tied a line to the end of this dock that stuck out beyond the tide mark. The guy comes out of his house, and he's yelling at me, 'Hey, is that legal?'

"I tell him, 'Yeah.'

"He says, 'Well, I'm going to call the harbormaster.'

"I tell him, 'Fine. I'll just work here in the meantime.'"

A certain irony attends these turf battles along what were once viewed as godforsaken shores. I remember that in November 1620, after failing to reach the Hudson River and nearly running aground on Monomoy Island, the *Mayflower* made its first landfall in America at what would later become Provincetown. When the Pilgrims "fell in with that land which is called Cape Cod," wrote William Bradford, the second governor of Plymouth, in his *History of Plymouth Plantation,* "the which being made and certainly known to be it, they were not a little joyful." A closer examination, however, tempered that first joy. They had arrived on the doorstep of winter and at a harbor near which they could find no source of fresh water. "Besides . . . what could they see but a hideous wilderness, full of wild beasts and wild men? and what multitudes there might be of them, they knew not. Neither could they, as it were, go up to ye top of Pisgah, to view from this wilderness a more goodly country to feed their hopes; for which way soever they turned their eyes (save upward to the heavens) they could have little solace or content in respect of outward objects."

Bradford's young wife found so little solace in that harbor that she committed suicide there. In December, finally, the Pilgrims abandoned Provincetown and found safe harbor, fresh water, and land ready for cultivation at Plymouth, on the other side of the bay. There, Patuxet Indians, ravaged by smallpox probably introduced by European fishermen, had abandoned their villages and left behind fields already cleared. When later considering this circumstance, Governor John Winthrop of the Massachusetts

Bay Colony drily concluded that "God [hath] thereby cleared our title to this place."

The Pilgrims intended to live more by fishing than by farming, but were ignorant of both and had brought hardly any fishing gear. In the midst of the richest fishing grounds ever recorded, they set about starving, raiding Indian food caches when opportunities arose but turning up their noses at the quahogs, soft-shell clams, and mussels the Indians relished. In 1622, William Bradford was ashamed to report that circumstances were so bad for him and his companions that the "only dish they could presente their friends with was a lobster." But gradually, with the help of English fishermen, they learned to catch cod, and finally began to enjoy good agricultural harvests once they learned to plow their cod waste into the ground as fertilizer.

By 1639 the Pilgrims had established fishing stations in Salem, Dorchester, Marblehead, and Penobscot Bay and were selling quantities of cod to merchants in Spain and the West Indies. That year in Plymouth, internal dissension and a shortage of good farmland compelled the younger and more vigorous half of that town's population to go not west but east, to Nauset, or present-day Eastham and Orleans, where the best soil in the colony was to be found. "But freedom from the vigilance and the restraint of the Old Colony seems to have gone to their heads," says Kittredge, ". . . for they hacked down the forests in such a wanton fashion that the generations which followed found the soil blown bare. The 'blackish and deep mould' which Bradford had seen there twenty years before had become dry and sandy waste, blown this way and that with no forests to break the force of the wind." Kittredge, however, sees in this a blessing in disguise, "for by ruining the soil, it drove the Cape farmer to the sea for a livelihood, and there he won fame such as no farmer can ever hope for."

That fame was at its peak when Henry David Thoreau walked the Cape two hundred years later. He found no vestige of those original forests of oak, pine, juniper, and sassafras, but along the beach he found persistent evidence of that "hideous wilderness" the Pilgrims had found so chilling:

> It is a wild, rank place, and there is no flattery in it. Strewn with crabs, horse-shoes, and razor-clams, and whatever the sea casts up, — a vast morgue, where famished dogs may range in packs, and crows come daily to glean the pittance the tide leaves them. The carcasses of men and beasts together lie stately upon its shelf, rotting and bleaching in the sun and waves, and each tide turns them in their beds, and tucks fresh sand under them. There is naked Nature, — inhumanly sincere, wasting no thought on man, nibbling at the cliffy shore where gulls wheel amid the spray.

And though the Cape at that time posted its sons to all the foreign ports of the world, it was not itself a destination. Such pioneer tourists as Thoreau and his companion were believed at first to be fugitive bank robbers. But Thoreau foresaw a different possibility for Cape Cod:

> The time must come when this coast will be a place of resort for those New-Englanders who really wish to visit the sea-side. At present it is wholly unknown to the fashionable world, and probably it will never be agreeable to them. If it is merely a ten-pin alley, or a circular railway, or an ocean of mint julep, that the visitor is in search of, — if he thinks more of the wine than the brine, as I suspect some do at Newport, — I trust that for a long time he will be disappointed here.

In fact, the visitor was disappointed for another hundred years or so. But from 1940 to 1990 — the period of the American wartime and postwar economic boom, the rise of a prosperous middle class, and the decay of that middle class and the retirement of the boom's first beneficiaries — Cape Cod's year-round population rose from 30,000 to 185,000. Its summer population now reaches a road-clogging 500,000, while over the past two decades an average of 5000 acres annually, an area approximately equal to the extent of Provincetown, have been developed on the Cape for commercial or residential uses.

In the process, bulldozers have filled in the cellar holes of the

Eldridges, the Nickersons, the Howeses. Century-old mulberry trees, some brought here as seedlings by clipper captains returned from China, have been plowed under. Pasturelands have disappeared entirely, along with open-field songbirds such as bluebirds, bobolinks, and meadowlarks. Orleans spreads at its circumference with shopping plazas thrown up in chain-store colonial style, while beyond, in subdivision after subdivision, row after row of eponymous Cape Cod cottages occupy parcels of land hardly bigger than the cottages themselves. "Outward objects" such as these remind Brian of the golden arches and car payments; they remind the Cape naturalist John Hay rather too much of anywhere else in America: "Abstracted, in the summer months especially, to the terms of the contemporary world, some of Cape Cod's more crowded areas have a familiar, continental look. They are covered with asphalt, cars, motels, cheap housing, shops full of grotesque souvenirs with no relation to the place they serve, and they amount, when you come right down to it, to receiving grounds for power, made by a conquering civilization. Will it be the same on the moon?"

Carl Johnston came to Cape Cod, arguably, as a child of that conquering civilization. But in becoming a fisherman, he became intimate with what Thoreau found so lacking in flattery; became divorced, in his fashion, from the fashionable world. Another Cape naturalist, Henry Beston, believed civilization to be "sick to its thin blood for lack of elemental things," and of course it is to treat this thinness that the fashionable world repairs here now. But once here, that world insists on its wine and mint juleps, its usual flattery and outward solace.

Fishermen feel this in ways both large and small. Nitrogen runoff from more and more septic systems has already smothered seven thousand acres of shellfish beds around the Cape. Skyrocketing shorefront property values push fishermen's homes away from the water and reduce their shorefront access. The business of fishing itself comes to be viewed as a "lifestyle choice" involving cruelty to animals, or at least nuisance smells. (When people tell Brian's friend Steve Smith that his truck smells bad, he replies, "Smells like money to me.") Considering the argument that transferring public resources to private ownership ensures that the resources will be protected and enhanced, Brian replies, "Whoever believes that should take a look at my front yard." This, with its stacks of drying lobster traps, its piles of buoys and warps and fish totes, its assorted pallets and sawhorses, its moldering skiff and spotty groundcover, its occasional nuisance smells, is no different from any other lobsterman's front yard. But I've heard reports on the radio that homeowners in Chatham are pushing for regulations against such yards.

Carl has more patience with this sort of thing than I do. He's working a narrow strip of hard-packed sand in twelve feet of water a few hundred yards offshore. This is a favorite spot of his, and he finds it by lining his boat up with certain landmarks onshore: an oak with its bark stripped off, a wooden staircase leading up from the spartina grass to a house hidden in the hardwoods. He keeps one eye on a white sign planted at the tide mark in the grass. He says, "If you paid $200,000 for a shorefront cottage down here, would you want to look out your window at a guy in an old boat and shitty clothes?"

Carl wipes his brow with a red handkerchief and then returns to his handle-pumping. Even now the sound of the quahogs' shells scrubbing against each other is faintly audible as the clams accumulate, one by one, in the dredge. Thoreau found heaps of clamshells in the fields of Nauset Plain, where the clams had been shucked as bait for codfish, "for Orleans is famous for its shellfish, especially clams. . . . The shores are more fertile than the dry land. The inhabitants measure their crops, not only by bushels of corn, but by barrels of clams."

Carl will not collect barrels of clams today, will not reach his limit, and he's not even sure that he'll find a buyer for them on a Saturday. He'll be able to grow his own barrels, however, and sell at his convenience, if that phone call comes from the feds next week. He can't help feeling that the stars are favorable, the planets in alignment. Last December, during the $2.5 million first-round distribution of federal fishing industry grants in New England, aquaculture projects were big winners. Other big winners were scientists at the University of New Hampshire, Northeastern University, and the Massachusetts Maritime Academy, whose pitch-perfect grant applications absorbed 15 percent of the available money. NMFS in fact took a lot of criticism for that from Representative Gerry Studds and other observers, and Carl expects to see fewer scientists and more fishermen on the receiving end this time around. That's good, but he particularly hopes that NMFS will find in his own application, with its research component, its collaborating biologist, and its part-time work for Chris Adams and Dan Howes, a pair of displaced groundfishermen, something that can answer to the virtues of both worlds and so pacify any of the critics.

In the meantime he hopes that Mark calls tonight so he can get out on the *Honi-Do* again and make some real money. A half-hour ago I saw a red fox flash through the spartina like a scarf being drawn through a loom. Now the flooding tide has risen almost to the crowns of the grass, and the bottom is hard to work. Carl cleans his dredge, breaks down the sections of his rake handle, stows the parts of the bullrake lengthwise in his skiff. The afternoon traffic is busier than ever on the other side of the Orleans Inn and the Goose Hummock, but this eastern side of the cove is silent and wooded, its conquering houses hidden like pillboxes along its slope. I see Carl moving like a target through the crosshairs of their windows as he cranks up his Evinrude and points the old skiff back to the landing.

8

Terminator Run

RONNIE HARRISON says he doesn't want any part of Brian's bushel of soft-shell clams. "I already got more than I can sell," he says. "I already turned Luther away."

"Okay," says Brian. Luther is another fisherman with whom Ronnie does business.

Ronnie blinks, surprised. Finding no resistance on this front, he seeks it on another. He stands toe to toe with the taller Brian, jabs a finger into his chest. "All right, I'll take the clams, but only on condition that you admit that Steven Seagal is the best tough-guy fag actor of them all."

Brian wants the evidence. Ronnie says okay, he took his kids to see *Under Siege II* the night before. There he saw Seagal beat dozens of bad guys bloody before killing them. As a law enforcement procedure, Ronnie says, this is far superior to Arnold Schwarzenegger's simply killing bad guys without requiring them to suffer first.

Brian is impressed but remains uncommitted. Ronnie is the owner of the Nauset Fish & Lobster Pool and Brian's regular buyer for lobsters and any other marketable species he might dig up or reel in. Ronnie steps back, achieving even in that a tough-guy swagger, a double-fisted rooster strut disarmed with a wink of irony, a nod of self-awareness. The noonday heat lifts like steam from the parking lot's black asphalt. Behind us the traffic on Route 6A sits nearly at rest, a military column extending unbroken from the heart of Orleans north to Eastham. Light jabs from its chrome and glass and makes Brian squint as, smiling, he regards Ronnie.

"Or if you deliver three lobster rolls over to Young's Fish Market in Rock Harbor," Ronnie suggests, his last and best offer. "I don't have anybody I can send over there right now."

The deal is struck. Brian lifts his bushel of clams from the bed of his pickup. We carry it by the handles into the back of Ronnie's store, where a scale weighs it out at seventy-six pounds, for which Ronnie will pay $1.30 per pound. The money will come at the end of the week. What Brian receives now are three lobster rolls and a tote of precious bluefish racks. The tote is slid on the end of a boat hook out of Ronnie's cooling unit by tall, brown-haired Esther, one of Ronnie's employees and Brian's ministering angel of bait. "Any cod in there?" Brian asks.

"Nobody's fishing for cod," replies Esther. "Why should they, with prices what they are?"

Today is Wednesday, and Brian is afraid that he may run out of good bait over the weekend unless he does a lot of scrounging now. Last night at around ten o'clock, we went down to Rock Harbor to empty the bait barrels there, and Brian was disappointed to come away with three totes of striped bass and nothing else. In his estimate, stripers are only a little better than nothing for attracting lobsters. Last night he slid the third tote into the bed of the pickup, raised and latched the gate, and said, "When I was a kid around my son Mike's age, striped bass was all I could think about. Then they disappeared, and now I'm probably the only guy on the East Coast not happy to have 'em back."

The bass went into a freezer in Brian's shed, next to one tote of cod that Esther had somehow scrounged for him the day before. Brian calls the freezer his oubliette, a conceit enhanced by his own shuffling step, the rope he uncleats from the cobwebbed wall, the creaking block and tackle that raises the freezer's heavy lid, and now the smell of abused flesh that escapes from it — the stench of bass edging toward a deshabille too advanced for even a lobster's palate.

Brian's strings off Wellfleet need hauling and rebaiting, but not for another day yet. This morning we went clamming, motoring in the skiff across Nauset Harbor to the tidal flats spreading from the back of the barrier beach. At a little after six o'clock the flats were still dripping and only partly abandoned by the tide. A gusting southwest breeze skipped with cat's paws across the harbor, chasing before it the brothy scent of the flats and a trenchant whiff of cigar from another clammer, rowing in a dinghy out to the flats. The skiff took us past the other clammer and then across shallow bottom stippled with sea stars and mussels. In the gin-clear waters of winter, fishermen pluck mussels off this bottom with pitchforks. Today a herring gull, swimming, dropped underwater like a loon at our approach and then emerged on foot, waddling up the slope of the flats with a crab struggling in its beak. The sky was a fibrous gray, the color of cigar ash, and the ground on which we beached was paved like Brian's driveway with a wrack of chalked and broken shell shards: quahog, steamer, razor clam, mussel, bay scallop, whelk, cockle. Brian anchored the skiff and then, sighing, lifted his shirt to lace a truss around his waist and lower back.

We dug meandering trenches across the flats, following concentrations of the chimney holes through which the clams extend their double-barreled siphons for feeding and waste. I learned from watching Brian how to use my rake to split the mud into even shelves, to extend my trench one shelf at a time, exposing the steamers while they were still half embedded in the mud, like pulling the wall away from an ant colony. Nonetheless, a number of the papery clams were pierced by the tines of my fork and pitched to the squabbling gulls who stood watching my work. Others were fished blindly out of the slurry of mud and seawater that covered the bottom of my trench. Many were seed clams, which slid like coins through the brass gauge that I had to fetch from the back pocket of my jeans each time I found one. I tossed these back into the trench.

Occasionally I brought up a razor clam. With its long, narrow shell in wood-grain hues of yellow and brown, this mollusk indeed looks exactly like an old-fashioned straight-edge razor. I know that Carl Johnston and others sometimes collect these in the winter, sprinkling a saltwater solution into their chimney holes to bring them to the surface. In winter they sell for anywhere from $1.00 to $1.80 per pound. That saltwater trick doesn't work so well in the summer, Carl says, and the clams are easier to catch anyway when the cold has slowed their metabolisms. Euell Gibbons, in *Stalking the Blue-Eyed Scallop,* says a razor clam is delicious in any form: in chowder, steamed, or fried in strips. Carl likes them raw, saying they taste better that way than steamers.

Today any exposed razors would have been easy prey for the gulls, since their shells don't entirely encase their bodies, and I kept them by my side for both their safety and my entertainment, so I could see a clam's pale, glaucous foot slowly extend from one end of its shell like a knifeblade gone limp in a Dali landscape and then slide wetly into the mud. Once extended, the foot's tip can be thickened and broadened by the clam into something like the helmet on a penis. With this for purchase and the foot retracting, the clam suddenly stands flagpole straight on the flats, an event that always looks like sorcery. Then, to the gulls' despair and my own enchantment, the clam descends into the mud like a driven nail.

Brian harvested three soft-shell clams to my one. Next to us, Nick Leroy, the muscular young clammer with the early-morning stogie, filled his bushel basket at a pace even more furious than that. Other clammers worked different sections of the flats as the clouds burned off and the sun showered light. Through the early morning the wind blew the greenhead flies away, though some found their way through my slipstream to claim pieces of my back and elbows. When the wind died at noon, the greenheads gathered in great numbers and worked the clammers as assiduously as the clammers worked the flats. Muscle aches made me abandon the preferred position for clam raking, squatting, and work less efficiently on my hands and knees. My sweat poured into the muck I kneeled in, and eventually what I took for miraculous quantities of sweat was shown to be the rising tide. We abandoned that flat for higher but less productive ground off Stony Island, on the other side of the channel leading out of the harbor. When that went under as well, Brian rinsed our bushel in the water and sorted through it, throwing away any seed clams and dead clams packed with mud. "Those must all be yours," he said.

Now, pleased with the price he got from Ronnie on the clams and having dutifully fought through the traffic to Rock Harbor in order to deliver those lobster rolls, Brian is on his way to Eastham, to a commercial smokehouse where he might be able to round up some more bait. In back of the smokehouse a bearded man in torn pants rewards him with a second tote full of bluefish. "I was saving these for another guy, but I don't know if he'll ever show up," he says. "You go ahead and take 'em. Hey, how are the lobsters running?"

"I was clamming today."

"With your back?"

Brian smiles. "My broker advises against retirement just now."

"Well, it's a day's pay, I guess. Prices are good enough."

"I wonder why they're so high, with all the clams that are coming in."

"Yeah, Monomoy's just stacked — clams right to the surface and three-deep down. But Maryland's not producing. They're down with warm water and an oil spill from New Jersey."

"Well, we could have a red tide here any time now."

"So how are the lobsters?"

Brian hesitates, blinks, considers. "I'm not complaining. They're okay now. But we're still way behind with that bad spring."

Buoyed by his payday for steamers and at least one tote of good bait, Brian drives home, where it only gets better. Suzanne is home as well and is playing Dylan on the CD player. Mike is at his summer job, working as a guide on the *Blue Heron,* the tour boat moored in the Town Cove. But he left his father a handwritten note on the kitchen table. Dylan prowls the alleys and byways of Desolation Row while Brian reads: "Dad, the Lower Cape CDC [Community Development Council] called. You guys got your FIG grant for $84,000."

"It makes you think."

"What does?"

Brian holds up a length of warp and points to the trap poised on the *Cap'n Toby*'s gunwale. "A trap with no lobsters in it and the mung all cleaned off the top, and a knot in the warp — like there were lobsters inside, but then somebody hauled it fast and then just threw it all overboard."

He pulls the old bait bag and replaces it with a fresh one of mixed bass and cod. Then he latches the door shut, untangles the warp, and lets the trap slide overboard, the warp paying evenly out. He stands at the wheel, letting his weight fall in with the momentum of the boat, staring through the salt-flecked windshield at a horizon bleached with light. The white cliffs of Wellfleet stand off to starboard. The wind is from the west-southwest, carrying with it the fugitive scent of meadows white with daisies and purple with vetch.

Brian still has nothing more than suspicions about who might be robbing his traps and how often. He hasn't found any scrapes of paint on his buoys that unequivocally match the bottom of the suspect's boat. He thinks for a minute more and finally says, "Well, I don't think there are lobsters here now anyway. Next time I haul, I'll get the gear out. Maybe moving it out a little bit will make a difference."

Suspicions are available wholesale. The council's Lobster Oversight Committee is meeting again tomorrow in Danvers to work further on reconciling the various EMT proposals ahead of the latest deadline, which comes later this month, but Brian is not the only lobsterman who suspects that the whole effort is now a waste of time. He said, "Smitty thinks the meeting's not worth going to. I called Dick Allen to ask if he thought it was worthwhile, and he never returned my call. I'll haul today and just go clamming again tomorrow."

A couple of talk-show deejays on WXTK radio out of Yarmouth this morning were teasing out the suspicions animating a matched pair of news stories. Put them together, said the deejays, and you have a tale of two cities. In Dennis disapproving selectmen had just denied a liquor and entertainment license to a family restaurant, while that same night in Provincetown a lesbian punk-rock band from San Francisco had played topless in a concert at the town hall. This is not to suggest that anything goes in Provincetown, said one deejay. Last week the *Mayflower II,* a replica of the Pilgrims' vessel, docked in Provincetown Harbor for a costumed reenactment of the signing of the Mayflower Compact. Two selectmen and a crowd of protesters gathered to object to the event as exclusionary to Native Americans. The selectmen also took exception to the accompanying hymns sung by the Joyful Life Gospel Band, charging that they represented "only a narrow spectrum of religious belief." "Whew! I sure wouldn't want my kids exposed to that," said the other deejay.

At Snow Shore, the tide was low and the flats were full of clammers. The white Lincoln Continental was parked on the beach again. Outside the harbor, bearing north to Wellfleet, we passed the *Kristine M,* with Greg Wade at the helm, heading out with a boatload of new traps. We hugged the shore, watching the bluffs rise from behind the beach, some with houses on top of them, some with the beach washed away and the houses now being undercut. We passed the red pulse of the Nauset Lighthouse, then the high-bluffed site of the pagoda from which Guglielmo Marconi's radio towers broadcast the first transatlantic radio message in 1903. The message was just a friendly greeting from President Teddy Roosevelt to England's King Edward VII; the immediate result was the U.S. Navy's cancellation of its standing order for carrier pigeons.

The towers are gone now, and only their concrete footings remain. The VHF radio hung from the *Cap'n Toby*'s wheelhouse ceiling hummed with a background buzz that occasionally congealed into voices. Beneath Marconi's station, the voices suggested a place like those in a Philip K. Dick novel, a nub in time and space wholly unprepossessing in itself, where in fact the present and the past are coterminous, as are — to ominous effect, and the vindication of the paranoid — other alternate realities. The voices were those of ghosts, clandestine government agents, or unimagined beings. To starboard, a cloud of gulls and terns hovered and keened frantically over a stretch of water where bluefish were feeding. The fish turned sideways as they fed, their silvery pectoral fins slicing the surface like sawblades.

Now we're hauling in forty-five feet of water and not doing very well, not even approaching the just-average days Brian has enjoyed recently. The warps and wooden traps come up frilled with red beard sponge. Brian says his traps don't fish well with that sort of growth. "Plus they're heavier. And they get *really* heavy if a storm blows silt into them. One of these traps will weigh seventy pounds dry, a hundred pounds in the water, and an extra forty pounds over that with silt in it. That's if you can even find your traps again. Five or six years ago, Hurricane Gabrielle blew by here in the fall, and I lost seventy to eighty."

A Coast Guard Jayhawk helicopter crosses noisily overhead, heading south toward Chatham with a litter of the sort used in medevac rescues at sea dangling on a line from its belly. Brian balances a trap on the rail, looks up, and says, "Wonder what that's for. Maybe Stallone or Schwarzenegger just saved the world and we're about to hear the explosion."

An opaque haze runs north to south, creeping like a caterpillar up the length of the beach from Chatham, while the air is clear, the water smooth and green and easy out to sea. For the most part, Brian's traps come up empty except for shorts and egg-bearing females, which are all pitched back. No other trap so far bears signs of tampering.

Brian throws another big egger over the side, muttering, "Piss poor." He watches the trap settle back into the water, the warp leap in wavelengths over the transom. "At least a bad day fishing beats anything else." When the next trap comes up entirely empty, he says, "But this does get old."

A flag on a floating buoy indicates the presence of a scuba diver. "We're not going to catch a damned thing with a diver around. He's down there collecting the lobsters attracted to the trap," Brian says. Another empty trap confirms this surmise.

The haze slowly dissipates in the afternoon. As Brian works his way through his strings, he thinks back to his days as a builder and the lunches he shared with bullrakers, who were already done with their day's work at noon and making three times as much as he was making pounding nails. "I don't have any regrets, really," he says. "This is still a better lifestyle for its independence and all the different things you can do. But the skills you develop are not really translatable to other fields, or for that matter recognized anywhere outside of fishing. The guys who stick with carpentry learn eventually to put out beautiful work that's there all the time for people to look at. My only recognition comes if I happen to be dropping a big eight-pounder into Ronnie's pool while some guy's wife in a tank top is in there looking at the lobsters. That's my brief flirtation with heroism."

We return to Nauset Harbor at three in the afternoon, dumping the stripped bones of the old bait outside the harbor and hosing the boat down. "Shorts 'n' eggers, shorts 'n' eggers," Brian repeats, savoring the action-hero sound of it. "We just completed what we call around here a terminator run — nothing but shorts 'n' eggers."

The water around the *Cap'n Toby* is a Mediterranean gray-green, and the beach a few hundred yards off is packed with sunbathers. The wind has come up, and the surf is heavy and foamed as it rolls over the sandbar at the harbor entrance. The boat rides the waves over the bar as though on horseback, rousing a pair of snowy egrets like puffs of froth from the surf. Brian stands at the wheel and reiterates that he's got to find more bait somewhere: "I need it bad."

He has been flirting with another sort of heroism, as successful fundraiser, since yesterday's phone call about money for the NFA's wild shellfish propagation project. His status takes a tumble, however, outside Ronnie's back door, where the proprietor begins by again recommending the new Steven Seagal movie: "Take your wife to it, Brian." Then Ronnie runs through a list of some of the other area fishermen who won FIG grants. He tosses out several names, Carl Johnston's not among them, his face lit with an appalled sort of wonder. "Jesus, what a bunch of frigging party animals," he ways. "Is this money all for drugs and booze? Can it be that easy to get those grants? Hey, I read in the *Cape Codder* that the grant to you guys and the Lower Cape CDC was for aquaculture."

"Aquaculture? Oh, shit. I gotta call that reporter."

At this moment Harry Hunt arrives, in the person of Fred Fulcher, another lobsterman who sells to Ronnie, who sees Ronnie and Brian together and gets out of his truck with his shoulders rolling and his legs pistoning up and down in imitation of the old lobsterman's gait, a combination of a sailor's roll and a linebacker's charge. The hands are held out in front and moving as though smoothing out wrinkles. "Everything's laying just like that," Fred growls, recalling Hunt's characterization of fair weather, good working weather. Then he becomes an angry Hunt, spewing malapropisms as Hunt did in that famous incident after a dragger had run over his traps. After threatening once more to prostitute the bastard, he grows calmer but still seethes with menace: "I wrote that guy a letter, you know, or I had a girl write it for me. Then I went and I even got the letter circumcised."

We might be beneath Marconi's station again, where Hunt perhaps nurses his grievances beyond the strictures of time or mortality, where at any hair-prickling moment his voice might be conjured out of the air by the VHF. Here outside the Nauset Fish & Lobster Pool, Ronnie and Brian laugh companionably, enjoying both the accuracy of Fred's routine and the truthfulness of Hunt's rage against those who break the rules, who cut the corners, who play the system, who steamroll their rivals and then get away with it. Behind Fred's pickup and the totes of lobsters in its bed, a row of European sedans are parked in front of Ronnie's store. One drives off, two more arrive; tanned men in polo shirts and cool, fragrant women in tank tops hurry in and out of the store. Fred throws up his hands, gives up the ghost, is himself again. Brian sighs and excuses himself, saying he has to find some more bait.

Down at Chatham Fish & Lobster, in the industrial park, he is pleased to come up with a couple of totes of cod and blackback flounder. He says that we'll clean out the barrels at Rock Harbor in the morning; Ronnie says the boats have been coming in with bluefish over there. Brian also runs into Mike Russo again in the parking lot, and Mike asks him if he heard about the accident off Monomoy Island today. Mike says a tuna boat moving at high speed in thick fog rammed the *Banshee,* a small Chatham charter boat. He says there were six aboard the *Banshee,* that two were critically injured.

The talk turns, Brian expressing his admiration for the big hauls of dogfish Mike and other long-liners have been bringing in recently. Mike snorts dismissively, saying that at least it's revenge for eating codfish. Brian says he remembers seeing boatloads of cod like that, and Mike replies that he's never really seen that kind of fishing. "But I'd like to — I hope to, someday. That's why I got into fishing."

Mike pauses a moment and then shakes his head, adding, "It's disturbing what's going on at the council these days. Amaru, shit, he's just a token in the middle of all those big-boat interests up there." He describes a conservation proposal brought to the council by the Cape Cod Hook Fisherman's Association, a sort of preemptive strike aimed at separating long-liners from whatever conservation measures might soon be forced on draggers. He says the CCHFA has proposed cutting its members back to a limit of eighty-eight days' fishing per year and a three-thousand-hook limit. "We can do that, I guess, but in our kind of boats, you know, we can't overpower the weather the way a dragger can. It's risky," he says.

"What if the rules we have now were enforced?" Brian asks, a species of Harry Hunt's prostitute-the-bastards fury beginning to bubble in his blood. "What if they just took the licenses away from violators? What if they *ever* did that?"

In the 1950s Robert Dow, the director of research for Maine's Department of Sea and Shore Fisheries, met with a group of Vinalhaven lobstermen. He arrived with a series of display charts illustrating his thoughts on the governing power of the market, as opposed to supply — to lobster population numbers — in determining harvest levels. The charts illustrated the relationship of price to such variables as water temperature, effort level, and landings, and the fishermen were keen enough to be able to call out changes in price accurately as soon as Dow changed a variable. Enormously pleased, Dow finished his presentation and asked for questions. The first man to raise his hand asked, "When are you people going to do something about increasing the supply of lobsters?"

Dow left that meeting disappointed, as no doubt the fishermen did. It is little more than a footnote in the history of the industry today that scientists such as Dow had in fact tried twice already to increase supply through hatchery-style lobster propagation projects. In 1888 — a year before Maine lobstermen landed 24.4 million pounds, a record that was not broken for 101 years — federal and state governments opened hatcheries in three states, using eggs from berried lobsters purchased from fishermen. During the next fifteen years 880 million artificially hatched lobster fry were placed in New England and Canadian coastal waters.

Through their first three molts and ten days or so of life, however, lobster fry are free-swimming, shrimplike organisms, half the size of a fingernail and vulnerable to predation. Efforts to raise fry at the hatcheries through their fourth molt, when they assume adult form and sink to the bottom to burrow and hide, proved at first futile and then expensive. The precise benefits of the program were impossible to measure, but they wrought no dramatic improvement in a regional fishery that by the early 1900s was declining rapidly, and in 1919 the hatcheries were closed. They were revived for a decade in the late 1930s, fared no better, and were closed again.

Also by the early 1900s, however, New England's rapacious canneries had been banished to the Canadian Maritimes (where they were then perpetrating the same mischief), and in their place was a series of landmark conservation laws intended to prevent history from repeating itself. Maine's statutes against taking berried females or lobsters below a minimum ten-and-a-half-inch total length were matched by similar laws in Massachusetts. It was widely expected that harvest levels would soon return to those seen in the 1880s.

But they did not. Instead, the character of the industry changed. When Maine legislated against the sale of berried females in 1872, and when Massachusetts passed its minimum-length law in 1874, lobstering became a managed and regulated fishery, its practices (at least as they were described in law) answerable no longer merely to fishermen, dealers, and consumers, but now also to state commissioners, game wardens, legislators, and biologists — people diverse enough in their goals and perspectives to ensure that frequently they would disappoint each other, and to ensure also that forging any political consensus would be difficult, especially in regard to an animal as incompletely understood (despite the best efforts of Herrick and others) as the American lobster.

The fishermen subject to these laws were of an independent cast of mind, and at that time they were largely unused to regulation. Herrick warned that such laws could be easily circumvented — that berried females could be brushed clean, that shorts could be sold anywhere for cash. It could hardly be otherwise at the time. A man struggling to meet expenses in a marginal fishery, in the midst of a terminator run, who found nothing but shorts 'n' eggers in his traps would consider not only how easy it was to cheat, not only how indifferent the regulators were to his own ambitions and desperation, but also the probability that his competitors were cheating. The dynamics of the tragedy of the commons take over. The herdsman resolves that positive utility lies in removing these animals from the commons. Into the dealers' pounds they go.

The laws looked good on the books and played well at the polls; committing the resources and the political will to enforcing them, however, was another thing. Game wardens were too few, too far between, and authorized only to seize illegal animals, not to punish their possessors. In 1920, repeated violations moved Maine's fisheries commissioner to suspend lobstering in the state's midcoast area. His decision withstood six days' simmering in the political pot at the state capitol and was then reversed.

Nor was it helpful that there was no consensus between state capitols in managing the industry; in particular, there was — and is — a rivalry between Maine and Massachusetts, the two chief lobster producers. Differences over the years in minimum length, in the size of escape vents built into traps, in licensing requirements, in trap numbers, in the wisdom of legislating a maximum legal size or of notching berried females' tails, built suspicion into the psyches of lobstermen of both states, convincing each that while they themselves might be sacrificing prosperity for the good of the stock, fishermen on the other side of the state line were grabbing the money on the run.

Herrick believed that since the biggest lobsters are also the biggest egg producers, this portion of the brood stock should be protected as firmly as immature lobsters. In 1915 he appeared before Maine's legislative committee on fisheries to argue his point. "The conservation of the lobster fishery is a purely economic problem, and if we strive to solve it upon the grounds of political bias or personal gain, we shall assuredly fail," he said. "If a tonic is indicated in this case, we must be willing to take our bitters."

Maine at that time declined to swallow this particular tonic, but for another twenty-five years the fishery there and in Massachusetts endured a malaise of lax compliance, lax enforcement, fluctuating demand, and a generally rising trend in fishermen and trap numbers. During the Depression, lobster dealers found themselves unable to market large lobsters anyway, and in 1932 the Maine legislature remembered Herrick's advice and established a legal maximum carapace size of four and three quarters inches, which in 1907 had been the state's minimum. Meanwhile, the advent of the four-cycle engine extended the lobsterman's range and allowed him to work earlier in the spring, deeper into the fall; winches powered by those engines could haul many more traps much faster than a lobsterman's aching back could.

A legal maximum did not turn the lobster fishery around, but World War II did, just as it did the groundfishery. Because lobster was deemed a luxury food, it was not rationed by the federal government. Consumers were hungry for protein, and full employment gave them the wherewithal in this instance to eat like William Randolph Hearst, who according to legend had founded the business of shipping of live lobsters by rail when he had some delivered in that manner to his Colorado mansion. Prices shot up from an average of 17.7 cents per pound in 1941 to 40.1 cents by 1945, and for the first time lobstering became at least marginally profitable as a full-time, year-round pursuit.

Unlike the case of the groundfishery, where demand and prices fell after 1945, this modest prosperity was carried into the postwar years. Consumers had developed a taste for lobster as a luxury food, a taste that was not displaced by the ready availability again of beef and poultry, and their own postwar prosperity preserved the means for them to keep buying it. But the means of catching a lobster in the first place became more competitive and more expensive as, one by one, lobstermen invested in the electronic depth sounders that had come out of the war and that did such a fine job of pinpointing the rocky rises and gullies where lobsters like to hide and feed. To pay for the depth sounders, more traps went into the water, and profit margins decreased.

So far depth sounders have been the last truly significant technical innovation in lobstering. They improved yield, and over the decades since the war harvest levels have generally climbed, though that has had more to do with radically increasing numbers of lobstermen and lobster traps. Depth sounders were simply a refinement, like the introduction of more foolproof parlor traps in the 1930s, to a methodology that had changed only peripherally from the dories and hoop traps of 150 years before. Scuba divers and draggers, meanwhile, were either outlawed or severely limited in their legal take of lobsters. There was no technical equivalent to the factory trawler in the lobster fishery, nor was there any such doomsday machine looming on the horizon.

Instead the fishery settled into a long tug-of-war between biologists, who urged a lengthening of minimum carapace sizes and warned of imminent stock collapse, and hard-pressed lobstermen who (along with dealers) typically resisted such increases and stood by the health of the stock. In 1957 as many legal-sized lobsters came out of the Gulf of Maine as had been harvested in 1888 — 24.4 million pounds — but they were harvested with 565,000 traps, nearly four times the number used in 1889. In the 1960s harvest levels dipped sharply. They climbed again through the next two decades, while more and more traps went into the water, even as some lobstermen geared up instead for groundfishing in the late 1970s and early 1980s.

By 1980 there were nearly two million traps in the Gulf of Maine, and biologists were estimating that over 90 percent of lobsters were being caught within one year of molting to legal size. Also by then all New England states were in accord on a single legal carapace size of three and three-sixteenths inches, though beyond that there remained substantial differences in regulation. Maine retained its legal maximum, required the notching of all berried females' tails before their return, forbade the sale of notched lobsters, and banned draggers. Massachusetts required none of this (believing the maximum to be ineffective and the notching to be ineffective and hazardous) and allowed some legal bycatch by draggers, but in 1981 — to Brian's dismay — went to a limited-entry system in licensing. In 1992, Massachusetts also imposed an eight-hundred-trap limit on its lobstermen.

Biologists still urged a much larger minimum size. Throughout the 1980s the Atlantic States Marine Fisheries Commission called for a standardization of states' regulations and a quantum leap to a three-and-a-half-inch gauge size. In 1987 the New England Fisheries Management Council tried to achieve just that and failed, though two years later Maine and Massachusetts did raise their minimum gauge sizes to three and a quarter inches. Now the council has been put in a throttlehold, or apparently so, by the federal government, and must try again to standardize and implement conservation measures. The council in turn has thrown the problem to the lobstermen themselves in the form of the Effort Management Teams. Proposed measures from those teams deal with trap limits, escape vents, hauling restrictions, closed seasons, and ITQs; none suggest any further increase in gauge size.

The lobster industry's intransigence on this issue has to do not only with a wish to retain market competitiveness with Canada's three-and-three-sixteenths-inch gauge size, but also with the consequences for individual lobstermen of the bitter, possibly fatal tonic of any gauge increase now in a fishery where nearly every lobster harvested (at least in the Gulf of Maine) weighs just one to one and a half pounds and bears a carapace only a hairbreadth over three and a quarter inches. That intransigence, however, also obscures a remarkable philosophical sea change that has occurred over the past few decades.

Maybe some small role was played by the example of the offshore canyons that Harry Hunt prowled, those dark trenches at the edge of the continental shelf — Oceanographer, Gilbert, Lydonia, Georges, Corsair — which, when first discovered in the 1950s, yielded great quantities of lobsters much too big to be sold in Maine. The area, however, had already reached the fourth stage of Herrick's model, general decrease, by the mid-1970s. A larger role probably has been played by the steadily rising costs of boats, traps, fuel, and bait, which put such a squeeze on lobstermen's profits that they have grown to appreciate the influence of conservation regulations in keeping prices up and quantities consistent. And then there is simply time, which brought into the industry a generation of fishermen raised on regulation and inhabiting a world more sadly informed by scarcity and finitude than their predecessors' had been.

Whatever the reason, while it remains relatively easy to circumvent the regulations of any state, lobstermen as a group are now much less prone to do so. "The compliance rate is much better now," Dick Allen told a panel of biologists convened to assess the state of the resource and industry. "Twenty years ago there was a big trade in short lobsters. People didn't look at it as such a terrible thing. There's a minimal amount of poaching now. Same thing with scrubbing eggers. It wasn't such a bad thing a few years back. Now it's minimal."

Violations still occur, just as trap robbing and gear-cutting wars still occur. Brian is scrupulous enough in his own fishing and vocal enough in his advocacy of conservation measures to be known as Dudley Do-Right among some of his more, shall we say, pragmatic peers. But the great majority of today's lobstermen are also fisheries conservationists of long standing, urging strict enforcement of current regulations and sometimes proposing or adopting stricter measures than their old adversaries the biologists deem necessary. Nevertheless, the issue of gauge length and the mystery of the great lobster harvests of the 1990s continue to divide the two camps.

Through the 1970s and 1980s, the fishery seemed to stabilize in something most like the third stage of Herrick's model, with fluctuating yields, multiplication of fishermen and traps, decreases in the size of lobsters caught, and increasing prices. To Herrick this was a "period of real decline, though often interpreted as one of increase," albeit it was not interpreted so in this instance. Rather, it presented to both camps the image of a fishery teetering on the verge of that fourth stage and ready to collapse at any moment, like the groundfishery on Georges Bank. Lobstermen saddled with boat payments and thin profit margins fought any gauge increases but generally agreed that something had to be done.

Yet in 1990 the Gulf of Maine's 101-year-old record for lobster harvests, which had been tied in 1957 — 24.4 million pounds — was shattered by almost 4 million pounds, and in the same year Massachusetts registered unprecedented landings of 16.6 million pounds. New England harvests as a whole topped 60 million pounds, and the following year that mark was topped again with landings of 63.4 million pounds. Overall landings declined somewhat in 1992 and 1993, though Maine by itself landed a record 31 million pounds in 1993. In 1994, however, records again fell or nearly fell all over the region in yields that far exceeded biologists' predictions: 38.9 million pounds out of Maine, 16.2 million from Massachusetts, close to 80 million for the whole of New England.

Last April, after the 1994 harvest totals were released, a cartoon in the *Commercial Fisheries News* showed two lobstermen reading the headlines on the record-breaking landings. One asks, "How are they going to make *this* sound like bad news?" The fishery seemed to have restored itself to Herrick's second stage of "greater supplies each year to meet a growing demand." But another characteristic of that stage is "lobsters of fair size." While Brian gets at least some fair-sized lobsters due to the migratory character of the Outer Cape's lobster population, Gulf of Maine lobsters consistently remain just barely legal.

Biologists shout in the face of the great harvests that the resource is overfished, that these record-breaking years are the result of record-breaking numbers of fishermen and traps, perhaps also warmer water temperatures and an accelerated molting rate — no doubt also record reporting, suggests Brian, now that lobstermen see quotas on the horizon — and that catch per unit of effort has actually been declining through the 1990s. With so much spawning stock removed from the fishery each year, they say, lobsters are in greater danger than ever before of being unable to sustain themselves.

Henry David Thoreau saw lobsters being caught off Cape Cod in the days of Herrick's "period of plenty": "About Long Point in the summer you commonly see them catching lobsters for the New York market, from small boats just off the shore, or rather, the lobsters catch themselves, for they cling to the netting on which the bait is placed of their own accord, and thus are drawn up. They sell them fresh for two cents apiece. Man needs to know but little more than a lobster to catch him in his traps." Since then respect for a lobster's wariness and intelligence has grown, while that for man's has generally decreased. "The more you know, the less you know," says Brian, summing up all that he has learned in twenty years of stalking the animals. For that very reason, silent springs of the sort that he's endured this year, and the terminator runs that still punctuate his summer fishing, all bad news, are not just hard on his profit margins: they play like coming attractions to an end-of-the-world economic scenario spun too far out of control for even Steven Seagal to head off.

There are no traps to haul today, but Brian needs to change the oil on the *Cap'n Toby,* and he can't find his oil pump. He limps perplexedly back and forth in drizzling rain between his basement and his shed, bending and lifting and peering and stooping and overturning. He mutters, "Everything I do is hurry up and stop." Finally he slows to a stop in the shed, gazing out the door through the drenched air at the abandoned cranberry bog that fronts his house, conceding that the archons of the intergalactic conspiracy have stolen his pump in the night, and now he's just going to have to buy a new one.

The archons have been busy with matters great and small. The rain came down in pikes and spears at Rock Harbor early this morning. The tide was out and the steel gangways down to the docks were steep and slick. A woman scrubbing the foredeck of her husband's charter boat, pouring rain or no, told Brian that one of the *Banshee*'s passengers had to have a leg amputated. Meanwhile I pulled racks of bass, not bluefish, out of one of Brian's barrels as a blonde in a lavender Eddie Bauer raincoat walked past me with her fingers clamped to her nose. Brian came up the other gangway hauling another tote filled with bass. "Didn't you hear Ronnie telling me yesterday that there were going to be a lot of blues here today?" he said. "That's why I only took one tote yesterday from Esther."

At the Nauset Fish & Lobster Pool, Brian picked up his check for the lobsters he took in last week. Esther wasn't around, and Ronnie was busy upstairs in his office and couldn't come down. Brian rang him up on the downstairs phone: "Hey, Ronnie, you remember what you told me about blues yesterday at Rock Harbor? Uh-huh. Do you know how I went out in the rain this morning and got two totes of nothing but bass? Well, maybe Esther can put me on her priority list. Just leave me a tote here. Okay. Yep. Now you can go back to whatever you're doing up there."

Searching afterward through the prodigality of the years in the shed and the basement and then stopping at Orleans Auto Supply for a new oil pump delayed our arrival at Snow Shore to a fashionable 11:30 A.M. Now the rain has been wrung out of the air, and the sky is breaking up into swatches of clean linen while crows rasp from the pitch pines. Behind the beach, the scrub's scurfy green is frothed with pinpricks of color: the butter of seaside goldenrod, the lavender of sea rocket, the coral pink of coast jointweed. In the bed of Brian's pickup are five totes of bass that are too far gone, banished even from the oubliette, and now Brian has to take them out to the boat and consign them to the mercy of the seagulls and amphipods.

Aboard the *Cap'n Toby,* Brian lifts his engine cover, fishes the wet black pad of oil-absorbent cloth from inside the milk canister that still hangs on a breather line from his crankcase, and uses a gaff to lift the bait bag packed with oily cloth from his bilge. The Volvo is still leaking two quarts of oil per day, and Brian has resigned himself to having to buy a new engine when this season is over. That's not to say he knows where the money is going to come from, given the way his paychecks from Ronnie have been going. He puts a fresh bundle of cloth into the bait bag, drops it into the bilge, and sets about attaching his new oil pump to the Volvo's sump.

Another lobsterman, Jeff Alberts, comes in from a morning's hauling while Brian cranks on the pump. He cruises by the *Cap'n Toby* and calls out, when he sees Brian, "A quarter pound per pot!"

Brian nods and waves, knowing how far short that is of a lobsterman's usual break-even point of a pound per trap, how poorly it compares to the typical yield of two or three pounds per trap that lobstermen got when he was starting out with Harry Hunt. His T-shirt today displays a *Far Side* cartoon: a worried lobster suspended in a chair over a pot of boiling water, while a line of chefs with baseballs prepare to pitch at a target linked mechanically to the chair. Brian has an idea now for a different T-shirt, one that refers to a conversation he's tired of having. "People are always coming up and asking you how you're doing this season," he says. "But they never ask lawyers how many people they're suing this year. They never ask doctors how many broken legs they've set or amputations they've performed. I'm going to have a T-shirt made up myself, one that just says, 'A quarter pound per pot — anything else?'"

He puts nine and a half new quarts of oil into the Volvo and replaces the engine cover. Then, one by one, we dump the five totes of bass over the port rail. The bass racks float, as bad bait will do, with ribbons of skin skeining, white in the black water, slicks of oil spreading to blossom-sized disks of indigo, vermilion, and gold. The gulls start picking among the racks before the first empty tote has rattled to the flooring.

On the beach, Jeff Alberts is loading gear and a few lobsters into his pickup. "I guess the big run is over," Brian says to him.

"Yep. Prices are down too."

"Oh yeah?"

"Four seventy-five for deuces over at Old Harbor, three-fifty for chicks."

On the way up to the truck I ask Brian what prices he's getting from Ronnie. He says he doesn't know. "Probably a little better than that. Ronnie's got a retail business, and the guy over at Old Harbor is more of a middleman. But Ronnie's funny about prices. He says he used to tell a guy he was paying three-fifty and then get all sorts of shit about why it wasn't three seventy-five. So finally he just stopped saying what the prices were. I just keep track of what I caught, divide it into my paycheck when I get it, and figure it out that way. Pricewise, I might do a little better playing the field now, but if I stick with Ronnie, he'll buy my stuff in September, when there's nothing but little chicks coming in and demand is down. A faithful buyer will do that."

I suggest to Brian that Ronnie needs a T-shirt: "The price is *fill in the blank* — anything else?"

Brian laughs. "No, he doesn't need that — not Ronnie."

Out through the windshield, just on the other side of the barrier beach, we see the *Blue Heron* float past, its sixty seats half full of tourists. Up in the bow, certainly, is Mike Gibbons, a microphone in his hand, his voice proceeding ghostlike from the boat's speakers as it informs the passengers about Champlain's doomed ship's carpenter, Nauset's early fleet of mackerel schooners, the long telegraph cable to France, and now the harbor's thriving lobster industry. Meanwhile the sun shines down bright and clean and the *Cap'n Toby* points jauntily out to sea.

9

Always Hopefully

PHIL COATES, the director of the Massachusetts Division of Marine Fisheries and the chairman of the New England Fisheries Management Council's groundfish committee, starts the hearing off with a spoonful of history and a dose of bitters. He briefly describes the events that have led up to Amendment Seven: the implementation in March 1994 of Amendment Five, with its limits on days at sea for groundfish vessels, its increase in minimum net mesh size, its area closures on parts of Georges Bank and the Great South Channel, its moratorium on new fishing licenses in federal waters, and its requirement for boat operators to keep detailed fishing logs; then, in August 1994, the announcement by a panel of NMFS scientists that Georges Bank cod were at a record low and were still being overfished, that yellowtail flounder stocks had entirely collapsed, and that fishing mortality rates "should be reduced to as low a level as possible, approaching zero"; then the emergency closure by Secretary of Commerce Ron Brown of an additional six thousand square miles of fishing grounds in December 1994; and now Amendment Seven, the goal of which will be to rebuild the spawning stock biomass of cod, haddock, and yellowtail flounder and to prevent other groundfish stocks from being overfished. "But there is no certainty that spawning stocks can come back," Coates cautions. "Levels this low have never been observed before."

With his trim white beard and inquisitive eyes, Coates looks like both a tenured professor and a soft-touch uncle. But there is no sugarcoating his remarks today. He speaks in a conference room at the Tara Inn in Hyannis, standing at a table he shares with two other council members, including a stone-faced Bill Amaru, and two staff members. Fishermen are still arriving, finally filling up nearly all of the 250 or so folding chairs that have been set up in the room. It's a room more crisply functional, less coyly opulent than the Marblehead Room of the Holiday Inn in Peabody. Its double doors are propped open, and gusts of fresh air blow in with the new arrivals, air flavorful with the mild, vaguely overripe tang of Nantucket Sound in September.

Mike Russo told me last week that he'd be coming to this public hearing, one of nine hosted this month by the council at various points in the Northeast. He also told me to watch for a gambit on the part of the Cape Cod Hook Fisherman's Association, through which he and other long-liners just might be able to dodge the Amendment Seven bullet. I scan the rows of seated fishermen and then take a seat in the rear, hoping to see Mike as he comes in.

Coates walks briefly through the amendment's several alternatives: first, and most infamous, the complete prohibition of fishing throughout the region with any gear capable of catching groundfish. "That would be the quickest rebuild," he remarks, "but it would come at the highest socioeconomic cost." Alternatives two through four involve various combinations of stricter possession limits, further reductions in days at sea, tighter gear restrictions, more area closures, and monitoring via quotas and total allowable catch figures, each fishery — cod, haddock, and yellowtail — to be shut down when its TAC is reached. Coates adds that a fifth alternative has lately been added to the list: status quo under the strictures of Amendment Five. He says that the latest scientific analysis indicates that some rebuilding may possibly occur that way, though at a much slower rate. He warns that projections also indicate a significant chance that stocks would continue to decline under Alternative Five.

None of this is news to anyone in the audience. All five alternatives have been spelled out in the newspapers and bruited about the docks, bars, supper tables, and wheelhouses. The assembly receives the first four plans with the enthusiasm of those granted a choice of their means of execution. Alternative One would be swift decapitation; the next three offer the tribulations of a more lingering demise, a death by a thousand cuts. Regarding the last alternative, Amendment Five is a devil in itself, but fishermen have lived with it for a year, and now, ironically enough, it is at least the devil they know. But Coates's manner plainly suggests the back-story to Alternative Five: that it was tossed in at the last minute as a matter of political expediency for the hearing process, and that Andy Rosenberg, the newly appointed Northeast regional director of NMFS, would never stand for it.

When the floor is thrown open for public comment, Mark Leach, the president of the Cape Cod Hook Fisherman's Association, is the first to advance to the microphone set up in front of the council panel. A television and VCR have been positioned to the panel's left, facing the audience. Leach says that the CCHFA is "admamantly opposed" to Alternative One because it would leave hook fishermen entirely without work. "Unlike draggers, we can't go out and catch whiting or shrimp instead," he says. Alternative Two or Three? Opposed. "With TACs, you have fishing derbies that favor larger vessels, whoever can catch the most fish the fastest. And once the TAC is reached, you have an Alternative One situation." Alternative Four, which would close the Nantucket Shoals for the first six months of the year? "That's almost as bad as Alternative One. For Cape hook fishermen, that's the only water you can work in small boats at that time of year." Alternative Five? "That's workable. We can support that. We would also support gear limits of thirty-six hundred hooks per boat and cuts in days at sea down to one hundred days per year."

That said, Leach moves into the substance of his comments: that hook fishing, whether it involves long-lining, jigging, or rod-and-reel fishing, is a profoundly different enterprise from the trawling that draggers do. For several reasons, he says, hook fishermen should be exempt from whatever measures finally make up Amendment Seven.

He asks that the lights be dimmed, moves to the VCR, and begins playing a segment from the Public Broadcasting Service's *Nova* series. Grainy black-and-white images appear of schooners in a fleet among heavy ocean swells, of men in oilskins bobbing in dories at the schooners' sides and unloading cod and halibut of extraordinary proportions. Until eighty years ago, the narrator says, schooners with baited hooks fished on the Grand Banks for century after century, their efforts always yielding good results. Another grainy image, this time of an old eastern-rigged side-trawler; this is followed by crisp color images of a factory trawler hauling the cod end of its net, packed to bursting with silvery fish, up the runway-sized width of its stern ramp. Then side-trawl vessels appeared on the Grand Banks, says the narrator, and were succeeded by huge stern-trawl draggers in the 1960s. Spawning cod that were once protected by sea ice were easy targets for these vessels, whose otter trawls are the most regressive fishing tools yet devised. Fish swimming ahead of them tire and turn back into the net. Next year's class of spawning fish are killed with this year's. As catches inevitably decrease, the nets grow in size and sophistication.

The *Nova* clip is followed by the appearance at the microphone of David Farrell, a Chatham lawyer retained by the CCHFA. He presents each council member with copies of an inch-thick packet of articles that has also been made available at the door to audience members. Articles range from such technical readings as "The Impacts of Mobile Fishing Gear on Low Topography Benthic Habitats in the Gulf of Maine: A Preliminary Assessment," to reprints from industry journals on the positive effects of trawl bans in other parts of the country, to personal statements from processors and restaurateurs on the superior quality of hooked fish to those caught by draggers.

Farrell reminds the panel that the Magnuson Act "requires the consideration of socioeconomic as well as ecological factors in the preparation of fishery management plans." He notes that Cape Cod is unique in the size of its small-boat hook fishery and defends that fishery's importance both to the summer tourist industry and to the health of the hardscrabble off-season economy. He echoes Leach on the absence of any alternatives to groundfish for hook fishermen, and quotes the U.S. Geological Survey from a 1992 study of Georges Bank: "Scarring of the bottom by groundfish trawls and scallop dredges . . . impedes colonization of the gravel bottom by attached organisms that flourish in untrawled areas, and thus reduces local biological roughness and overall species diversity and abundance."

Farrell's presentation takes better than an hour, though he speaks only briefly. He pauses often for the testimony of jiggers, long-liners, dealers, and fishing gear technologists summoned from the audience to throw the weight of their authority behind his observations. They speak of familiar bottom features in areas such as the Mussels or the Figs that over the years have been altered or destroyed by otter trawls. They speak of the bait crews and shanty workers their boats support, and the impossibility, given real estate prices, that that shanty economy would ever come back again if it were shut down. They speak of the relatively negligible amount of groundfish they take each year compared to draggers — 7.3 million pounds in 1993, for example, versus the draggers' 71.3 million pounds — and emphasize, repeatedly, "We didn't cause this crisis."

Veteran long-liner Fred Bennett, one of the acknowledged highliners whom Mike admires, a fisherman who was once a draggerman himself, offers a sort of demonstration of gear impact on bottom features. He spreads a tarpaulin on the floor in front of the panel and places a ceramic vase full of fresh carnations on it. Then he jigs a short hook-rigged groundline among the blossoms. He draws out a single carnation on the hook and hands the carnation to Bill Amaru. "I could still put it in my lapel," Amaru says, smiling.

Then Bennett, straining, lifts a twenty-one-inch trawl roller above the vase. "This weighs ninety-two pounds," he says. "I had five of these in one section on my old stern trawler. The sweep alone weighed thousands of pounds on that, and then you also had the weight of all the spacers, washers, chains." He drops the roller, and the bouquet bursts apart as the vase shatters into bits. The carnations lie scattered on the tarpaulin amid a rubble of white china shards. Laughter and applause ring through a room largely dominated by hook fishermen.

The few draggermen present at the hearing seem momentarily stunned, though the hook fishermen's tactics cannot have been unanticipated. The CCHFA has had a longstanding proposal on the table suggesting that hook fishermen be exempted from any closures on Georges Bank and that the council provide a way for trawl gear to be converted to hooks. Hook fishermen hold 53 percent of the groundfish permits in New England, the CCHFA has observed, and yet in Massachusetts they account for only 5.5 percent of cod landings. Last March, NMFS scientists, at the behest of the council, developed computer models displaying the effects on groundfish populations of completely shutting down various fisheries. Hook fishermen were elated when NMFS announced that according to their models, closing the trawl fishery alone would leave enough cod to sustain hook fishermen and gill-netters and still meet the reduced mortality rates necessary to rebuild the stock.

Five of the council's eight voting members on the groundfish committee, however, are draggermen, and last spring the committee supported council president Joe Brancaleone's suggestion of emergency measures that pointedly ignored that computer model: larger mesh sizes for trawl nets and a six-month closure on Georges Bank during the only time smaller boats can work those waters. "This was nothing but a turf war today," complained Bill Chaparales, a Harwich long-liner and tuna fisherman, to the *Cape Codder* after Brancaleone's proposal hit the floor. "We've had our plan before the council for months, without a vote or an offer to do any research, and Joe just pulls this out of his back pocket over a fish sandwich at lunch, and they bend over backwards to accommodate him."

Finally the CCHFA found a sympathetic ear in Congressman Gerry Studds, who criticized the council's reluctance to rein in big dragger interests over the years and urged Governor Weld to provide for some representation of Cape Cod's small owner-operators in his nominations to fill the about-to-be-vacated council seats of Dick Allen and Tom Hill, a Gloucester recreational boat owner. Bill Amaru, then working on the council's groundfish advisory committee, applied for Tom Hill's seat, as did Mark Leach. Hook fishermen were not gladdened by the eventual appointment of Amaru, who was once a long-liner but who now runs a dragger, albeit a small one, no bigger than Mark Farnham's *Honi-Do.*

The draggermen at the Tara Inn are all small owner-operators, the friends and relatives and neighbors of Mark Leach and Mike Russo, men who have been there to help if a hook fisherman gets in trouble out at the Figs, and probably they recognize that they are not the primary targets of this withering fire from the CCHFA. But whether friendly fire or not, it is directed at their gear and boats, and I suspect they are surprised by the power, range, and calculated precision of these antitrawl salvos. Finally, in the prickly pause that follows the end of the CCHFA's presentation, a draggerman rises to say that he's glad he's wearing his camouflage hat, "because I feel like prey." He concedes that the use of rollers on the bottoms of the nets of the big trawlers may cause habitat destruction, but the sort of flat-net trawling that his boat does will not. "The problem is the bigger boats," he says. "If they wiped out the fish, then they should pay. But if you're going to shut down this fishery anyway, then just do it. Hang me up, slit my throat, but just do it. Don't let me die slowly, just bleeding to death."

He is followed by Jim Kendall, the executive director of the New Bedford Seafood Coalition, a man who does business with big draggers and who thought he was "at a public hearing, not a hanging." He says, "This is like the family at the funeral — everybody's trying to grab a piece of the pie. Over in New Bedford we've tried to bring in everybody, hook fishermen included, to combat the same evil that is befalling all of us — Amendment Seven. We support Alternative Five, and we support fishermen working in partnership, not one fishery dictating to another. There may be bottom damage from our boats. There may not be. But you're going the wrong way, pitting fisherman against fisherman. If you think taking food off my guys' plates is going to help you, you're wrong."

Other angry draggermen follow, and then other interests. While you're at it, suggest gill-netters, you can distinguish our habitat-friendly, size-selective gear from that of draggers and give *us* an exemption. Sport fishermen point out that for the number of dollars of state revenue generated per pound of fish, no fishery is as lucrative as theirs, and they should therefore be granted, if not an exemption, at least a priority. Members of all camps observe that in complaining to this panel about the big boats out of Gloucester and New Bedford, they are complaining to the foxes who are plundering the chicken coop. One gill-netter finds the size of the panel telling: "I'm sorry that there are only three council members here to listen to our ideas on the preservation of the small-boat sector of the industry. I guess to them it really doesn't mean shit."

The infighting pauses as one of the few women at the hearing walks to the microphone. Certainly most of the fishermen expect a suck-it-up endorsement of Alternative One from Niaz Dorry, who represents Greenpeace. Surprise ripples visibly across the aisles and rows as she announces instead that Greenpeace cannot support any of the alternatives and is disappointed in the council's failure to address in any way the issues that might promote the viability of sustainable small-scale fisheries. She says that among the elements Greenpeace considers essential to an effective groundfish plan are, first, the sort of gear that minimizes bycatch (hooks or gill nets); second, a management approach that treats fisheries as integrated ecosystems, not separate spheres of stock; and third, priority to the small owner-operators whose ties to local communities promote standards favorable to sustainable fisheries.

Dorry returns to her seat in the back of the room under the umbrella of a long, puzzled hush. Finally a gill-netter rises to say that everybody here, by the way, is falling into a trap the federal government has laid for them: "This is the way the government tries to beat somebody, by dividing them up — hook fishermen versus draggers, sport fishermen versus commercial, lobstermen versus everybody." The jab at lobstermen draws laughter that for an instant unites all the camps. Then, through subsequent speakers, they resume their rivalries, fall into the trap. No one rises either to support or to object to Dorry's testimony. Each speaker steers around the elements of Greenpeace's position here, as though Dorry's statement simply marked out potholes in a road.

Last to speak this afternoon is Mark Farnham. The big draggerman rises out of the right side of the audience in his trademark red suspenders and advances to the microphone as though he were about to run it down. He says he agrees with a number of previous speakers that Amendment Five is working, that all that's really needed is for the council to give it a chance and stick to the timelines it attached to it. He wonders if scientists have considered all the reduction in effort happening anyway as fishermen abandon the cod and haddock fisheries. He believes that the council is now simply asking too much of them. "I couldn't live under Amendment Seven," he says. He looks at Phil Coates and says, "You couldn't live on twenty percent of your income," rousing quick laughter when he adds, "Though it's a pretty big income, I'm sure."

Coates smiles back. "Are you hiring deckhands? I work for the state."

"Not the feds?"

Coates allows that he stands corrected — at least today he's here on federal business. He glances at his watch, says thank you, that should wrap it up. He thanks us for our input, thanks us for coming, says that he hopes the council will have "finalization of action" on the issues raised today by December. "That's the hope."

Amid the clatter of chairs, the shuffle of paper and feet, and the clipped squelch of the microphone, Mark goes back to his seat. I seek him out there to say hello, sit down, and am surprised a moment later to find Mike Russo seated next to me. Mike says he came early and was sitting over on the left side all the time. I sit between the two, a little uneasy, talking to both, my head swinging back and forth like a gate in the wind because the draggerman and the long-liner have nothing to say to each other. In the hallway outside the conference room, meanwhile, where the air all afternoon was fresh and sweet, an argument breaks out.

The Coast Guard's Chatham lighthouse station has been flying its storm warning flag, a black square against diagonal red stripes across a white field, but by now the flag is redundant. Fifty-knot winds out of the southeast have planed it as flat and stiff as sheetrock, with only a thin margin of its leeward edge buzzing like a tuning fork. The flag, its colors crisp and bright in the pale light of October, buzzes atop its pole against layers of granitic clouds broken into shelves, trenches, wind tunnels. Here at the very nub of the Cape's elbow, the horizon out to sea spans nearly 270 degrees and is smoking from the spume of clouds knocked out of the sky and down to its rim. Meanwhile the wind comes barreling out of Nantucket Sound, rounds the slippery corner of Morris Island, and then slingshots up the length of Chatham Harbor, Pleasant Bay, Pochet Neck, the northern reaches of Nauset Beach, and out to sea.

Mike Russo and I have stopped for a moment on Shore Road, directly beneath the tower of the lighthouse, and have piled out of the pickup to lean against the rail overlooking the harbor channel and beyond that at the nearly submerged length of sand, Nauset Beach's dwindling southern tail, that stretches out toward Monomoy. The island is still stacked with clams. A moment ago we were at a marine supply store off Main Street in Chatham, where Mike bought some seven-sixteenths-inch shackles and a length of three-eighths-inch chain for his scallop dredge. There a magazine photograph of a boat overloaded with Vietnamese refugees had been tacked to a bulletin board with a hand-lettered caption: "Due to new NMFS regulations, clammers are forced to carpool to Monomoy."

Mike is grateful to have survived another summer of the regulatory groundfish wars. He bought bait all through August and September, gambling each time that the fishing would last long enough for him to use it, and each time it did. In August, though the prices were still low, he went out for cod every day he possibly could, fishing alone, despite his back pain, and fishing hard: partly because he was used to having a crewman and so used to fishing long hours, and partly because he was never sure when the council might decide that this particular cod would be his last.

That was the month that Andy Rosenberg was named NMFS's new Northeast regional director. Only forty but an experienced biologist and administrator, Rosenberg said his top priorities would be getting Amendment Seven defined and implemented in the groundfish industry, and also finally getting the state and federal plan in place for reducing effort in the lobster industry. Meanwhile, Lobster Oversight Committee Chairman Eric Smith still hoped for a diplomatic solution to the impasse over trap numbers with the Gulf of Maine EMT. "Every step that we take away from what the EMTs proposed, we lose credibility, so we're always balancing what we think we need to do with keeping the industry's support," he told the *Commercial Fisheries News* in July. "As long as we're making substantial progress and don't delay for an inappropriate purpose, the world is not going to come to an end the morning of July 21."

The world did not come to an end after that deadline for a council-approved public hearing document slipped by; nor did it come to an end after August 10, the next missed deadline. But finally the council's, and the EMTs', involvement with the issue may have come to an end. Rosenberg has begun soliciting comments on two proposed federal actions: either a withdrawal of the current federal lobster plan, which would also open U.S. markets to undersized Canadian lobsters, or preparation of a secretarial amendment to that plan to be written by NMFS instead of the council — a plan that might open the door to ITQs as a management measure in the lobster fishery.

Hurricane Felix cuffed the Cape in August, doing much less damage than feared, though it further weakened the stone revetments protecting shoreline properties in Chatham and made the actual route of Thoreau's tour of Nauset Beach — a path that currently lies a hundred yards out to sea — just a tad more inaccessible. Brian was glad to get out of the hurricane with the loss of just seven traps, though he regretted the enforced time ashore. He was glad to get out of the EMT process too: "It sounds good — we all get together to talk and compromise. Then Hitler rolls his tanks into Poland. It's tough to walk away from your business for a day and then go negotiate with people who are very well schooled in the adversarial process. It's the flower children meet Jimmy Hoffa. And the whole thing has made me a reviled person around here just for being involved with it. People who don't even go to the meetings come after me with all these critiques. No good deed goes unpunished, you know. I'm just burned out on it."

In the salt marshes, meanwhile, the spartinas bloomed and went past while male blue crabs, their claws the color of butane flames, feinted and parried and boiled into combat with other males. Mike saw small rafts of common eiders floating offshore, though there was no telling if these were year-rounders or early arrivals from the Canadian arctic. At dusk the twitterings of the chimney swifts faded away as the birds adjourned to their wintering grounds in Peru, and at night the Perseid meteor showers drew chalk lines against the sky.

Then, on Labor Day, in road-jamming, horn-blowing, white-line-straddling numbers, the fashionable world summarily abandoned Cape Cod. Roads and cottages and businesses went empty in a parody of the plagues that emptied the Wampanoag and Nauset Indians from these lands; over that weekend and the next few weeks, the Cape lost more than 60 percent of its summer population. The weather turned, its meteorological gears sighing and whistling, as the moist, placid air masses that had blown up the coast from the southwest all summer gave way to more volatile affairs pushing in directly from the continental west. On clear nights Saturn, sailing past with its rings on edge, shone with a silvery light from the midst of Aquarius in the southern sky.

All summer common terns had nested by the thousands on Tern Island, the sandy hummock that provides Chatham Harbor with its only securely protected anchorage, where Mike keeps the *Susan Lee*. Henry Beston, often harried by these birds as he walked Nauset Beach, admired them as essentially spiritual creatures, and wrote, "There are crowded days when I live in a cloud of their wings and the clamour of their cries." Thoreau marveled at their delicacy, deciding similarly that they were adapted to the Cape's "boisterous shore" more by spirit than physique, concluding, "Theirs must be an essentially wilder, that is, less human, nature than that of larks and robins. Their note was the sound of some vibrating metal, and harmonized well with the scenery and the roar of the surf, as if one had rudely touched the strings of the lyre, which ever lies on the shore; a ragged shred of ocean music tossed aloft on the spray."

One night, as a mass of cool air moved in from the west, the edge of the front tearing swaths as it advanced from a rolling surge of storm clouds, and as the setting sun shot bolts of light through the breaches in that surge, the terns rose in a slashing maelstrom into the sky, the light smoldering on their breasts, their wings as white and sharp as cut paper against the dark smoke of the clouds. The next day, in mid-September, Tern Island lay silent and dispirited.

Ten days after that, after a hearing in Orleans District Court, a complaint for negligent operation was issued against the West Hyannisport captain of the *Hawkeye* for his collision with the *Banshee* off Monomoy. The Coast Guard determined that the *Hawkeye* had been racing at fifteen knots through that day's heavy fog, and eyewitnesses reported that the forty-five-foot tuna boat struck the *Banshee*'s port side and then cleared its deck and stern. Chatham summer resident Charles W. "Johnny" Johnson — the first Allied soldier to enter Dachau, said the *Cape Codder,* and a former military attaché to the United Nations — died nine days after the collision, and his son lost his left leg. The *Banshee*'s four other passengers, including the skipper, Ron McVickar, were not seriously hurt.

That was the month that Mike had to fish a few days even beyond his fishing-too-many-days schedule in order to complete the gear selectivity experiment he had promised to help the state conduct. The theory was that different hook sizes would catch different sizes of cod, and that using larger hooks — size 11-0 and 13-0, both full-circle designs, rather than the 10-0 semicircles that Mike tends to favor — would mean that fewer undersized cod were caught and released. Don't bet on it, said Mike, and over eight trips with a state biologist aboard and alternate bundles of his gear set with larger or smaller hooks, Mike's prediction was borne out: small hooks as often caught big fish, or "markets," as large hooks, and while big hooks did allow fishermen to catch bigger fish than otherwise, they did nothing to reduce the numbers of small fish, or scrod, that were caught. Ultimately, there was no useful correlation between gear and catch size.

The price for dogfish came up to twenty cents per pound, then rose another nickel, and when Mike heard that some of the big gill-netters were landing fifteen to twenty-five thousand pounds per day, he went out for dogs himself. Sometimes he made two trips in a day, working in six fathoms of water less than a mile off the beach and landing as much as five thousand pounds. "There's no glory in it, but plenty of dough," he said. "Makes me think of buying a gill-netter myself." At the same time, the Commerce Department designated the Northeast a "fisheries disaster" area and earmarked $25 million for the implementation of a fishing vessel buyout program. This followed in the wake of the National Oceanic and Atmospheric Administration's successful $2 million pilot program, which last year received sale offers on more than a hundred vessels and which in August bought thirteen, one of them a thirty-eight-foot Chatham gill-netter. But the additional $25 million in buyout funding is contingent, warned John Bullard, the director of NOAA's Office of Sustainable Development, on "evidence that the New England Fisheries Management Council is serious about rebuilding groundfish stocks."

The price of Mike's bread-and-butter catch, market cod, finally snapped out of its doldrums, climbing steeply to $1.40 per pound during the first week of October. Mike joyously left the dogfish grounds and steamed back out into the Great South Channel, returning usually with loads between a thousand and sixteen hundred pounds. He was also elated to find an experienced crewman for the winter: Ben Bergquist, the son of a Chatham lobsterman, working now as a sternman for his father through the fall. "It was basically just luck," Mike explained. "We were both just hanging around the dock asking each other the same question: 'What are you doing this winter?'"

Greenpeace was in the news when activists in Seattle temporarily seized the *Pacific Scout*, a 238-foot factory trawler owned by American Seafoods. Protesters chained themselves to stations around the vessel at its dock in Seattle and hung banners from its rails. A crab fisherman from Alaska handed out bumper stickers — SAVE OUR SEAS, BOYCOTT TYSON — and told the media, "I have no desire to be a sharecropper for a big company." Representatives from the American Factory Trawler Association, meanwhile, handed out copies of a *Wall Street Journal* editorial describing the sort of plantation fishing that the crabber feared as the laudable and inevitable result of economic efficiency and "commercial consolidation."

At midnight last week Orion loomed over the horizon, the great stars Betelgeuse and Bellatrix glittering like frost on the hunter's shoulders. Two weeks ago noisy flocks of grackles — Cape naturalist Robert Finch calls them "thieves of summer" — arrived with redwings and starlings, moving in squabbling storms through the oaks, picking off acorns as they went. This week Brian is bringing his gear in for the winter. By now he has eighty traps stacked like cordwood in his yard, the laths green with seawater or bearded red with mung. Thirty more traps lie stacked at Snow Shore, if the wind hasn't knocked them over.

Here the wind and the ebbing tide have succeeded in knocking a sailboat over, a thirty-foot fiberglass sloop, its sails furled, its sheets and halliards snapping in whip-cracks against its mast, its keel nakedly exposed as the vessel lies on its port side in mud flats far from the harbor's main channel. En route from Gloucester to Nantucket, the sloop's captain tried to sneak into Chatham before dawn in breakers that were already cresting at eight to ten feet. He ran aground, and finally the Coast Guard had to get him off with a motorized inflatable raft.

Mike stares at the stranded vessel for a moment and then says the sandbars are moving around out there even now — they're always moving around, but particularly so today. In my mind I see them heaving beneath the breakers, see them as the lumbering coils of some tentacled leviathan. The channel is booming with the same waves that Henry Beston heard pounding Nauset Beach from his cottage, his outermost house, during the winter he lived there in 1926 — waves "monstrous with a sense of purpose and elemental will" — while the bars and flats that run to the foot of this point are raked by shallow rollers boiling with smoke and spume. The sloop pitches shallowly from bow to stern on the point of its hull, like a horse down on its flank and trying to rise. Meanwhile that leviathan is moving.

The *Susan Lee* isn't bobbing in the chop behind Tern Island today. Yesterday Mike moored it in Ryder Cove, just up the shore around Allen Point, at the southern extremity of Pleasant Bay. We climb back into the pickup, drive north up the deserted Shore

Road to Route 28 and then the cove. The harbor here is triply protected — by the barrier beach, the sentry bulk of Strong Island at the mouth of Bassing Harbor, which opens into the cove, and then the encircling arms of North Chatham and Nickerson's Neck. But even in this triple-wrapped teapot the tempest is blowing, the milky water pricked into short, agitated waves.

The *Susan Lee,* thirty-two feet long, white above the spray rail and slate-blue at the waterline, its wheelhouse spiked with antennas, is dwarfed next to the big gill-netter moored beside it. "Which one of 'em is yours? That little dinghy there?" teases the harbormaster as he shuttles us to the boat in his motor launch.

Mike smiles. "Yeah, that's mine."

"You hauling out today?"

This is the boundary between Mike's summer and winter seasons, a time he routinely uses for one of his twice-yearly haulouts of the *Susan Lee.* He cleans the boat from top to bottom, mends what needs mending, improves what bears improving. He knows prices are due to tumble again as the Thanksgiving season depresses demand for groundfish. And the weather makes it hard for him to get out at this time of year anyway. "Yep," he says. "I'll do better scalloping."

"Hopefully."

"Always hopefully."

Abigail wept when the *Susan Lee* arrived. Susan brought her out into the front yard when the trailer truck that was towing the boat appeared. The truck was hard put to maneuver in the narrow lane outside Mike's driveway, and the vessel that looked like a dinghy next to that gill-netter in Ryder Cove loomed like a tanker among these ranch houses and young trees. The boat was backed bow first into the driveway, the trailer's wheels hissing and skidding as the truck pushed it up the incline, and Abigail flinched in the midst of her tears at the popping of the trailer's hydraulic valves, the slow settling of the *Susan Lee*'s keel onto blocks Mike had set up in the driveway, her hull onto a set of padded, three-legged shores. Susan passed the child to Mike, who tried without success to calm her. In the end Susan had to take her inside.

At Ryder Cove, Mike said that when he first bought the boat, Susan had applied something of a woman's touch to her cabin beneath the foredeck. No such considerations are in evidence now: the *Susan Lee* is a working boat, musseled and algaed from six months at sea, slimed and flecked and blooded from all the fish she has caught, part killing machine and part survival capsule, with nothing of elegance aboard her now, nor any breath of comfort save Mike's sleeping bag rumpled across one berth and the two immersion suits stowed up in the bow.

The *Susan Lee* rocked slightly in the wind-clipped chop as we boarded her at the cove. Then we stripped the antennas off the wheelhouse roof — the loran, the radar, the VHF radio, the cellular phone — and scrubbed the cockpit and decks with a mixture of hydrochloric and oxalic acids shot from a plastic garden sprayer. Finally Mike untied the boat from her mooring and piloted her into the loading ramp, where the truck's trailer unit had already been extended deep into the water. When the trailer, with the boat aboard, was pulled back to the truck like a caterpillar contracting, the *Susan Lee* emerged dripping from the cove, looking at once monumental and vulnerable, like a racehorse dangling from a harness. The algae below the waterline bloomed like gray lichens. Mussels clumped in nests of black eggs about the base of the propeller shaft, and the keel was stippled with barnacles. Mike loaded the sprayer with bleach and began hosing the hull.

Now, stranded in Mike's driveway, towering from the blocks up to the level of the house's rooftop, with the wind whistling through the trees around her wheelhouse, the *Susan Lee* looks like the furniture of a dream — not only for her size, but also for her flowing lines, the sheer and sweep of her hull: curves so sure and perfect and liquid against the squared-off corners of the house as to seem the product of a higher mathematics. For her size the *Susan Lee* is a heavy boat, and her narrow eleven-foot beam leaves her with relatively little carrying capacity. But her grace here in the open air suggests all those propitious handling characteristics that Mike has already described to me: her talent for nosing her way down a swell in a following sea and then bobbing like foam over the next one; her flair for staying secure on the top of the water, no matter its antics — what Mike calls her sea-keeping ability. She rests now with her transom to the road — rests in a fidgeting sort of way, with a sense of barely arrested motion. She looks as if Mike might be able to climb into her wheelhouse tomorrow morning, turn the key in the ignition, back her down into the road, and drive her away to work.

A moment ago Mike leaned a ladder against the port side and climbed up to the cockpit. Now he's passing items down for me to stack at the opposite side of the driveway: fish totes, twenty-pound trawl anchors, round rubber buoys the size of medicine balls and as red as beach plums, and finally the high-fliers, buoys with six-foot flagpoles and aluminum polyhedron radar reflectors, which Mike uses to mark the ends of his lines. Today's task is to clean the boat out and scour her bottom. Over the next few weeks Mike means to replace the bent propeller shaft, replace the stuffing boxes that keep the hull watertight around that shaft, and rebuild the starboard side of the wheelhouse. At the same time he'll dredge for bay scallops in Pleasant Bay using a small skiff he built last fall.

Regarding the future, Mike feels a lot better now than he did this summer. Partly this has to do simply with what he refers to as "the fate factor," the interplay of desire and destiny that turns many men into fatalists. In Mike Russo's instance, however, fate has made him a Cape Codder, given him a calling, and buoyed him with a conviction that eventually things work out for the best. His grandfather was an Italian merchant marine sailor who emigrated to America and wound up working a farm in eastern Pennsylvania's Pocono Mountains. His father grew up there, did a hitch in the air force, married, and then moved to the Cape to be near his wife's sister, who lived in Eastham. Mario Russo worked as a firefighter and took his only child out bass fishing. In time Mike went to work for the Rock Harbor charter captains, and eventually for Stu Finlay. "He yelled a lot, I guess, but I didn't let it bother me," Mike told me. "I learned a lot about boat maintenance from him, and I think I also got a lot of my competitiveness from him. You're always matching yourself up against other boats in that fleet, just like in this one."

Mario wanted his son to become an engineer and encouraged him to work on his math at school, Mike said, "So of course I dug in my heels, and that's exactly what I didn't do." Then his parents wanted him to go to Virginia with them after his father began selling fire-protection systems down there. That was in 1983, just after Mike had graduated from high school. He said no thanks, he'd stay and go to work as a sternman for a lobsterman out of Nauset Harbor. No matter what happened, Mike figured, there was always some way you could make money in the fisheries here. If all else failed, there would still be the bay scallops and steamers and quahogs and their promise of a reliable $80 for half a day's work on a Sunday morning. Just as it was when Brian was starting to fish, and just as it was a century ago on the Cape, those open-access shellfisheries existed — and still exist — as a sort of community bank account: for retirement income, for unemployment insurance, and as a sort of safety net for youngsters breaking into the business.

Mike worked three years for a pair of Nauset lobstermen and then spent a year dragging for groundfish as Bill Amaru's mate on the *Joanne A*. He left that site for reasons he says are nobody's business and with somewhat prickly feelings toward his former skipper. In 1989, Cecil Newcomb, now a lobsterman, hired Mike to run the *Del-Hy,* a small long-liner out of Nauset. "By then I already had some experience long-lining. I knew how to handle the gear, how to find hard bottom," Mike said. He learned more the following year, working with the veteran Charlie Melbye on the *Shenandoah,* a forty-four-foot Chatham long-liner.

During those first ten years he also ran a twenty-one-foot skiff that he used for bullraking quahogs and also, eventually, for offshore lobstering, working just beyond the pale of state waters, which by then had gone limited-entry. Slowly, putting in twelve-hour days on both his own boat and others', he paid off the money he owed on his outboard and lobster gear and began to build some savings. In the winter of 1992, after he had sold his lobster gear and turned in his federal permit, he rigged the skiff for long-lining, fished the inshore waters, did well, and decided that he had worked long enough for other skippers and other boats. In Dennis he found a down-at-heel lobster boat, the *Jim-Dandy,* that with a new engine and a lot of work could become the sort of vessel to take him long-lining into the waters off Chatham. The one-time reluctant math student drew up a three-year business plan and went into a local bank to ask for a loan. "I got laughed at," he remembered. Instead he put up the down payment himself and arranged with the owner to pay off the balance in five years. He is about to wrap up his payments on the *Susan Lee* two and a half years ahead of schedule. "And I've held to about ninety percent of that business plan I wrote," he added.

Also in 1992, after a three-year courtship, Mike married Susan Lee Collins, the daughter of a Chatham funeral home director and the chief secretary of the Orleans police department. Susan is three years older than Mike and as a senior had been in a government class at Nauset Regional High School that included her husband, just a freshman then. She paid him no attention. After high school she worked two jobs, clerking and waitressing, and traveled every year, to California, Oregon, Washington, Alaska, Australia, Norway, Sweden, Denmark, the Bahamas. She met Mike again through a friend's recommendation and a more-or-less chance encounter at the Land-Ho, an Orleans bar and restaurant. After Abigail was born last year, Susan quit both her jobs, directing the energy that used to go into those days and nights at work and those interludes of travel into the baby and, during the quiet moments, into her spinning and knitting.

These events are all part of the same sense of directed happenstance that has begun to take on something of a malicious aspect for Brian but that to Mike remains, even through all his recent trials, essentially serendipitous. His aunt no longer lives in Eastham, having moved away shortly after his parents moved to the Cape to be near her, and Mike regrets that Abigail sees so little of his side of the family, regrets that he has twenty-three first cousins, mostly in Pennsylvania, whom he hardly ever sees himself. But he says there are good things as well about being the lost relation, and that he didn't have any choice about the matter anyway: "I feel like I was born to be here." Lobstering was all right, but he prefers being able to bring his gear home at night, clear of easterly winds, draggers, and the character flaws of his competitors. And he likes the American-dream faithfulness of cod: "There's nothing like groundfishing. The cod still pays the bills. If you put in the time, if you work hard, you're going to get reimbursed. Sooner or later you're going to have a big month or two. And with my wife not working and staying home with Abigail, I've got to make good money. I've got to hit those big months."

Now he has the expectation of more of those months to come, at least in the foreseeable future. He thinks the hearing in Hyannis went well, all in all, that the hook fishermen got their point across and in the end might be treated a little differently from the draggermen by Amendment Seven. If not, then at least he has some time. If the council is pointing to a decision on that by December, says Mike, then in reality the decision will be later — maybe a lot later. "I don't believe Rosenberg will go for that fifth option, the status quo. If he does and everything stays the same, then I can stay in business, no sweat. So I got my fingers crossed, but I doubt that's going to ride. If they go with one of the other four, though, it's going to take them another year anyway to get it all in place. They're just going to fiddle for a while. If they moved any faster, they'd have to go back to real work. They've got it figured out."

He admits to being impressed by all the dogs that big gill-netters were bringing in earlier this month. By now the cockpit is clear, and Mike is back on the ground, helping me to scrub the hull with a stiff long-handled brush and a mixture of soapy water and sand, speculating as he does so about what it might be like to own a gill-netter: "It's twice the overhead, twice the cost, but then you get more than twice the profits with dogs the way they are now." He adds that he doesn't really like gill-netting, that he is uneasy with that fishery's higher levels of bycatch and discard, at least as compared to hook fishing. "You've got to keep your integrity," he says. "But then you get that dollar sign on your forehead, you know? I'd feel lower than whale crap doing it, but it pays real good. I guess gill-netting's a clean fishery if you use a large mesh and stay on top of your net. You do what it takes. Never say never."

He pauses, steps back, the wheels turning in his head, the angles being calculated, the future in its various guises swimming before his eyes. At the same time he gazes up and admires the *Susan Lee*, saying, "I don't know. I'll probably stay put. I've got so much put into this boat now."

By late afternoon the wind has died back and the hull is clean. We put the boat to bed for the night by covering it from bow to stern, from wheelhouse to pavement, with an enormous blue polyethylene tarpaulin. The last remnants of the storm nuzzle at the tarp's edges, billow along its length, come whistling out through holes nibbled by mice.

In appearance this is no less surreal than the boat itself, this roiling edifice of blue plastic. "It's landscape art," I tell Mike.

"There go the property values. But I'm not moving."

10

A Sea of Troubles

CARL SAID THIS MORNING, as his skiff, towing a scallop dredge, carved the edge of a sandbar near Little Pleasant Bay's Barley Neck, that he wished the *Honi-Do* could turn like the skiff could. "But then, if I were in the *Honi-Do,* I'd be warm," he added.

He's in the *Honi-Do* now, in the wheelhouse and out of the wind and only just warming up from the chill of the November breeze across the bay. When we pulled out of Carl's driveway at 5:45 this morning, the radio said twenty-eight degrees in Chatham and gale warnings posted for later in the day. A trio of ghosts floated and rattled in the bare apple tree in Carl's front yard, paper cutouts hung there before Halloween by his children, Susie and Peter. The dirt road was still rutted and channeled from the rains of Hurricane Felix. A fox stared boldly into the sweep of Carl's headlights as he turned toward Baker's Pond. A deer slipped into the woods, the white flag of its tail popping like a flashbulb in the glare.

We met Dave Slack, in wire-rimmed glasses and a heavy Nauset Fishermen's Association sweatshirt, at the town landing at Quanset Pond, which is perched like a raindrop on the western lobe of Pleasant Bay in south Orleans. In the winter, on days when the *Honi-Do* can't get out or Mark doesn't want to fish, Carl picks up a little money working for Dave, who lays bathroom and kitchen tile; in return, whenever Carl is running the *Honi-Do* and needs a crewman, Dave is first on his call list. It's a good arrangement for both of them, and Carl is grateful for the work he can get with Dave during the lean winter season, though sometimes he chafes under the ribbing he gets from the carpenters and other contractors who work alongside Dave. "I thought you were supposed to be a fisherman," someone always says to him. "How come you're not out fishing?" Screw you, thinks Carl, who has been tickled too many times by that particular needle to laugh anymore. You do whatever's necessary.

It was necessary this morning to plant the seed quahogs that Carl and Dave have been cultivating in bottom trays in grants positioned side by side near Pochet Neck. Dave was already at the landing, loading gear into his own skiff. The boats pushed through doughy panes of ice in clearing the pond. A pair of black ducks whistled by overhead, their wings flashing white and windmilling like a weathervane's, while herring gulls cried from off toward the beach. Carl led the way north, up into Little Pleasant Bay, past Hog and Sampson Islands, and then toward the curling northern terminus of the bay between Barley Neck and Pochet Island. The sun had lifted by then into a flawless ice-blue ether. In front of the islands the light chipped uncut diamonds off the crests of low, coal-colored waves.

In the shallows at the neck, a hundred yards or so off the shore, Carl's and Dave's bottom trays — seven for Carl, five for Dave — were dark rectangular shadows in the water. Carl anchored his skiff near the westernmost shadow, closest to the shore, and pulled on orange rubber gloves with diaphanous sleeves that reached above his elbows. In these and his corduroy hat, his gray sweatshirt with the hood turned up, his chest-high insulated brown waders, he assumed a mailed, warlike look. Finally he splashed noisily into water that was hip deep, taking arms against the sea of troubles that had dogged his aquaculture work this fall and that made his presence here this morning, so late in the season, such a heavy-hearted exercise.

The first blow had been the failure of his application for a FIG grant. In August the application had been returned to him with no explanation other than the attached comments of the three anonymous readers who had reviewed his packet at NMFS. Two had been impressed. "Honest hard-working project manager who seeks to change his profession from hunting to gathering. Good consultant and graduate students who should add a lot to the results," said the first. The second found Carl's intended use of clam tents, fine-mesh arrangements of netting whose purpose would have been to collect and protect drifting soft-shell clam and quahog larvae, to be "an interesting application because it helps to determine if the development of local stock is limited primarily by predation. . . . It's very gratifying to see a proposal that comes directly from industry in a partnership with the regulatory agencies and academia." But the waspish response of the third reader no doubt doomed his application: "The proposal is not adequately described to decide what the real benefits are. Background information is limited. Anticipated benefits are small. What is a clam tent? What is the purpose of using it? . . . This is a project that might employ one or two individuals on a limited basis."

Carl spent a long time staring at that response, getting used to its finality, its absence of recourse or appeal. The next day he looked at the names in the *Cape Codder* of those fishermen who did receive grants for aquaculture projects, and like Ronnie Harrison, he could only wonder, and be reminded once again of that ornery element of surprise in the way things work themselves out.

A second reminder was provided when the new seed quahogs he ordered from New York for this year's fall planting never showed up. Carl had joined Dave Slack, Mike Russo, and several dozen other Cape grant-holders in a Cape Cod Community Resource initiative to order seed from a Long Island operation promising swift-growing stock at lower prices than they could get from the Aquaculture Research Corporation or other local suppliers. Carl needed his 56,000 seed quahogs in September, the same month that Director Phil Coates happened to take a vacation from his desk at the Massachusetts Department of Marine Fisheries. For two weeks the papers necessary for the seed to enter Massachusetts had sat on Coates's desk, since no one else at the DMF was willing to take responsibility for them. By the time Coates was back at work and the papers were moving again, the supplier had been forced to plant half the ordered stock — the swifter-growing half, the best seed — in flats down in Long Island. With nothing available now but reduced quantities of short, slow-growing seed, many fishermen, including Carl, canceled out. Carl placed double orders with three other suppliers, including Aquaculture Research. But none, finally, could provide anything for him at that late date.

The whole affair made Carl furious at the time. A seed quahog takes three years to grow to market size, and he estimates that the loss of his fall planting will take a 20 percent bite out of his income from this grant three years from now. But the passing weeks have made him more philosophical. He told me the mix-up was nothing unusual, really, in a business where planters, suppliers, and the bureaucratic middlemen are all new to what they are doing. At least he was better off than some. Mike Russo, for example, is in his first year of being a grant-holder, and his order represented all the stock that he intended to plant this year. "So now guys like Mike are a whole year behind on their grants," Carl explained, "and if it's their first year, they've got to go through a progress review by the town. If they don't have anything planted, the town could pull their leases. But they probably won't. It's a small town, and people will understand what happened."

Living in a small town, however, also means that rivalries over shellfish grounds, especially these newly privatized ones, can get hot and personal. I remember Brian's receiving a knock on his door from his friend Jay Harrington one evening last July. Using a skiff borrowed from Brian, Jay had just harvested and sold to scientists at Woods Hole a quantity of horseshoe crabs, whose copper-based blood is used by the pharmaceutical industry to detect endotoxins. Jay is also a jazz aficionado and hosts a weekly jazz show on WOMR radio in Provincetown. That night he brought a gift of appreciation for Brian's loan of the skiff: a boxed seven-CD set of Miles Davis playing live at the Plugged Nickel.

Jay stayed for some coffee and confessed that he was stung by a letter written to the Orleans selectmen accusing him and other wild shellfish license-holders of constituting a "good ol' boy network" dedicated to excluding newcomers from the shellfisheries. This was because at the last selectmen's meeting Jay had argued (in vain) against setting aside more acreage for aquaculture grants in Orleans. What particularly hurt, he said, was that the letter was written by someone he considered a friend: a prosperous new arrival, a professional, whom Jay had helped get started in aquaculture through loans of his own boat, gear, and expertise.

Jay himself is a grant-holder in Pleasant Bay, but he shares Brian's convictions about the primary importance of the wild fisheries and worked with Brian on the Nauset Fishermen's Association FIG grant application. Brian was chairman of the town shellfish advisory committee which several years before had drawn up the regulations governing the awarding and development of the town's aquaculture grants. "We drew up those regulations as volunteers, and afterward we were congratulated by the selectmen for the job we had done on them," he said. "The next thing you know, people start driving their sport utility vehicles through all the loopholes left in the regs by us nonlawyers."

Jay was disturbed by shellfish aquaculture's runaway growth and the manner in which the Orleans selectmen had compromised the limits established for that growth by increasing the number of grants available and grandfathering in certain notorious loopholes. Brian was disturbed that the primary beneficiaries of these extra grants were often not the out-of-work fishermen whom his advisory committee had intended to help but well-heeled professionals and retirees, a new breed of gentleman farmer, or intertidal squire, perhaps — archons of that conquering civilization John Hay describes.

Jay's skin was shoe-leather brown beneath a close-cropped beard bleached prematurely white. He and Brian sat at a table strewn with mail and newspapers and books, listening to Davis play on the CD player in the living room. Brian said that a lot of the talk you hear about aquaculture now is from people who really know very little about it, "but they're schmoozing with committee members down at the yacht club and spewing all this bumper-sticker philosophy about how the fishermen ruined the fishing industry, and now it's up to the Lands' End crowd to save it through aquaculture. It's another good example of how this crisis in the commons is being used as a lever for privatization."

Jay sipped his coffee, glancing at Brian, his pupils creeping sideways in his sun-blinked eyes like winkles lurking under rocks. Tony Williams's drums came snapping out of the other room, their rhythms struck square on the beat, deep in the pocket. Davis's horn playing was forceful, spare, unhurried. "We're commonists," Jay said, and smiled.

Brian does not deny the role that fishermen have played in bringing about their current woes, and objects when others in the industry are disingenuous about that. At the same time, he is uneasier than ever about the gathering speed and momentum of the aquaculture bandwagon.

The Nauset Fishermen's Association's FIG-funded wild shellfishery enhancement project is scheduled to begin next spring, but two days ago, in a sort of rehearsal for the work on that project, I went with Brian to Asa's Landing on the east side of the Town Cove. There Jay Harrington, Mike Russo, and some other volunteers gathered to plant seed quahogs that had been cultured in bottom boxes since last spring as part of a town-funded wild shellfish propagation project. That day, under pewter skies, with a hard wind blowing out of the northeast and the tide running out, some eight or nine rubber-booted fishermen walked out onto the wet flats of the landing. Twelve three-by-ten-foot bottom boxes lay recessed in the mud — enough to make that stretch of the cove look like a graveyard for basketball players. Brian and the rest pulled the olive-colored mesh netting off the tops of the boxes and shoveled out each box to a depth of several inches. They ran the excavated sand and mud through a hand-held sifter that separated the seed quahogs, a few marauding green crabs, and the general detritus of shell fragments the fishermen call shack from the sand. The seed quahogs were only half the size of a fingernail.

Discussion ran on recent stories in the *Cape Cod Times,* one of which announced, "Fish Landings Plunge 18 Percent in Massachusetts." NMFS had just released its 1994 statistics, and this particular number now was being used at the New England Fisheries Management Council as a boost to the more stringent versions of Amendment Seven. Recently a *Times* fisheries columnist had suggested that this statistic was misleading, writing that since NMFS was counting landings and not fish in the sea, things really weren't as bad as the agency claimed them to be and NMFS was exaggerating the scope of the crisis to suit its own political purposes.

This provoked a letter to the editor from Brian, one that reminded readers that NMFS's population assessments were based not only on landings but on extensive and statistically consistent trawl surveys. "Blaming NMFS for the cold, hard facts is not going to change those facts," he wrote. "An honest recollection of the politics that have brought the fisheries to the current state would recognize our industry's resistance to the many suggested and mandated conservation measures over the years since the inception of the Magnuson Act. We might have required a few fillings in 1978 or a root canal in 1986. Now we're facing major surgery, and the politicians are touting a new set of 'choppers' called aquaculture. NMFS cannot be blamed for this. We shouldn't focus on shooting the messengers. . . . We need to learn how to deal constructively with the problem."

Bill Amaru, a council member for three months now, has been

preaching tirelessly on behalf of stock rebuilding and stern conservation measures and has expressed his firm support for Amendment Seven. For that reason he has been taking a lot of flak from many of Chatham's long-liners and gill-netters and even his fellow draggers, some of whom interpret his position as another way of advancing the interests of big boats at the expense of small. "I expected that," Amaru said to a *Times* reporter. "When you are in the front of the line, you get booted in the rear end more often, but that [anonymous criticism] hurts because I feel I'm doing the right thing."

Brian knows that Mike Russo nurses some sort of grievance against Amaru, and he couldn't resist asking Mike if he'd seen Amaru's picture in the newspaper that day. "Shit, do I have to see his face there every day now?" Mike said. He stripped off his rubber gloves, stood up, and swabbed his forehead with the back of his hand. "He's got sensible stuff to say, really. I just would have liked to see Mark Leach get that council seat instead of him."

That morning the seed quahogs were dispersed into new bottom boxes that were then packed with mud. Those boxes, each only a few inches deep, were sealed with a fine-mesh screen and carried out to deeper water, hip deep on Brian and well beyond the extreme low-tide mark. There, throughout the winter, at least a foot of water would protect the young quahogs from freezing, and the new screens would protect them from predation. Eventually they would be distributed throughout the Town Cove and dispersed into the shellfish commons, becoming fair game for any hunter-gatherer with an Orleans shellfish license.

Carl's and Dave Slack's quahog grants are similarly protected by high water, though ice shifts around on the tide when the bay freezes over, and sometimes the moving ice tears the top mesh off bottom boxes. This morning the mesh was clean and tight, so the owners had to remove only modest amounts of seaweed as they pulled the mesh from the boxes, only occasional crabs as they worked the boxes — four by eight and screened top and bottom, like the new boxes at Asa's Landing — free of the mud. The crabs were no more numerous than at the landing, but were more cosmopolitan: green crabs in leaf-pattern hues of forest and lime, rock crabs in buffed shoe-leather brown, skeletal spider crabs with skull-like shells alive with an infernal landscape of algae, hydroids, byrozoans, and sponges.

Carl said that a crab will insert the point of one leg into a quahog's shell while the mollusk is feeding with its siphon extended, and then leave the leg there, unharmed, when the quahog clamps its shell shut. Slowly the crab will work a second leg into the seam maintained by the first, then a third, and eventually the quahog will tire and fall open. The thin, hook-shaped legs and claws of spider crabs, he said, are particularly deft at this sort of siege. Carl and Dave exact the quahogs' revenge, impaling every crab they find on the tines of their rakes.

Carl was disappointed with the size of his seed. He and Dave worked their boxes out of the bottom one by one, each of them heavy. Then they turned each box over, removed the bottom scrim, propped the box on chunks of Styrofoam to keep it off the sand, and raised one corner, allowing the seed and shack to collect in the opposite corner while the mud and water drained out through the mesh. The seed here, planted last June, were hardly bigger than grains of sand, and Carl regarded them with a look that plainly suggested he had hoped for more growth, that certainly he would have liked to have added to these that lost shipment from Long Island. "Damn, these are small," said Dave, no less disappointed, when ten emptied boxes had filled only a little more than five one-gallon pails with seed.

Earlier Dave had warned Carl not to put on his sunglasses or it would get cloudy, but Carl had done so anyway, and sure enough, clouds were being laid like sheets of wet newspaper from one horizon to another, and the wind was picking up from the northeast. The men stacked the bottom boxes in Dave's skiff, scooped the seed in handfuls from the pails and scattered them like wheat directly onto the bar in long, narrow strips, and staked long swaths of three-eighths-inch netting, its edges hung with lead-line, at the corners of the strips above the seed. Finally they hammered the lead-line into the bar using a water pump run off an engine propped between the thwarts of Dave's skiff. Carl, hunched, his nose almost to the water, held the nozzle of the pump's hose close to the bottom and moved along the length of the bar like someone vacuuming pennies off the bottom of a wishing well.

The Atlantic bay scallop, *Aequipecten irradians,* is the Roman candle of the bivalve genus: peripatetic, agile, neurologically complex, swift to grow and die. In comparison to it, the hunkered-down quahog is more like a smoldering coal: stodgy and unimaginative, barely giving off smoke, much less light, a sitting duck for a crab or starfish if uncovered, but generally much longer-lived. The scallops, about three dozen of them from Carl's last tow, were wafer-thin compared to quahogs. They lay among the wrack in gritty shades of khaki and slate gray and lavender lit with highlights of orange or cream. In their shells' perfect symmetry, their curtainlike folds, their matched marginal wings, they look more like artifacts than animals. In fact the shells once served as badges for Christian pilgrims on their way to the Holy Land. Now the scallop is the badge of Shell Oil, whose product conducts pilgrims today to all the sacred places of the car-payment culture, as well as out to Georges Bank.

The scallops lay clamped tight on the culling board, grimly concealing their astonishing blue eyes: thirty to forty in an ink-drop necklace around the mantle, each with its own cornea, lens, and retina. A scallop's eyes probably cannot form distinct images, but are sensitive to light and shadow and work in combination with its chemical receptors to detect reliably the approach of a starfish, skate, or scallop dredge. This provokes sharp contractions of the scallop's Schwarzeneggerian adductor muscle, which opens and closes the shell and forces water out of the mantle cavity in pulsing jets that may be sent in any direction. The only swimming bivalve, a scallop flaps its shells as a butterfly flaps its wings, achieving the submarine equivalent of a butterfly's darting, bouncing flight.

But this was the undoing of the scallops Dave now culls. In front of his dredge Carl has attached a tickler, a small concrete block towed ahead of the dredge and rigged so that it sweeps back and forth across the dredge's path, knocking eelgrass to the side and moving the scallops off the bottom, where the dredge would harmlessly roll over them. The wise or the fortunate dart to one side at the approach of the tickler. Most retreat, and are finally overtaken and pulled into the dredge like blown leaves into a street sweeper.

Of the three dozen that lay on the culling board a moment ago, a little more than a third have already been pushed over the side with the shack. A bay scallop lives only two years (compared to a quahog's twenty-five or so, or a black clam's century-long lifespan) and spawns but once, extruding its cargo of eggs and sperm early in the summer of its second year. Scallops possess both male and female organs, though they discharge at different times and so avoid self-fertilization. Juvenile scallops by this time of year are indistinguishable from adults in size, but lack the usually conspicuous growth ring that undulates across the width of an adult's shell. This year the growth ring is close to the hinge on most scallops and often hard to see. But scallops lacking that ring are thrown back to provide next year's harvest, while the adults, spawned out and certain to die this winter in any event, are taken on what might be described as a guilt-free basis.

If a "clean fishery" is defined as one whose operations have little impact on the populations of other plants or animals and as little impact as possible on the reproductive potential of the target species, then this modest bay scallop fishery is easily Cape Cod's cleanest. It is also one of the few surviving scallop fisheries on the East Coast, the rest having fallen victim not to overfishing but to various forms of environmental distress: eelgrass blights, tides of poisonous phytoplanktons, and frequently the nitrogen-loaded coastal runoff that follows commercial and residential development. The slack in the domestic market has been taken up by cheaper but much less flavorful farm-raised scallops from China. Meanwhile the remaining wild fisheries grow less and less productive.

Dave's face was red from the cold and pinched with concern once the dredge was back in the water. His own skiff was just visible, resting at anchor near the grants at the northern end of the bay. A five-eighths-inch polypropylene line strained and shivered and snapped off water along its exposed length between the bay and a cleat on the skiff's starboard rail. Another line rose limply from the water to a block and tackle hung like a tassel from the top of the ten-foot mast Carl had stepped into his skiff. The first tow of the morning had brought up about four dozen scallops, counting both seeds and legals, but subsequent tows ranged between one and three dozen and were dropping. "The dredge is hauling awfully clean," Dave suggested. "Maybe we should shorten the towline."

Carl took some length out of the towline, cleated it off, scanned the width of this narrow channel between Pochet Island and Barley Neck and then the spread of Little Pleasant Bay to the south. There were only four other boats in sight. "Look who isn't out here today — Mike Russo, Freddie Fulcher, Chris Adams, Brian Gibbons," he said. "That tells you something."

He tightened the drawstring on the hood of his sweatshirt and laughed, remembering an incident during the 1982 scallop boom in these waters. The whole town turned out that year with dredges, rakes, or just bare hands, and Orleans's shellfish constable was adamant that the fish totes that the scallopers brought in be "gently rounded" at the top, not heaped to slippery, skittering heights in an effort to pack the maximum number of shellfish into each picker's five-bushel limit. "But we're all just heaping 'em high anyway," he said. "So when Dan Howes comes in with some totes all piled up like that, the guy stops him right there and is reading him the riot act. But right at the same time, one berth over, Bindy's got some baskets that are even higher than that, and she's getting help unloading from the guy's assistant."

Now every scallop season opens in hopes of a gold strike like that, but every season since that year and the next has been a disappointment. This one is no different. "Scallops scarce on opening day," said the *Cape Cod Times* on November 2. "Gray skies, cool morning temperatures, and a smattering of scallops greeted about a dozen shellfishermen who ventured into Little Pleasant Bay yesterday for the opening of the scallop season. 'Beats poverty, but there ain't much here,' said Michael Russo, as he bent over the culling board . . . to find the tasty bay scallops that pass legal muster, and may bring nine to ten dollars per pound for fishermen at the market."

Today didn't even beat poverty. The price on scallops dropped in the nine days after that, to around $7.50 per pound, shucked, and nobody found any secret stashes. With its purselike steel frame and chain-link bag, the dredge rose against the clouds like a piece of medieval armor. The last tow, along with its customary freight of shack stippled with tiny oyster and slipper shells, also brought up a sea-wash ball, a white, spongy, pincushion-sized item that produces its own soapy lather if scrubbed and is actually the egg mass of the waved whelk. Only a handful of scallops lay like corroded doubloons on the culling board. Dave looked at those, and at the single tote, about six pounds of meat, that he and Carl had managed to fill in three hours. "That's not a day's pay," he said.

The *Honi-Do* rests fitfully at its mooring in Aunt Lydia's Cove, just inside the lifeless hump of Tern Island. Mark is up in New Brunswick, deer hunting, and this week and the next is one of those treasured times when the boat is Carl's to run. But with gale warnings posted for the afternoon, Carl decided today would be better for working on his grant and trying some scalloping with Dave. Now the wind has grown claws, blowing a cold thirty-five knots out of the south, and the waves are as thick as harrow's teeth on the other side of the island and in the channel out of the harbor.

Carl is here to replace the valve on the boat's big hydraulic reel, but before he does that he gives me a rundown on the *Honi-Do*'s electronic capabilities: its loran, radar, backup loran, backup radar, VHF radio, CB radio, cellular phone, satellite-link black-and-white television (suspended from the ceiling on the companionway side of the wheelhouse), depth sounder, and video plotter. This last, a Furumo GD 3100, is the newest addition to the boat's information arsenal. Its monitor, bigger than the television, dominates the console in front of the windshield just to the left of the wheel. Carl powers it up, punches in the loran coordinates of the Tuna Grounds, an area fourteen miles east of Chatham that he and Mark often work. Readings come up on the margins of the screen: recommended heading, bearing, distance to that spot, steam time at any given speed. A series of maps outlined in green show bottom topography, the channel out of the harbor, and then the course to the coordinates. A number of yellow lines that cross or run parallel like arrows in a quiver show the precise routes of previous trawls in the grounds. Blue squares mark either the beginnings or endings of trawls, or else sites where the *Honi-Do*'s net got rimracked, or hung up.

For all its video-game panache, the electronic plotter is in one respect simply a notebook, showing in visible form the range of Mark Farnham's (or Carl's) previous solutions to the algebra of weather, season, species behavior, price, and fisheries regulation underlying each decision as to where and when they trawl — an algebra whose scratchwork is done in dock at the cove, where the *Honi-Do*'s gear is changed, adjusted, and repaired, and also in Mark's head, as he weighs, in his solitary fashion, which days to go out and which days to stay home. Yellow lines etched in August, after the new engine was put in, represent trawls for blackback flounder, plaice, and sand dabs, and those in September, after the gear was changed, trawls for whiting out to the Tuna Grounds or another spot six miles east of Nauset.

Whiting, or silver hake, is a small groundfish similar to the cod. The fish is plentiful off New England and is not, like cod and haddock, a regulated species. But because a great many cod have historically been caught as bycatch by fishermen geared for whiting, the whiting *fishery* is heavily regulated. This year, to go for whiting, fishermen have to sign up in advance with NMFS, receive identification and permit numbers and call in to NMFS when leaving port, remain in the fishery a minimum of seven days, use nets with a mesh of two and a quarter inches and no such ground gear as the cookie disks that allow the rigs to roll over small bottom features, and use a Nordmore grate, which is a rectangular piece of aluminum tubing positioned in front of the net's cod end, through the bars of which only fish the size of a whiting or smaller can fit. Theoretically, larger fish such as cod will hit the grate and then escape through an opening cut into the net above it, while the little whiting sift through.

Carl is no admirer of Nordmore grates, which indeed reduce bycatch to near zero, he says, and the catch of whiting to numbers just slightly above that. "The whiting see the grate and then just swim out through the hole. Or else skates plaster themselves against it so that nothing else gets in." Previous efforts to minimize bycatch in this fishery have prompted other citations of the law of unintended consequences. "Two years ago I was running the boat with Dave Slack on board and we did our biggest trip of the year with whiting, the week before Thanksgiving," Carl explains. "We didn't have to use those grates then. Then last year — to conserve cod, they said — they shut down the small-mesh fishery entirely, and there went twenty-five to thirty percent of our income. So we went back to larger mesh, to six-inch squares in the cod end, and we went from catching four hundred pounds of groundfish bycatch per day in the whiting fishery to taking fourteen hundred pounds of flounder and cod per day with that large mesh. So that's how they went about conserving regulated species. That made no sense, closing down that small-mesh fishery."

In October the *Honi-Do* went back to flounder, working around Hurricane Felix and getting out for three trips in the first two weeks, each of two or three days' duration. Toward the end of the month the flounder began to fall off, and when the net began bringing up more dogfish than anything else, Mark quit going out, though he gave Carl permission to run the boat with Dave. During the last week of October the wind blew hard from the north, and the tides had rolled around so that the boat could only get out at four or five in the morning. Dogfish then were selling for twenty-two cents per pound, and on one trip, after fighting steady twenty-five-knot winds all day, moving into the beach for shelter, and then beating five miles out to work again, Carl came in to the dock with an eight-thousand-pound load of dogs, only to find that Dave Carnes of Chatham Fish & Lobster, Mark's regular buyer, had sent nearly all his help home already. There was only one dock worker to help pack out four tons of fish, and only one truck to ship it in. Under those circumstances, Carnes didn't want to buy.

"Usually we don't land that much," Carl says, "and Carnes was complaining he'd been trying to call that day, trying to find out if we were coming in or not. But if you're trying to steer the boat in hard weather and at the same time trying to work the deck in a crowd of fish, what are you going to do? You're sure as hell not going to answer the phone. I told 'em before I left that I was coming in on Tuesday and that I wanted a decent check for the week. The hell with them if they don't listen."

Carl walked down to the other end of the dock, to the offices of Chatham's Finest, another buyer, and started working out a deal in which he would throw in a small load of flounder that he had if they would buy the dogs. Then he went back to move the boat from its berth at Chatham Fish & Lobster. "That was when Ray, that one fish packer, came out and said he'd talked to Dave and that Dave was going to buy the dogs after all."

Carl finishes installing the reel valve, and we row ashore in the dinghy that Mark keeps beached on the north side of the docks.

He says that he had a couple of good trips running the boat himself last week and yesterday left a check for $5500 at Mark's house. "Twenty-five percent of that is for Dave Slack. I wanted him to get at least a thousand dollars for the week, and he will. I'll get my thirty-five percent — I get an extra ten percent when I run the boat — and Mark'll get the rest. Before he left, Mark was telling me, 'Make me lots of money.' So he ought to be happy. He called from up there yesterday, and he hasn't even seen a deer."

Carl has; I remember the blaze of that deer's tail near Baker's Pond this morning. He says that we're going to have to swing by Mark's house on the way home: "I got to ask his wife to write me a check for some grocery money." Behind us the *Honi-Do* and two other boats, Bill Amaru's *Joanne A,* and Chris Armstrong's *Overdraft,* are blown in a line back from their moorings, rattling together in the wind like Susie and Peter's paper ghosts.

11

Another Fire Drill

IN THE UNCERTAIN HOUR before sunrise, beneath a dark new moon, the lights at Sesuit Harbor are the lanterns of madness. They stare out of the gloom like the stars that stared back at van Gogh. They cast their chloroform light across chasms of time and space: on racks of striped bass simmering in the summer bait barrels at Rock Harbor; on the protests that greeted the arrival of the *Mayflower II* in Provincetown Harbor; on the empty concrete footings at the site of the Marconi radio station in Wellfleet; on the fragments of a ceramic vase scattered like fish bones across the floor of the Hyannis Inn; on the high prow of the speeding *Hawkeye* as it looms out of the fog off Monomoy; on Jimmy Hoffa's luncheon with the flower children, an event catered by the archons of the intergalactic conspiracy; on Miles Davis sweating in the spotlight as he made his horn sing at the Plugged Nickel; on the miraculous resurrection of Harry Hunt one afternoon outside the Nauset Fish & Lobster Pool. Across the asphalt parking lot I saw a shadow I thought I recognized, and I wanted to stop. Instead I hurried down to the docks, padding behind Dan Howes, and the shadow disappeared with the blowing of a horn from somewhere else in the harbor.

Now Dan descends into the cabin of the *Last Resort* with a flashlight corded into his dashboard and powered by his marine batteries. The first sweep of the light revealed the black ribs of the boat's quahog dredge lying collapsed on the culling board, the rocky battlements of a breakwater opposite the boat, and water that looked silty and claylike in the crisp yellow beam. Now the light dances like a fairy about the cabin, winking at the companionway and the portholes. Then the engine groans, turns, fires. The boat shudders and settles into a swift, living pulse as the engine runs. Dan comes up the companionway and douses the flashlight. Samuel Eliot Morison defined a Yankee as "a new Nordic amalgam on an English Puritan base." In the dim glow of the lights along the quay, Dan's face betrays a Puritan's passion for righteousness married to a Squarehead's desire to get on in the world, whatever its madness and disarray.

Sesuit Harbor is a crescent-shaped cut in East Dennis's Quivett Neck, about eight miles west of Rock Harbor and close to the Brewster flats where Everett Howes and Tom Johnston first took their sons fishing. Dan moved his boat here earlier this month in anticipation of Rock Harbor's icing up. Now, at the end of November, he expects at most another month of quahogging before Sesuit ices up as well. Before that happens, he'll haul his boat out and find a different way to get on in the world until spring.

Still in darkness at six A.M., he noses the boat out the harbor and into calm seas, the temperature in the low thirties and the wind blowing five to ten knots out of the northeast. Beyond the breakwater the *Last Resort* is encircled by what Robert Finch describes as "the lighted necklace of the Bay": the ruddy glow of Plymouth and the eastern Massachusetts mainland, their lights blending to create a false dawn in the west; the green blinking beacon at the entrance to Sesuit and the red one at the entrance to Rock Harbor; the scattered household lights of Sandwich, Barnstable, Brewster, Orleans, Eastham, like campfires along the beach; to the north the steady white twinkle of the Pilgrim Monument in Provincetown and the light at the tip of Long Point, at the entrance to the harbor; to the northwest a winking buoy light that Dan says marks the grave of a wreck, and in the north-northeast the pulsing sweep of North Truro's great Highland Light, where Thoreau stayed some nights and listened to lightkeepers' complaints about the quality of oil issued them by the government. The cumulative fluorescence rises and seeps into the seams of the cloud cover, and the bay lies spread like a petri dish beneath its dim refrigerated refulgence. This dish where America was spawned and cultured has the vaguely portentous look now of an ongoing and possibly dangerous experiment.

It's a forty-five-minute steam to the island, by which Dan means Billingsgate, which was once inhabited but now is only a barren swath of sand and rock at low tide. This morning, with the tide nearly at ebb, the sunrise reveals the shoal poking up like an outcrop of slag amid a field of sooty two- to three-foot waves. A mile to the east the waves lick at the rusting hulk of the SS *Longstreet,* and to the north — less than a mile — at the whittled point of Wellfleet's Great Island. Tom Smith steamed out of Sesuit just ahead of us in the *Sea Wolf II,* a handsome converted gill-netter with a stern track for its quahog dredge, and is now nosing around off Dan's port beam, looking for a place to drop his dredge. Dan studies the video monitor of his depth sounder, surprised by the expanse of yellow tints on the bottom, indicating bare sand bottom only six feet down, and the relative lack of the yellow and red spikes that indicate eelgrass. "I guess a lot of grass let go last night with the big tides we're having," he says.

Satisfied with the bottom below him now, Dan hits the handle on his hauler. The boom snaps the line taut on the empty dredge and lifts it from the culling board. The clouds show through its ribs as it twists for an instant above the water over starboard, and then the cage settles, gurgling, into the waves. Almost at the same instant Tom Smith drops his dredge off the stern of the *Sea Wolf* like a dirt-track race car being dropped from a trailer. Dan lets about seven fathoms of towline run over the stern, cleats it off, and returns to the wheel. "I love machinery," he says, smiling, "when it works."

He loves these inshore state waters as well, and feels a Puritan's prickle of outraged righteousness on the rare occasions that his work calls him into federal waters, as it did in September for tuna fishing. With the *Sea Biscuit* still unsold, with a lot of tuna being caught off the Cape in July and August, and with a favorable exchange rate between the yen and the dollar, he hoped to land a bluefin himself in September or October, when the great fish were at their fattest, and knock a chunk out of his debt for the *Last Resort* and for the two blown outboards that preceded it. He stripped the quahog gear off his boat and fished for four days amid a fleet of seventy to eighty other boats, many of them employing spotter planes, and didn't see a thing. Only two fish were taken those days (Mike Russo hooked a bluefin that was too small), but Dan wasn't worried. There was time.

And then suddenly there wasn't. On September 12, NMFS closed the commercial bluefin fishery for the year, announcing to stunned fishermen that 1995's total allowable catch for tuna had been exceeded. Fishermen had questions. What about NMFS's promise to run the fishery through October by breaking the TAC into monthly quotas and closing down early-season fishing temporarily for the sake of time in the fall? Well, the fishing was good near the ends of July and August, NMFS replied, and the agency was reluctant to disrupt that, the result being severe overages in those monthly quotas. What about the 145 metric tons of TAC kept in reserve to cover such overages? Not enough, said NMFS.

Dan saw the expended overages as another example of "a basic level of incompetence with the feds." But there was nothing for it except to submit to the tedium of regearing his boat for quahogs much sooner than he had wanted. As he did so, he considered how narrow his niche in the Cape fisheries had become. In 1993, after a lot of debate and hesitation, the New England Fisheries Management Council had instituted limited entry in the groundfisheries of Georges Bank and other federal waters. Thanks to Dan's documented work in those waters between the control dates of December 1990 and February 1991, he was issued and still possesses a permit — he thinks — "but it depends on the amount of bullshit I have to put up with whether I keep it or not."

The permit was issued to him as the skipper of the *Sea Biscuit*, and he has since written a letter to NMFS stating that the *Sea Biscuit* is no longer working, is up for sale, but that the permit itself is not for sale and will remain with him as the skipper of the *Last Resort*. "I hope that's enough," he says. He keeps the logbook required of groundfishermen by NMFS, and each month simply states in it, "Did not fish." His own official assessment of the status of his groundfish permit is "in limbo," and he has no desire to wrestle with the regulation, paperwork, and dim prospects of that fishery. He did not go to the Hyannis public hearing on Amendment Seven, deeming it inconsequential, and is not encouraged by the recent change of management at NMFS. His logic in this is of the deductive kind Brian would like, though Brian might contest its major premise: all scientists are prima donnas; Andy Rosenberg is a scientist; therefore . . . Nevertheless, Dan hopes to keep his permit current and active, in case of in case of.

He did not qualify for a lobster permit when that fishery went limited-entry in state waters in 1977, but he wishes he had, noting that "if I had five to seven thousand dollars lying around, I might want to pick up one of those." Officially, lobster permits cannot be transferred through sale, but fishermen are driving their Chevy S-10s through the loopholes, and Brian, for one, is impressed by the speed and sophistication with which the market has been brought to bear on these supposedly nontransferable commodities. "Guys who can't even write their names," he says, "can figure a way around these regulations in about two minutes." He is no less impressed, and angered, by the degree to which entry into the fishery has become a commodity, available only to those who can meet its price, and by the walls now surrounding what was once a true commons, a public resource. "You look at Dan Howes, who is a native Cape Codder, and he no longer has the right to catch lobsters in front of the house where he grew up," he told me. "Same thing with the groundfishery now. If I catch two hundred thousand pounds of cod one year, does that mean I have the exclusive right to catch two hundred thousand pounds of cod henceforth, year after year, and other people don't? Are those the rights our forefathers died for? I don't think so."

Dan, who doesn't like worrying about what other guys are doing, is not sure that he has the temperament for lobstering, particularly in Cape Cod Bay, where lobstermen are said to be more territorial than on the Outer Cape. Still, the lobsters have been coming in. After the silent spring, after the June slump and the July drought, Brian and other Cape lobstermen were rewarded with high prices and relatively good landings in the fall — a combination good enough, said Brian, to transform a terrible season into at least an average one. In either event, Dan could only watch.

The first tow today comes up slim, with half a bushel of quahogs, maybe an eighth of a bushel of littlenecks and cherries. Dan lengthens the towline, allowing the dredge to take a deeper bite, saying that the clams burrow down this time of year. He says that in the spring a short little tow in this area will cram the dredge to bursting. Apparently Tom Smith has also had a lean first tow. He calls Dan on the VHF and suggests they try the Horseshoe, a section of the shoal that was a cove on the island before it submerged. Dan discourages that, saying there is a lot of eelgrass there now.

The wind is freshening, and the clouds are layered in puffy lines of rouge and charcoal. Succeeding tows are better as Dan homes in on the edges of the grass banks, trying to finish each tow in the midst of a bank and so beat the grass down for towing there tomorrow. A startled bay scallop comes up in one tow, its valves clacking like castanets, its necklace of blue eyes blinking in rapid-fire peekaboo. The ambulatory scallop is part fish, it would seem, and the same tow yields a fish that is part mollusk. The sand lance is a five-inch darning needle of a fish, bluish silver with a white belly and a faint green stripe down the length of its flank.

Preyed on by nearly everything, the fish schools near the bottom and uses its pointed snout to burrow like a quahog into the sand, there to hide, rest, and even hibernate after its spawning in December or January. A later tow brings up several sea grapes, animals that don't look much like anything — like small, cowhide-covered water balloons, maybe. Like quahogs, these are siphon-equipped plankton feeders. Dan says the squirts are fun to throw at shipmates, who think they're rocks.

Some quahogs come up sliced into neat zoology-textbook cross-sections by the dredge's knife. Dan throws these overboard, where they're retrieved off the bottom by the single horned grebe that shadows the *Last Resort* all day. A few come up thoroughly crumbled, the shards of the valves held together only by the pale muscle and viscera of the animal, answering graphically to Wellfleet resident Marge Piercy's description of shellfish as "bits of fog caught in armor," "chill clots of mortality and come." Others show up merely cracked, and these are thrown back as well, since the clams will be able to repair the cracks by growing a new layer of shell beneath them. In certain spots some quahogs come up with their shells moldy and frayed: stippled with barnacles, ulcered with oysters and limpets, rotted by the sulfuric acid of boring sponges. "There's so many of 'em here," Dan says, "that there isn't room for them all to get down into the sand."

Still others, though not many, come up with their shells sprayed with the tan and umber shadings I saw in the seed quahogs at Carl's aquaculture grant in Pleasant Bay. "Those are 'hogs escaped from a grant," Dan explains. "They'll interbreed with wild quahogs and keep their markings for three or four generations." He holds one in his gloved hand, studying it. "That's a pretty color. Must look nice in a supermarket display case."

By 9:30 Dan has tied off his fifth bag of 'hogs, his second of cherries and 'necks. The wind has come up to fifteen knots and the boat is working in ten feet of water, treading in a dream over what was once dry land attached to the mainland, where

Wellfleet farmers used to pasture their horses. It was here that an early form of privatization was modeled. Reports Kittredge: "At Nauset the settlers showed themselves even more avaricious than usual, and the Indians more than usually compliant; for when they were asked who owns Billingsgate, they replied that it belonged to nobody. 'In that case,' said the settlers, 'it is ours'; and the Indians believed them."

Later Billingsgate was cut off from the mainland, though Great Island was (and remains) joined to it. Fishermen raised houses on Billingsgate and lived there from the spring to the fall, and a school was built even while the wind and tides pared the new island down. Warren S. Darling, a quahogger who worked out of Rock Harbor in the first decades of the century, described in his memoirs the eventual abandonment of the island, and the day the tides first met over its salt meadows:

> In the late summer of 1927 there was a stretch of stormy weather that kept the quahog fleet idle. Finally there came a day when the weather seemed to have moderated. After watching the sky and the surface of the Bay all morning, two of the impatient men decided to go out on the noon high tide.
>
> The tides were such that they normally would have raked back-of-the-flats for a few more days. But the inaction and the no-pay days of the previous week were too much for Reed Walker and Herman Taylor, who took his teenage son along. They decided to go to the Channel, rake until sunset, spend the night in the Horseshoe and work eight or ten hours the next day before heading for the Crick [the entrance to Rock Harbor].
>
> Unfortunately the weather had a different plan in mind! In the late afternoon the nor'wester breeze, which they hoped would die with the sun, did just the opposite and became a real blow. The two men headed for the lee of Billingsgate Island as it was much too late to get to the Crick.
>
> This time the Island offered little protection from the Gale! The men anchored their boats in the deep water just off the Horseshoe with both anchors at the bow to try to keep them from dragging. Even then they stayed up all night to be ready to start their engines if

their anchors did not hold. In the morning light they were amazed to see white water where the Island should have been. This was the first time this had ever happened to anyone's knowledge!

Quite a crowd of concerned people were on hand at Rock Harbor as the two boats attempted a crossing from Billingsgate. Their progress was continually monitored through Art Smith's big telescope, which had been used for the same purpose many times before. A cry of, "There they are" went up from the watches each time a boat rose into view on the crest of a wave.

After they finally made the Crick, a relieved Cap'n Benny Bangs Nickerson was heard to mutter, "Them dang fools better believe Rideout's weather report next time!"

Dan says there were still empty houses on the island until the 1938 hurricane swept them away. There were rumors as well of pirate treasure buried there. Now the *Last Resort* plucks clams in nickels and dimes from its salt-meadow pastures. Dan ties off another bag of 'hogs and says, "Some people bitch about all the 'hogs they get out here around Billingsgate, as compared to cherries and 'necks, which pay about five times better. But at twelve dollars a bag for 'hogs, I'm not going to complain."

By midmorning the wind is gusting to twenty and the boat is lightly rocking on its beam. Long, throbbing skeins of eider ducks are flying low over the water. A small raft of them have joined the horned grebe still following the boat. They arrive singly or in pairs, their breasts thrown back, their wings churning, their webbed feet thrust out in landing like toes about to test the bathwater. They float sturdily atop the waves like dories, then wink out of sight as they prospect the bottom, soaring over undersea pastures and old foundations. In the northwest the cloud cover is starting to wear thin, and light is seeping through its sheets like oil through paper. At the same time a spattering rain begins to fall. The wind is still building.

All day long the *Last Resort* and the *Sea Wolf II* have been running in concentric circles, evenly spaced, like two sprockets on a bicycle, Tom a few hundred yards to the west. At one o'clock, with the tide still coming and the water up to seventeen feet, a third vessel arrives, the *Worker's Pride,* in blue and gray with boxy, bargelike lines. This boat drops its dredge inside the radius of one of Tom's circles, and we hear Tom on the radio to the new boat: "Just give me fifty feet. I can't work like this."

Dan shakes his head. "Even out here. Well, that guy's famous for that stuff. We'll let 'em fight it out. It's good over here. If I make a day's pay, I don't care what anybody else is doing."

Within another hour Dan decides he's got a day's pay, at least for this kind of weather: twelve and a half bushel bags of quahogs, five and a quarter bags of cherries and 'necks. He lifts the dredge clear of the water, coils the towline, cleans the culling board, and drops the dredge back on the board. He points the boat through the slapping waves to Sesuit, but then notices that the *Sea Wolf* is lying dead in the water behind him. He tries unsuccessfully to raise Tom on the radio and finally circles back to the shoal.

Tom is only doing some repair work on his dredge, but he points to the *Worker's Pride* and wonders aloud across the waves why that guy thinks this is the only fucking spot in the ocean. Dan shouts back that he did better the farther east he went, and then points for home again. "Maybe Tom didn't answer so there wouldn't be any witnesses around when he rams him," he muses.

I look back over the transom and see that Tom has thrown out a yellow buoy to mark his section of the shoal, and the skipper of the *Worker's Pride* has thrown out a red buoy to mark his. "Hague Line," I suggest.

Carl said that for the third year in a row the *Honi-Do* could not make money on whiting and Mark had gone back to groundfishing again. But while prices on cod and flounder usually climb around Thanksgiving, they dropped this year, back down to just fifty cents per pound, dampening Mark's enthusiasm for running the boat. "He always says, 'Don't call me, I'll call you,' whenever I start bugging him for us to get out there," Carl said. He would be happy to run the boat himself, with Dave Slack and one or two other experienced hands, any one of whom would be happy to leave on a moment's notice. "But, 'No, we'll go,' says Mark. 'We'll get out there.' Well, it's his boat, I guess." Instead Carl has been doing a lot of tile work with Dave, and in spare moments he is teaching Dave to mend net: "The more I learn about tiling, the more he has to learn about net mending."

Early in December, Carl ran into Mike Russo on the docks at Aunt Lydia's Cove, where they stood together for half an hour to "piss and moan," said Carl. Dan Howes was right about the Hyannis public hearing — inconsequential, at least regarding any special dispensation for hook fishermen or gill-netters. All groundfishermen, it appears, will be in the same closely monitored boat on May 1, when the new regulations are scheduled to go into effect. The final form of Amendment Seven will probably not be as draconian as Alternative One, nor as relatively palatable as Alternative Five. Instead, its regulations will be a combination of measures from the middle alternatives: a 50 percent reduction in allowable days at sea for groundfishermen, to be phased in over two years and now to include small-boat fishermen as well, who formerly were exempt; area closures, as yet unspecified; total allowable catches, to be treated as targets, with any overages to be deducted from the following year's TAC; and limited entry in the hook fishery, which currently has open access, along with a limit on the number of hooks a vessel may carry.

"Cape fishermen have long bristled," reported the *Cape Codder,* "at having to pay for what they see as the sins of the big draggers out of New Bedford and Gloucester, which catch the overwhelming majority of fish. 'We hear the argument from everyone, "I didn't impact the resource," but if you went fishing, you did,' said Mr. Rosenberg. 'It's a real unproductive argument to say, "I didn't do it." You did it.'"

Though disappointed by the inclusion of the hook fishery in these regulations and surprised by the speed with which the council is moving on them, Mike Russo still has his piss and vinegar. The *Susan Lee* is back in the water with a new propeller shaft and a rebuilt wheelhouse, and has been doing pretty well with cod down in the channel, despite the low prices. "What they're doing with TACs is one step away from ITQs," Mike told me, "so I got to keep landing as much as I can in case of ITQs, where your individual quota is determined by your landing history."

The probability of limits on his days at sea has him thinking harder than ever about buying a bigger boat, something that would be able to force its way out of the harbor on at least some of the windy days that keep the *Susan Lee* at home. "If I had a forty-five-footer, I could get out twenty percent more often than I do. Of course, then my operating costs would double. Well, it doesn't matter anyway. I've already been to the banks looking for money, and nobody's lending for boats now."

Nevertheless, Mike was much encouraged by the success of NOAA's $2.5 million vessel buy-back pilot program, which drew 114 bids, more than 80 of them from fishermen willing to undervalue their boats in order to get out of fishing. The pilot program pulled thirteen boats out of the New England fleet, representing 2.6 percent of the revenues generated by groundfish landings last year. At that rate, the proposed $25 million buy-back program could reduce the fleet's capacity by more than a fourth. In that event, provided he gets enough days at sea, Mike thinks having a small boat with low operating costs and a lot of owner equity could work to his advantage. "With reduced days at sea and hook numbers, those larger operations aren't going to be able to meet their payments and expenses with cod," he figures. "So if ninety percent of the boats that are left are chasing dogs to pay the bills, I can go my merry way and catch more cod."

Whatever happens, Mike looks forward to finding a way to turn it to his advantage, to gainsay the cynics. It may be the end of the world on some calendars, but he finds himself almost eager for the arrival of May 1. "I got a buddy who's taking a land job this winter, he's so spooked. But all this crap about everything going down the tubes here is just that — crap. If you keep your chin up and your eyes open, there are more opportunities out there than ever before. You look at the Depression. Some people came out of that as millionaires," he notes.

Amendment Seven won't make Brian a millionaire, or any poorer either, unless he can't find cod for bait, or unless the exodus from the groundfishery results in more inactive lobster permits coming out of mothballs. His chief interest has been his conversations with Bill Amaru, comparing notes on the dubious pleasures of public service. Brian remembers talking with Amaru shortly after the council representative had a particularly tough meeting with some angry Chatham long-liners.

"Amaru's afraid that if he hardballs it with the hookers against the draggers, with no concessions from the hook side, then the whole thing's going to get knotted up, and then Rosenberg will just throw it out the window," he told me. "He'll reject the whole amendment process and write his own secretarial amendment with a low total allowable catch and a complete shutdown once that's reached. So they've got a problem there. But when you bring bad news or problems back to your friends and colleagues, well, they're no longer your friends. Smitty and me found that out with the lobster Effort Management Team process."

Brian calls these presentations of evil tidings "hyena sessions." He believes, however, that Amendment Seven has too much momentum now for the hyenas to stop it: "The council seems to have a pretty good sense these days of what the regional director will accept. This is all stuff that at least stands a chance."

And it is just the sort of bitter tonic that Brian would like to see administered to the lobster fishery. Andy Rosenberg remains adamant that just such a tonic is needed, that effort must be immediately reduced. But in the wake of 1994's record landings and a current year that finished much better than it started, more lobstermen than ever are starting to question whether the drastic cuts required by NMFS are really necessary. In response, Rosenberg has chosen neither to withdraw the federal lobster plan from offshore waters nor to write his own secretarial amendment, but rather to shift authority in this process from the council to the Atlantic States Marine Fisheries Commission, a grassroots agency that oversaw the remarkable recovery of striped bass stocks in Atlantic waters through the 1980s.

The ASMFC has already set up a fourteen-member Lobster Advisory Panel, though some of its members are experiencing déjà vu all over again. "Dick Allen, the ASMFC's governor's appointee from Rhode Island," reported the *Commercial Fisheries News,* "noted that many of the advisers nominated had served as EMT members under the council process, and that they were naturally likely to continue their deliberations in a similar vein."

They'll have to do so without Brian, who thinks it "curious" that neither he nor Steve Smith was nominated to the panel's Massachusetts delegation ("But I wouldn't want to be on that anyway"). Instead the state is represented by men whom Brian finds more philosophically akin to the Maine EMT membership: "They're industry lobbyists whose first commandment is regulate everybody else, not yourself. They're happy to be catching lots of lobsters and are adversarial to the science that says it can't last. People on that panel would be shocked if anybody brought up the sort of reductions in effort the Outer Cape was willing to try."

I saw Brian a week before Christmas, showing up at his house in the early dusk to help lift a replacement window frame into the north side of his porch. The *Domino,* the skiff that Brian fished from before he bought the *Cap'n Toby,* was up on blocks in the driveway, a small fir tree fixed into its step in place of a mast, and strings of colored lights circled the tree and ran twinkling down the stays from its peak to the bow and stern. His lobster traps had all been brought in by then, and they were stacked as I remembered them from the spring, running in lathed and hoary walls up and down the length of his yard, like battlements raised against the archons.

Brian was grateful that his fall had been better than his spring, though the weather had been rough. In October a trip with his son to New York City for a taping of the David Letterman show was followed by four straight days of thirty-knot winds, a frosted-up freezer, and twelve spoiled totes of bait, all of it prime bluefish and cod. When he could haul his traps, he was rewarded with harvests that he described as "average, nothing to get revved up about." But all over New England prices were 20 to 50 percent higher than they had been the previous fall, and the handsome paychecks that Brian drew from Ronnie have given him at least a little cash toward what he didn't want to do this winter but must: replace the leaky Volvo in the *Cap'n Toby* with a brand-new engine. He has his eye on a 260-horsepower Mercury Cruiser V-8, a gasoline engine that will be more expensive to run than the diesel but cheaper to buy and install, and will give the *Cap'n Toby* about a five-knot speed boost. "The price tag on that, all installed, down at the yard is eleven thousand dollars. They'll only give me a thousand dollars trade-in on the old engine. They know what's wrong with it," he noted glumly.

Brian also knows what's wrong with Amendment Seven, as much as he applauds its premise and goals. He doesn't like its intended use of total allowable catches, which can be easily converted into ITQs. "John D. Rockefeller once said, 'The age of the individual is over.' And it's a mortal sin what he did to independent oil producers," he said. "That's what we'll be facing with ITQs around here, if it ever comes to that."

Nor is he cheered by the imminent loss of another open-access fishery, this time the hook fishery. "The great thing about this small-boat inshore fleet is its versatility — guys moving into one fishery and out of another as prices fluctuate or as stocks decline or recover, like what Mike's got going with dogfish. Being versatile, being able to zig or zag as needed, is the way small-boat fishermen have always dodged the bullets that come their way. But with all these limited-entry permits, you're seeing the fisheries becoming stratified and specialized, and everybody being corralled into one box or another."

He didn't say it, but his previous rhetorical question hung in the air: are these the rights our forefathers died for? More than anything else, Brian loves the New World promise of fishing, its classlessness and virgin space, the breadth of opportunity it offered to him as a young man with debt and a family and a sense of place, and the purity of its relationship between skill and reward. The forefather to whom this love answers best is Thomas Jefferson, whose vision of an American yeoman husbandry finds its best expression today in the lives of men like Brian — men who make a living, if not on their own small farms, then on their own small ships, and who therefore cannot be forced into the wage-labor relationships Jefferson viewed as exploitative; whose ability to produce food helps to guarantee their independence, supporting a society that is itself independent, resilient, and not hostage to internal or external commercial interests; who participate directly in political processes of local self-rule; and who conduct a community life, finally, defined by relatives and neighbors, associations and clubs, congregations and guilds, rather than anonymous buyers and sellers.

In his day, Jefferson feared the nascent stirrings of a powerful centralized government dominated by big capital. He feared the urgings of those who saw large commercial farming enterprises, such as plantations, as more economically efficient than and therefore preferable to small family farms. He feared a society in which laborers and wage-earners would effectively resign from the processes of government, concerning themselves only with their own self-interest in an economy in which they had no choice, he wrote, but to "eat . . . one another."

In 1946, an anthropologist named Walter Goldschmidt, working at the behest of the U.S. Department of Agriculture, put Jefferson's suspicions of the costs of centralized capital and the alienating effects of wage labor to a sort of test. Goldschmidt examined two towns in California's Central Valley that were similar in the size and nature of their agricultural economies but quite different in the character of the farms that surrounded each town. Dinuba was girdled by small, independent farms worked chiefly by the families who owned them, while Arvin's lands were given over to fewer and much larger farms worked largely by seasonal workers. The total volumes of agricultural production in the two towns were very similar, but Goldschmidt was struck by profound differences in the towns' social fabric.

He found, for example, that Dinuba had more institutions than Arvin for democratic decision-making and broader participation in such decision-making by its people; that the small farms in Dinuba supported about 20 percent more people at a higher standard of living; that most citizens of Dinuba were independent entrepreneurs, while two thirds of the population of Arvin were wage laborers; that Dinuba had better community facilities — more schools, parks, newspapers, churches, and civic organizations; that Dinuba had twice as many business establishments, which did 61 percent more retail business, particularly in household goods and building equipment; that such public facilities and services as paved streets, sidewalks, garbage disposal, and sewage disposal were more available in Dinuba, whereas in some areas they were entirely lacking in Arvin.

These were not facts, actually, that the U.S. Department of Agriculture wished to hear. It canceled Goldschmidt's research, invoked a clause in his contract forbidding him to discuss his findings, and refused to publish his report. The anthropologist finally published the report himself years later, and in 1972 he was called to testify before a Senate committee investigating land monopolies. "In the quarter century since the publication of that study," Goldschmidt said,

> corporate farming has spread to other parts of the country, particularly to the American agricultural heartland, which has always been the scene of family-sized commercial farmers. This development has, like so many other events of the period, been assumed to be natural, inevitable, and progressive, and little attention has been paid to the costs that have been incurred. I do not mean the costs in money, or in subventions [subsidies] inequitably distributed to large farmers.

I mean the costs in the traditions of our society and its rural institutions.

Ultimately Goldschmidt, like Jefferson, went unheeded. The family farm is now as quaint a notion as Jefferson's yeoman husbandry. Its passing has not slowed the march of the American economy. The decline as well of the sort of small-town manufacturing in which Carl Johnston's father worked has been balanced in the gross domestic product by the growth of a monetized service sector. But the journalists Clifford Cobb, Ted Halstead, and Jonathan Rowe question the accuracy of the gross domestic product as an economic measuring stick, and suggest that this growth comes at the expense of American families and small towns, where services were once performed for reasons other than money. This is a shift, I believe, that has been felt nowhere more profoundly than on Cape Cod since World War II. Cobb and his colleagues speak in terms that stretch in the 1990s from sea to shining sea but resonate with a particular sense of loss through the gridlocked summer streets of Chatham and Hyannis:

> Parenting becomes child care, visits on the porch become psychiatry and VCRs, the watchful eyes of neighbors become alarm systems and police officers, the kitchen table becomes McDonald's — up and down the line, the things people used to do for and with one another turn into things they have to buy. Day care adds more than $4 billion to the GDP; VCRs and kindred entertainment gear add almost $60 billion. Politicians generally see this decay through a well-worn ideological lens: conservatives root for the market, liberals for the government. But in fact these two "sectors" are, in this respect at least, merely different sides of the same coin: both government and the private market grow by cannibalizing the family and community realms that ultimately nurture and sustain us.

It all starts to resemble what happens in a lobster trap if it's left too long to soak: the residents start to eat each other. With its fishermen corralled into ever smaller boxes, with the big trucks of Tyson and ConAgra poised to come rumbling over the Bourne

Bridge, with the monetized service sector becoming the eight-hundred-pound gorilla of the Cape economy, this brave new world of self-absorbed wage-earners starts to look more and more to Brian like what Harry Hunt always claimed it to be. It makes the memory of such gentle and guileless men as Toby Vig, the Cap'n Toby after whom Brian has named his boat, all the more poignant.

I remember Brian and Ronnie Harrison reminiscing about Toby Vig one afternoon last May in the parking lot outside the Nauset Fish & Lobster Pool. Brian told me that Cap'n Toby was a transplanted Squarehead, a man who had dredged sea scallops off the New Jersey coast in the 1920s and 1930s and then run draggers out of New Bedford to the Outer Cape and Georges Bank for the next three decades. Even now some of the runs made off Nauset Beach in the winter by the *Honi-Do* and other draggers are known as Toby Tows. Ronnie, a fish buyer, enjoyed Toby's stories about his battles with the New Bedford buyers. "Toby always said, whenever he didn't like the prices they offered him, 'I'm just gonna go dump 'em all on Butler's Flat,'" Ronnie said, laughing. "'Just go dump 'em.'"

Brian laughed too. "Which he did, more than once."

That sort of resolve in his commerce and insistence on fairness was coupled in his personal life with a spirit whose sweetness was undismayed by either tragedy or betrayal. Toby came to Orleans to live out a retirement devoted to driving a truck for Ronnie and helping Chatham's draggermen and purse seiners build and mend their nets. Carl learned his net-mending skills from Toby, and told me how once, under Toby's tutelage, he made the mistake of counting squares of mesh using both hands. Toby slapped one hand away, growling, "One hand. You take one big handful, like when you piss over the side."

Brian said that Toby had had a daughter who had died at the age of eighteen of leukemia, and a wife who had died a year or two later of a broken heart. "At least that's what Toby said. Then he had a second wife who cleaned him out," he added.

Ronnie said, "He'd come into the Lobster Pool in the morning saying, 'Yeah, she threw me out, and then she took all my money out of the bank. But what are you gonna do? Can't do nothin'.'"

Then he had a boat partner, said Brian, who took whatever money was left via accounting tricks, until finally Toby had nothing to live on except his social security checks and his pension from the Seafarers' Union. Brian remembered Toby driving an old car up near his house early every morning and positioning it so that he could pull in the daily fish prices broadcast on an AM station out of New Bedford and then take them back to Ronnie. I imagined him there fiddling with the knobs and dial on Marconi's machine, the only man in Orleans, a penniless saint or mystic, who knew what a fisherman's work was worth that day.

"We used to tease the hell out of him," Brian recalled. "I went over to his house once in the middle of a blizzard so bad that a D-9 bulldozer had gotten stuck on the mid-Cape. While I was there, he gets this phone call from Ronnie. 'Hey, Toby, can you drive down to New Bedford to pick up some fish for me today?' Toby says, 'Today? You want me to go today?' Ronnie says, 'Sure.'" The laughter of both men rose into the blue sky above the back door of the Lobster Pool. "He would've gone, too, if I hadn't been there to stop him."

Brian said Toby suffered a stroke in 1985 and died three years later. "I don't know how much money they made in those days when Toby was fishing," Ronnie observed, "but they had fun. Then the money came in, and it all got too serious."

Now Brian checks his carpenter's level and decides that it's time to nail the window frame into the porch wall. His ranks of lobster traps have disappeared into the winter gloom. All that is visible through the panes of the new window are the *Domino*'s holiday lights, radiant with the threadbare majesty of a stable floor, a humble skiff, the promise of good will toward men.

The day before yesterday, only a week before Christmas, the temperature was forty degrees in the early morning at Sesuit Harbor, but predicted to drop through the day; the wind was dead calm, but predicted to blow thirty knots out of the northwest by the afternoon. The *Last Resort* was working the island again, dredging in seventeen feet of water over humpy bottom amid low, rippled waves. Schools of sand lance puffed up in white clouds from iodine-orange bottom strata on the monitor of the depth sounder. The hauler sang a guttural *yadda-yadda-yadda* when Dan hauled in. One quahog balanced improbably on the latch handle outside the cage of the dredge. "Trying to lead an escape," Dan surmised, tossing it down to the culling board and then emptying the dredge on top of it.

The quahogs spilled out in company with a broken whelk, a small rock crab, a handful of scallops, only a few strands of eelgrass, and a rank, sour smell. Dan held a scallop to his nose and said they were starting to die already. He checked the depth sounder, dropped the dredge back into the water, felt the boat wince, then advance again as the dredge took hold. "The grass is really going now too," he said. "I put that dredge right into the middle of a bed and it dug right in."

The quahogs were dug right in too, plumbing deep, beyond the reach of the cold water temperatures, often beyond the reach of Dan's dredge. The *Last Resort* was working in tandem again with the *Sea Wolf II,* and at 11:30, with the wind freshening and the waves kicking up, Tom was on the radio to say it was getting scary out here. Dan hung on for another hour and a half, by which time he reckoned that a load amounting to just 50 percent of what he was getting a week ago would have to do for the day. The waves had grown to six feet; their flanks were carved with tunnels, and they ran in north–south columns across the bay. The *Last Resort* bore southwest to Sesuit, rolling like a bobsled in the trough of each wave and then rearing skyward as its bow broke out of the trough and arced over the wave's crest. Sleet was coming down, and the wheelhouse windshield ran with tears. Dan looked over his shoulder at the low piles of bushel bags in the wheelhouse and under the culling board and said, "Another exciting day on the water."

In Chatham's industrial park we stopped at a windowless warehouse, in front of which was parked a truck from Harvester, a Massachusetts seafood distributor. A heavyset man in Champion sweats told Dan to attach one of his bags to the clam-sorting machine that dominated the little office inside the door. With its electronic counter and slanted banks of paired parallel rollers, this is a gadget that would have delighted Rube Goldberg. Dan hooked a bag of cherries and 'necks to an automatic intake unit on one end of the machine, and then the rollers, dripping water, were put in motion, rotating so that the quahogs danced and jigged in single file, like a mollusk's dream of chorus girls, along the length of the top pair. These gradually opened up to the width of big chowders, and nearly all of the smaller quahogs quickly fell through to the next pair of rollers, which opened to the width of cherrystones. About two out of three fell through these as well, clacking into a trough for collecting littlenecks. The rest bounced and jigged down the length of the second rollers into a separate trough for cherries.

A teenage boy came in with a bushel basket of steamers as Dan's catch was being sorted. He stared in bug-eyed amazement at the eight bags of quahogs stacked behind Dan, which would total about 450 pounds of chowders at sixteen cents per pound, 480 cherries at twelve cents each, 700 'necks at sixteen cents each — the whole load worth about $240. "Shit, have some quahogs," he said at last. "You guys long-raking? What? Dredging? No shit? Is that legal?"

The questions were eloquent of the degree to which Dan's fishery has become so small as to be almost invisible on the Cape. In Warren S. Darling's time, the waters around Billingsgate and elsewhere in the bay were peppered with twenty-eight-foot sloop-rigged catboats, their masts set nearly flush to the bow, and the sort of bullraking that Carl did in the Town Cove last October was done in water as deep as forty feet with rakes attached to cunningly fashioned pine handles. Darling loved the grace and workmanship united in those rake handles:

> One of the most beautiful sights I can imagine is that of a quahog boat anchored in the distance with the quahoger pulling up his rake.

> The thing that made it so beautiful was the graceful curve of his well made 56-foot pole arching across the boat just before the tee [the crossbar on the upper end] touched the water 20 or more feet away. This smooth arch was an indication of a well proportioned pole. It was the result of great care taken in matching the natural direction of the bend of each piece that had been shaped and joined to make the finished pole.

Kittredge described the punishing labor of harvesting quahogs with those rakes from catboats anchored at both ends:

> Beginning well aft, he hitches his way forward, pulling his heavy drag after him by its twenty-foot handle with a series of jerks that give him the back and shoulders of a Hercules. When he has reached the bow, he hauls his rake up and culls out the quahaugs from the mixture of sand and eel-grass with which its net-basket is filled. Then he shuffles aft again and repeats the process again for hours on end, coming up a boat's length on his hawser as soon as he has exhausted all the bottom within reach of his rake.

Darling says that frequently a throw of the rake would result in a "squeaker": the sound of a tooth on the rake scraping the inside of an empty shell. This meant that the rest of the rake would be unable to penetrate the bottom, and the rig would have to be arduously hauled up and thrown again. Nevertheless, Kittredge says, a man could dredge up thirty or forty bushels per day at the turn of the century; Carleton Nickerson's ship, the *Luella Nickerson,* carried eight-hundred-bushel loads of quahogs to New London, Connecticut, as fast as the schooner could make the trip.

In the late 1920s, gasoline engines powerful enough to haul a dredge became available and the pretty catboats disappeared, though the sturdy quahog didn't. Dan tells me he's seen a picture of Rock Harbor in the 1930s with as many as thirty gas-powered quahoggers in the harbor. Then they harvested only the big 'hogs for Snow's Clam Chowder, and tossed the cherries and littlenecks back over the side. After World War II, however, Snow's switched to the larger, tougher, but cheaper surf clams found off the coast of New Jersey, and then it was the second fleet's turn to disappear, succeeded by Rock Harbor's present complement of charter boats and sport fishermen. The development of a restaurant trade for cherries and 'necks in the 1970s is all that makes dry-dredge quahogging in the bay a paying proposition now, and only for very few boats. These high-priced smaller clams go to markets in Rhode Island, Dan says, while Harvester buys the 'hogs for its chowders and stuffed clams. He says that maybe there are two or three boats out of Rock Harbor now that have been in this fishery full-time "for years and years." Like Dan, Tom Smith is a recent entrant, though Dan wouldn't call Tom a full-time quahogger, since he also does a lot of tuna fishing and jigging for cod. Brian would call it zigging and zagging.

Today the northwest wind that blew up Sunday afternoon and then blew all day Monday and all last night is still blowing at 6:45 A.M., and Dan says it's definitely a wait-till-daylight situation. He's not sure whether to zig today or just stay home. The seas are rough, and at twenty-eight degrees the temperature is marginal: any colder and quahogging would be illegal, since seed quahogs could freeze on the culling board. We walk to the docks through an elephant's graveyard of blue shrink-wrapped yachts and charter boats reared up on blocks against the sky, and we come to rest finally in the lee of the *Annie L*, a forty-five-foot charter boat with a long bowsprit poking out of its wrap. Tom Smith, shorter and huskier than Dan, his face looking vaguely elfin within the hood of a fleece jacket pulled up against the cold, joins us there in a few minutes. "Well, it's not windy here," Dan offers.

Tom looks up at a wind gauge spinning on top of the marina building beyond the boat. He says, "That thing's moving so fast it looks like it's frozen."

They wait and talk about the fall: good for quahogs through October and November but not December, with the weather shitty every day. They say that all the Chatham gill-netters have steamed down to Maryland to chase the migratory dogfish, though most aren't doing well there so far. Dan sighs, reckoning that it's nearly time to haul the *Last Resort* out for the winter. He might go to Florida, where the federal mackerel quota still hasn't been filled, "but the rents really go up down there after New Year's. If I don't go south, I might look for a site on a long-liner, I guess." His tone reflects the scant enthusiasm he holds for working for someone else, particularly on a long-liner.

Tom says that he's definitely going to Florida for three weeks in January and he'll start quahogging again with the *Sea Wolf* in February. "Then I might go jigging in March and April, but I don't know. Cod prices are so low, just a dollar a pound now. Usually they're twice that much this time of year. I hear New England landings have been so sporadic, buyers have turned to other markets."

By 7:15 the wait-till-daylight is over. Sesuit Harbor is bathed in wintry, chrome-colored light, and the bay beyond the harbor's stone jetties is a sea of wild horses, their manes and flanks blowing foam. Tom looks up at the wind gauge. "It's coming northeast now. Getting raw."

"It feels colder than twenty-eight," says Dan. "I can't take the cold like I used to." He hitches up his coat and contemplates the bay. "I don't know. I'm not sure it's worth getting the snot beat out of you for what you can get now. If it was like this a few weeks ago, I'd go."

"You can catch a little bit of lee from that island at low tide," Tom notes.

"Yeah, but by then you're not catching anything in the way of clams."

"Well, I promised a bushel of 'hogs to Mark for Christmas. I'll probably go out for a little bit, anyway."

Mark Nichols is Tom's mate for tuna fishing. Dan has made no such promises, and at 7:30 we watch from the dock as the *Sea Wolf* advances, bobbing, into the bay, then veers to the east as it heads up into the seas and climbs the breakers. "That looks pretty shitty," Dan says.

He walks back to where the *Last Resort* is berthed and calls Tom up on the VHF radio. "Is it as bad as it looks out there?"

"It's not too good."

"That wind blowing northeast now?"

"Well, north-northeast."

"Looks like you're gonna catch 'em all today."

"Probably won't catch anything today. Don't know how long I'm gonna be out here. Not long."

By the time Dan signs off, the *Sea Wolf* is out of sight. We walk back to the pickup with the wind running a rat's maze between the hulls of the muffled vessels. This is the second day in a row of no work for Dan, and the third this week. "Another fire drill," he says. "Well, that's what winter fishing is like."

On Route 6A, in the empty parking lot in front of the storefronts that make up the Orleans Marketplace, a pair of crows have pecked through a puddle's pane of ice and are sipping the cold water beneath.

≈ 12

The True Atlantic House

THE RUINED HOUSE may have been an omen. I went with Carl to Aunt Lydia's Cove at 1:30 yesterday morning. As we drove past Ryder Cove on Route 28 from Orleans, the Chatham Light just up ahead extended a whitewashed spoke into the darkness. The wind was blowing hard out of the southeast, and the spoke of light looked as if it were being spun by its force like a turnstile.

Dave Reed, Carl's mate for this trip, the quahogger who took me out to the *Last Resort* on the day I missed my first departure with Dan Howes last summer, was waiting in the parking lot for us. The February air was as thick and wet as raw meat. The lamps overhead threw a freezer light's thin sheen across the piers, the pilings, the scattered pickups in the lot, the low, hunkered roof of Chatham Fish & Lobster. From the other side of Tern Island, on the eastern flank of the barrier beach, we heard the froth and hiss of the waves heaving themselves against the sand and then boiling away like acid. Carl took off his hat and scratched his bare pate.

"The buoy report from offshore says it's blowing twenty-five knots, but it's supposed to die down by morning," he said. "We'll try again around eight-thirty and then just stay out all night, I guess. Bring your sleeping bags."

The *Honi-Do* was geared for shrimp, and Carl wanted to get out and fish, something he had not been able to do much this winter. "Don't call me, I'll call you," Mark has been telling him, and whenever Carl hears that, he knows — even when the forecast is good, the tides are favorable, the nights are calm and clear — that he'll be doing a lot of stone and tile work with Dave Slack, or clamming if it's warm enough to do so legally. Once, at the beginning of this month, he thought he had it all arranged to take the *Honi-Do* out himself on a day on which Mark had jury duty. He had two men lined up to go with him, only to learn that Mark was tired of the short loads of groundfish they were getting when they did go out, and also of the volatility of cod prices (one day $1.00 per pound, the next $2.00, the next $1.25, and so on), and had signed the boat up with NMFS for the winter shrimp fishery. The result that day was that they couldn't go out with the gear they had on.

Carl didn't mind gearing for shrimp; he just wished they could have started some other day. Cod and flounder are always hard to find this time of year, and northern shrimp, spawning now in the inshore waters off the Cape and in the Gulf of Maine, have been modest but reliable midwinter moneymakers over the past few years, commanding a steady eighty to ninety cents per pound and yielding regular, if unspectacular, loads of one thousand to fourteen hundred pounds on Mark's short day trips. But shrimp fishermen only work five-day weeks, stormy winter weather permitting. This year NMFS is closing the fishery every Sunday, and buyers on the Cape won't take shrimp on Fridays, since they can't be auctioned off in the fish markets until Monday. That makes days of good weather all the more precious, and last week Carl was antsy enough to try something reckless: simply telling Mark that he'd take the boat out if the weather was decent on a certain day and then steering the conversation in other directions before this motion was subject to discussion or amendment. Last Thursday the weather was decent, Carl took the *Honi-Do* out with Dave Slack aboard, and eventually he got a call from Mark on the boat's cellular phone. "What are you doing out there?" Mark said.

"Well, I'm out here fishing," said Carl.

Pause. "You catching anything?"

"Yep. Hey, I told you I was going Thursday."

"No, I thought you said the thirteenth."

"Well, I'm out here."

Carl said it was "funky" hearing Mark's voice on that phone asking him what he was doing. He came back with a big load of shrimp, covering expenses and making some good money for Mark, who chalked it all up to a misunderstanding and appreciated the extra change. Carl gambled, won, and then decided maybe he'd better not push it anymore. He went back to waiting for Mark's calls.

This week, however, Mark is out of town again, this time on a ten-day vacation to Florida with his family, and the *Honi-Do* once more is Carl's to run. Yesterday Carl wanted to get out on the 1:30 A.M. tide, fish all through the morning and into the next evening, and get back home that night. When that didn't work, he decided to catch the 8:30 tide and make an overnight of it after all. The important thing was to bring in a good load, not only for the extra money it put in his and Dave Reed's pockets, but for whatever nudge one more big check might give Mark toward letting Carl run the boat more often, or at least going out more often himself, and staying out longer.

So we met again in the morning in the parking lot at Aunt Lydia's Cove and rowed in the dinghy kept on the beach out to the *Honi-Do*. The wind had veered to the southwest, but it was only a light breeze now, and the temperature was up into the high teens, with light, airbrushed clouds riding high and clean overhead. The tide poured in quickly through the half-mile sluiceway between the Chatham headland and the tail end of Nauset Beach, and congealed pads of ice — hard white floes and rubbery mats of slush — sped past the cove entrance like targets in a shooting gallery. A harbor seal had hauled out onto the beach at Tern Island. Two Coast Guard men, standing on the pier where the Coast Guard's forty-four-foot inflatable rubber-and-steel surf boat was tied up, were gazing at the seal through binoculars.

As Dave Reed climbed aboard the *Honi-Do,* his face was stiff with distress, his blue eyes pinched beneath the weight of his bunched eyebrows, squeezed by the swelling of his left cheek. He said he had a toothache, was taking an over-the-counter pain-killer, would manage all right. He had news of Dan Howes and Tom Smith: Dan had driven down to Maryland to see if he could catch a site aboard one of the Chatham gill-netters fishing for dogs down there; Tom had hauled his boat out of Sesuit and already put it back in again, hoping to get a jump on the rest of the quahoggers, but the boat was frozen into the harbor now. Then Dave climbed up the mast to clear a wire hung up on a spreader while Carl cranked the *Honi-Do*'s motor and booted up the electronics.

The monitors on the wheelhouse console rose from a jumble of rubber and cotton gloves, tide charts, coffee mugs, paperback novels, a notebook, a calculator, a flashlight, a roll of paper towels, some random tools. The boat's immediate course — out of the cove, down the length of the harbor past the sheared-off ends of Holloway Street and Andrew Harding's Lane in Chatham, past the Chatham Light, and finally out beyond the end of North Beach — appeared on the video plotter as a series of yellow angles between the sinuous topographic lines of the beach, the headland, and the sandbars. Meanwhile the radar scope showed the headland as a solid green mass, the waves outside the cove as a pointillistic clutter, both of these elements emerging in brush-strokes out of the scope's darkness as the spoke of the radar pulsed over them. Outside the wheelhouse, four-foot waves were just beginning to break in curlers over the crest of a bar. Once under way, the *Honi-Do,* with its bargelike breadth and bulk, seemed immune to these waves. The boat eased its way through them like a float in a parade.

But the headland was not immune, nor the houses that lined the narrow beach on the approach to the lighthouse. A piece of the beach had been stripped away in the night, and one summer house below Holloway Street was slowly settling into the sea. Its shingles stripped, its angles crimped, its foundation eaten partly away, with the waves gnawing at it again with the incoming tide, the house was sagging fatally at one end into what Thoreau described as "that seashore where man's works are wrecks . . . the true Atlantic House, where the ocean is the land-lord, as well as the sea-lord."

Dave asked Carl if he wanted to try the shoal waters six miles out. With Carl's assent, Dave punched a set of loran coordinates into the video plotter, whose monitor marked the *Honi-Do*'s speed at 7.7 knots, 8 miles to destination, 68-degree compass bearing, east-northeast. Outside the harbor, beyond the tidal bottleneck of the beach and the headland, the sea was clean of ice and syrupy beneath a five-knot breeze that sailed over its surface without finding any purchase. Chris Armstrong had followed us out in the *Overdraft* and steamed behind us several hundred yards off the port quarter. Another boat lay two miles ahead to the northeast, a green blip on the radar screen. "That's probably Amaru," Carl surmised.

He grabbed a novel and went below to read in a cabin berth until we reached the grounds. At eleven he came up again to reset the video plotter for the set of tows he wanted to run across this section of bottom. The *Overdraft* was about to work this same area, and Chris called Carl on the VHF to give him the loran coordinates of a snag on which he had torn his net here recently. The breeze had sharpened and begun to cut chips and wrinkles, the veiny ridges that are called capillary waves, into the sides of the slow swells moving under the boat. "Maybe we'll be in at eleven tonight," Carl said. "The wind's picking up already."

Bearing north over soft and silty bottom fifty-seven fathoms below, with a hairline of land, the Outer Cape, showing over the western horizon, Carl took the wheel and dropped the *Honi-Do*'s outriggers. These unfolded like a pair of wings from around the boat's mast. Each outrigger had on its end a paravane, otherwise known simply as a bird — a glider-shaped float made of flaked steel and carrying eighty-five pounds of lead ballast. These stabilizing devices nestled onto the waves like storm petrels — Mother Carey's chickens — twenty feet from the boat on either side.

By then Dave and I had taken positions at the stern on opposite sides of the gallows frame and net reel. The cod end of the net, made of a flexible blend of nylon and Dacron, hung like the tip of a green mesh stocking cap off the stern side of the net reel. I watched as Dave took a rubber hammer, stepped into the cockpit beneath the reel, and banged the cod end up, closing it off by driving a pin through the middle of the cowbell, a bronze assembly that hangs from the nub of the cap. Then, with the engine idled down to 900 rpm and the net reel slowly unwinding, he took a shovel to the descending cod end and pushed its tip over the removable plywood panel — the pen board — that served as the boat's transom. Once the cod end began sinking into the *Honi-Do*'s backwash, the Nordmore grate in front of it descended from the reel and followed it into the sea, as did the various sections of the rest of the net: the extension in front of the grate, the long belly of the net, the floats, and then the top and bottom wings, all of the sections meshed in different-colored twine to facilitate repair. An occasional square of mesh went over still plugged with the remains of a gap-jawed whiting or a desiccated sand dab.

The legs and ground cables came off the reel last. The legs are the four steel lines that tie the net at each corner into the two ground cables, and if Carl were fishing for groundfish today, the legs would be fifteen fathoms wide and would carry Styrofoam floats up top and rubber bobbins on the bottom. For shrimp, only bare wires are allowed, and the legs are shorter — five fathoms — to maximize the gear's vertical (rather than horizontal) bite. At the end of each ground cable is a steel pair link clipped to two heavy idler chains secured to the net reel. Carl stopped the reel when the pair links and idlers appeared, and then Dave and I took equally heavy chains called backstraps off the iron-slabbed doors hung on either side of the stern and snapped them with g-hooks into the pair links. Finally Carl let the reel run a little farther, just enough for the weight of the net to pull the backstraps taut and allow the idler chains to go slack. Then we disconnected the idler chains, freeing the net from the reel, and watched as Carl dropped the doors off the stern.

These fell like tombstones into the foam. Each door was shackled onto a half-inch steel towline spooling out of a great winch behind the wheelhouse, and Carl let them run out to the three-fathom mark on the towlines, then throttled the boat up to 1300 rpm. With this the doors turned kitty-cornered to the boat's northern heading and strained outward, east and west, spreading the ground cables and the legs and finally the wings of the net. At last the whole assembly filled like a windsock and descended silently to the bottom, down through the darkness.

Setting the net took a little more than ten minutes. Carl ran out about nine hundred feet of towline as the *Honi-Do* moved north at a stately three knots, steadied from any roll by its paravanes. It steamed with the wind behind it so the net would sit more evenly on the bottom, running a route that Carl had selected on the video plotter, negotiating a path between the yellow squares on the monitor that identified hangups or junkpiles, including the hangup whose location Chris Armstrong had just shared. Above us the mackerel skies of the early morning were beginning to congeal into lower and thicker clouds. Dave went up to the wheelhouse, stripped off his gloves, rubbed his tender jaw, told Carl that he couldn't wait until the middle of March, when he expected he could go quahogging again.

"I've been thinking about that myself," said Carl, who has grown ever more appreciative of steady and reliable work.

"Can you catch much shrimp in the dark?"

"No, we'll probably tow till dark and then steam in to the beach for the night, and start again at four-thirty tomorrow morning. High water at Chatham is around noon, so we'll head in around ten or eleven A.M. — if the weather holds."

Three hundred and fifty feet below us, the doors kicked up storms of turbulence as they bumped along the bottom, as did the steel ground cables.

If the *Honi-Do* had been geared for cod or flounder or whiting or any other groundfish, these twin walls of turbulence would have created what has been described as the herding effect: the fish flee in a line ahead of the net, unwilling to try the turbid, silty water on either side. Scuba divers have seen groundfish grazing on the bottom dart away at right angles to the wall of silt as it approaches them, only to resume feeding again and finally to swim calmly, without apparent fright, away from the mouth of the net as the walls of silt finally converge on it. At three knots, however, a cod can swim for only ten minutes or so before tiring and slipping back into the belly of the net, and then down the extension, and finally into the close confines of the cod end.

But northern shrimp — small and clawless arthropods, gentle cousins to the lobster, growing up to seven inches long — appear to be immune to this effect. From December until April they migrate inshore, congregating by the millions to spawn in frenzied, swirling pods. Then they disperse again into the offshore waters, where their populations are too diffuse to be of any benefit to fishermen. While spawning, they respond not at all to clouds of silt and gravel, but go about their orgiastic business unconcerned by the advance of a dragnet. The trick for the draggerman is to find a large pod and then clear a path through its center. Neither the doors nor the sweep cables will compel any sort of herding behavior.

At noon a school of harbor porpoises appeared off the port rail. Their dorsal fins arced into sight above water nearly as black as those fins and then disappeared. The Greeks believed that porpoises and dolphins were the embodiments of *thalassa,* the vital force of the sea, the source of all life, and occasionally one of the porpoises leapt clear of the water, gone lighter than air with the thrill of that force. Birds followed us as well: herring gulls, a couple of great black-backed gulls, and a big northern gannet, its black-tipped wings spread wide, its chiseled beak duplicating the tapering spearpoint of its tail. Toward the end of that first tow, Bill Amaru in the *Joanne A* steamed past us off to starboard, his boat trailing its own entourage of birds like a white scarf in the wind.

The water behind the *Honi-Do* bubbled and boiled when Carl began hauling in the net at 12:30. The gulls hovered, swooped, jockeyed, and screamed as the doors came up and were chained again to the stern, as the wings and belly and extension of the net, groaning, were wound around the reel, as the cod end rose and hung suspended over the dumping pen. Dave hunched down and took the rubber hammer to the cowbell. The pin gave way, and the shrimp and a few stray finfish poured into the dumping pen like pudding.

With their long, wiry antennas, many times their body length, as well as the pink plastic translucency of their shells and the segmented redundancy of their limbs, the shrimp looked less like animals than like several gross of standardized electronic components. Their eyes were tiny wafers of silicon. Their roe were dusky lines of circuitry along the flanks of those hermaphroditic shrimp reproducing as females this time. Dave and I reattached the backstraps to the doors, and then Dave reinserted the pin he had taken from the cod end's cowbell. He pushed the cod end over the pen board again, and while the net unwound from the reel and drained into the sea, he used the shovel to lift piles of twitching shrimp onto the culling board adjacent to the dumping pen.

I was surprised to see any bycatch at all, given that the Nordmore grate would admit nothing bigger than a shrimp into the cod end. But sometimes finfish get tangled in the squares of mesh preceding the grate, Dave said, or a school of skates bunches up and plugs the escape opening above the grate. We found among the shrimp a few small cod, some herring, a pair of sculpins, a butterfish the size and shape of a silver dollar, and eight small whiting. We saved the cod and the whiting for our dinners, or those of friends, and shoveled the rest over the rail. Then the shrimp were spilled into fish totes and hosed down with seawater. They filled a little less than three totes — about two hundred and fifty pounds.

"That's not good," Dave said. "We should get nine or ten totes on a good tow."

Carl came out of the wheelhouse and stood with his arms folded across the bib of his oilskins. "Last week Amaru got six or seven thousand pounds working out here. But Mark wouldn't go out that day — said it was too rough."

Ten minutes later Carl got a call on the VHF radio from Amaru, who had just been in contact with Chatham Fish & Lobster. "Ray wants to know if we can come in tonight so they can ship tonight," Amaru said. "I said I probably could, but I wasn't sure if you guys could manage it with the way the tides are now. Ray's standing by on eighteen."

Carl signed off and stood at the radio, hesitating. "They want us to come in, and we want to stay out. Well, I know they don't want to wait till ten o'clock tonight." He flicked the radio to channel eighteen and raised Ray Durkey, one of the fish packers at Chatham Fish. "You don't want 'em tomorrow?" Carl said into the microphone. "Gotta be tonight? Well, we can't get in till nine, nine-thirty. How late you staying open?"

"Dave wants the shrimp by seven-thirty," Durkey said.

"Can't do it. Maybe by nine. Can you wait that long?"

Ray said he didn't know, that he was just a middleman, and that Carl would have to talk to Dave Carnes directly. Outside the wheelhouse window, beyond the *America's Funniest Home Videos* reruns playing on the suspended TV, among the troughs between the paravanes and the hull, the wave crests were curling into glassy white streaks of foam.

We towed throughout the afternoon, running the net for ninety minutes and then hauling out. On the second tow, hauled at 2:15, we found among the shrimp a few more herring, pretty fish with their greenish blue backs and silvery white flanks, their bellies brushed with mauve, their gill covers glossed with brass; some sleek and steel-blue mackerel with tiger stripes and lunate tails and scales so small their skin was velvety to the touch; a five-inch worm called a sea mouse, shaped more like a football than a worm and covered with bristles as soft as fur along both flanks; and a fifteen-inch, three-pound haddock.

The haddock is a smaller relative of the cod and is also a pretty fish, though while the cod is a golden ingot, the haddock is silver: a purplish silvery gray with pink reflections above, an opaque white along the belly. The fish is pretty as well in the filmy apparel of its three dorsal fins and two anal, these being slightly larger and more prominent on the haddock than on the cod. Like the cod, the haddock wears a lateral stripe along its flank, one that is actually a series of sensory pores to detect local disturbances in the water. In the cod this stripe is pale ivory, in the haddock a charcoal gray, running from the neck of the tail to the back of the gills and just touching the sooty patch behind the pectoral fins that is the haddock's most distinctive feature. New England fishermen, the sons of Puritans, have called this patch the Devil's mark. According to folklore, the cod was the fish that Christ used to feed the masses; when Satan tried to do the same with the haddock, the fish wriggled away from his burning fingers, acquiring in this manner its shoulder patch.

The meat of the haddock, like the cod's, is lean and white and unfolds in firm white leaves when cooked. Its historical abundance on Georges Bank and in the Gulf of Maine made it cheap, however, and so it was considered poor man's food, at least until the advent of quick-freeze technologies and the first swooning decline of the cod in the late 1920s. Then haddock ascended to the cod's culinary status, and soon afterward — much more swiftly than the cod — plummeted into decline and scarcity. In fact the abrupt crash of haddock stocks and their slowness to recover were major factors in the founding in 1950 of the International Commission of Northwest Atlantic Fisheries, an organization that made an attempt, at least, to manage and conserve fisheries resources on a multinational basis. But the inability of the ICNAF to control the factory trawlers of the distant-water fleets and, most conspicuously, the success of those trawlers in fishing Georges Bank haddock down nearly to nothing led in a direct line to the two-hundred-mile limit and the Magnuson Act of 1976. American fishermen effected a very modest surge in haddock landings in the first decade of the Magnuson Act, but since the middle of the 1980s the haddock has been the rarest of New England groundfish, and today it is still considered commercially extinct.

I find it hard to believe now that in 1913, during the early days of steam-driven otter trawl fishing, between 60 and 70 percent of the fish taken by trawlers on Georges Bank were haddock, while only 10 percent were cod. In its peculiar sensitivity to fishing pressure, the haddock has been the canary in the coal mines of New England fishing, wilting away well in advance of hard times and being difficult to find throughout their duration. This haddock was below the nineteen-inch minimum for commercial take, but no matter — no haddock of any size may be legally taken now as bycatch by a dragger. Dave pushed the struggling fish down the length of the culling board and over the rail, returning it and its sooty patch to the coal mines.

By midafternoon, horizons on either side had disappeared and the sun was a swatch of white gauze stitched to a sky of cadaverous gray. Around five o'clock the gauze moldered to a purple bruise in the west, and we finished our fourth tow in the dark, hauling the net in the hard glare of the *Honi-Do*'s cockpit light, the gulls hovering in silence at the frayed edges of the darkness. "We need a nine-box tow," Dave said as the winches began to grind, the reel to turn, and a drowned gull came up attached to a sweep cable, one foot tangled between the cable and a chain link.

In quantity the fourth tow was like the previous three: just a little more or less than three boxes of shrimp. But this time there was plenty of bycatch: dozens and dozens of herring, seemingly outnumbering the shrimp. Pretty or not, a herring's scales come off at only a touch, and many of the fish were already as deshingled as that house below Holloway Street, many of the pink shrimp as sequined as circus performers with stipples of loose herring scales. Dave looked at the mess and swore, anticipating all his weather gear getting stippled in handling the fish and then stinking as the scales dried, stuck fast, and rotted. Carl came out of the wheelhouse again, and Dave said, "You're fired. I'd rather pull my tooth out with pliers than clean all this shit."

Carl nodded, accepting this demotion without complaint. He said he had talked to Dave Carnes, who wouldn't take anything after 7:30, but he had also talked with Chris King, a Provincetown buyer who had a truck in Chatham. King said he could hold his truck a little later into the evening if Carl could come in now and get into the back moorings at Aunt Lydia's Cove, where the water should be deep enough even at dead low for the *Honi-Do*'s seven-and-a-half-foot draft. "Then we can have Billy Amaru offload the shrimp for us. He's going in too," Carl said, smiling at Dave and folding his hands again into the bib of his oilskins. "I get along fine with Billy," he said. "Now if Mark was here, those two would be fighting like cats and dogs." He turned to go back to the wheelhouse, calling over his shoulder, "The forecast is for twenty knots and gusting southeast tomorrow anyway — twenty-five to thirty knots offshore."

Carl pulled up the paravanes and wheeled the boat south at 7:15, with the stars still blotted out overhead but the horizons clear enough for the lights of Chatham to glow like fireflies through the window on the wheelhouse's starboard side. The *Honi-Do* moved evenly through moderate seas in winds blowing at fifteen knots. The blue waters, pearly white sands, and pretty red maillots of *Baywatch* flickered across the TV above the portside console in the wheelhouse.

The *Joanne A* and the *Overdraft* had preceded us into the harbor. Outside the channel, Carl contemplated with visible unease the floodlit vista of whitecaps marching across the bars there. Finally he hailed the two other boats on the VHF. "Where's the tide?" he wondered.

"It's just a little past dead low," Bill Amaru replied. "Just go from flag to flag down the channel. We're getting a skiff ready here to help you unload. Just come in from flag to flag, and you should have no problem."

"You gotta go about two hundred yards past that flag buoy, Carl," added Chris Armstrong. "Then make a straight shot. Come right up the beach to the red flag. You gotta go further around that one."

We were still some three miles out from Aunt Lydia's Cove, working our way into the half-mile strait between Chatham's South Beach and the feathered tail of North Beach, sniffing out the wave-raked channel snaking between the outer bars that spread from the shores of each. The *Honi-Do* moved circumspectly toward the channel, breasting breakers that sprouted like dragon's teeth from either bar and, for all its weight and bulk, pitching beneath our feet like a skiff the moment those breakers closed about it. All was darkness around the boat except for the waves leaping in the spotlight and the lamps of Chatham to starboard, which were yawing above the wheelhouse window and then plummeting below it like film slipping out of its frame. Carl said between clenched teeth, "And they wanted us in here an hour earlier."

The flag that should have marked the entrance to the channel was gone, blown away in high winds over the weekend. Carl found the entrance nonetheless, and then the flag buoy, which was missing its radar reflector. He moved past this, though not two hundred yards, choosing to stick close to the flags, as Amaru advised, and then bore northeast, heading on a straight shot for the red flag, and then for a short run along the beach to the third flag, and the dogleg to the west that it marked.

But a straight shot was hard to run, as was a crisp turn around a buoy. In the wind and surf, and in a running tide that required the sluggish *Honi-Do* to crab-walk across its current, Carl found it difficult to keep a bearing. After the third flag he fought with the wheel as the depth sounder began a giddy slide from fourteen fathoms down to four, and then quickly down to zero, while the boat got pushed to the western edge of the channel. Either the *Honi-Do* was then able to plow back into deep water or else the bar just fell away, since three and then six fathoms of water filled in under the boat. But just as quickly the sounder was falling again: five fathoms, three, one, zero. "What's that?" Dave said as a shiver ran through the boat, something like a tuning fork's sympathetic vibration. The second grounding was more defined, lifting us to the balls of our feet. The swooning moan of thwarted steel rose to our ears.

Finally, with one more shuddering sigh, the *Honi-Do* seemed to hit, skid, float briefly again, and fall for a third and final time into the zero-fathom breast of the bar. The wheel spun uselessly in Carl's hands. He killed the engine and turned toward Dave with a hard-edged smile, saying maybe they'd better get an anchor down; probably dropping the port door would be the best thing.

Bill Amaru's voice crackled on the VHF. "Carl, how are you doing?"

Carl picked up the mike. "I got a funny feeling I just lost my rudder." He clicked the mike off, speaking mostly to himself: "We wouldn't have this problem if Carnes had just been willing to take the shrimp tomorrow." Then he flicked the radio to channel sixteen, which the Coast Guard monitors. "Chatham Station, this is the *Honi-Do*. Chatham Station, this is the *Honi-Do*."

"Roger, *Honi-Do*, this is Chatham Station."

"Yeah, I'm hung up on the outer bar here, and it's looking like I lost my steering."

"What's your weather right now?"

"Wind's twenty knots out of the southeast with a good chop out here."

"Are you taking on water?"

Dave had just gotten back from unclipping the chains on the port door. He pointed to the lazarette light above the hydraulic controls on the wheelhouse bulkhead opposite the wheel. The light shone red as the bilge pump ran in the stern compartment where the steering mechanism is housed. Then it shut off as the pump went dry. "No," Carl said.

"If you've got life preservers, please put them on."

"We got survival suits."

"Make them available. We'll have a boat out there as soon as possible, and we'll call you at fifteen-minute intervals. We can't tow you just at present."

Bill Amaru broke in. "We'll wait for you in here, Carl, and help you offload."

"Do you have an anchor?" the Coast Guard operator asked.

"Well, we're hard aground. We dropped one door, just in case."

"Our instruments indicate that you're drifting."

Carl checked his video plotter. There the *Honi-Do* was a cursor blinking west of the green line marking the route into Aunt Lydia's Cove and west of the yellow line marking the edge of the bar we were now stranded on. "My plotter says we've only moved two tenths of a degree in twenty minutes. But the sooner you get out here, the better."

"We're doing our best."

"Well, we're not going anywhere."

Carl kept his spotlight focused on that fourth flag, the one he had nearly fetched, almost as if at any moment the *Honi-Do* might break free of its rut on the bar and round that buoy with steerage intact after all, steaming into Aunt Lydia's Cove just in time to call the truck back from the exit of the parking lot.

The flag and its radar reflector swung back and forth in the surf like a metronome. Forty-five minutes later, once the tide had risen enough to make pulling the *Honi-Do* off the bar at least plausible, the Coast Guard's twenty-eight-foot inflatable rescue craft appeared at the edge of the spume-flecked circle of the spotlight. The rescue boat was little more than a rubber raft with a jet unit mounted in back and a doghouse amidships, manned by a coxswain and three blue-uniformed, orange-life-jacketed crew members. The air temperature was around thirty degrees, water temperature perhaps fifty. Aboard the *Honi-Do,* the cold had settled in, seeping like moisture into both steel and bones. On the Coast Guard craft, it had whitened the faces of the crewmen, who clung to its gunnels like ghosts.

The coxswain hailed Carl on the radio at the craft's first appearance in the spotlight, saying he would need a man in a life preserver up on the *Honi-Do*'s bow in order to catch the rescue craft's towline. Dave stripped off his raincoat, his fleece jacket, his shirt, and finally a back brace, the consequence of an injury from a car crash. He dressed again without the brace and struggled out onto the foredeck through the wheelhouse's starboard door. He became a black silhouette in the glare of the spotlight, a ghost himself, clinging stiffly and painfully to the boat's forestay while the inflatable disappeared beneath the *Honi-Do*'s bow, while the chop slashed back and forth in the light ahead of us, while a rope snaked up into the light near his feet. He wrapped the line about the mooring bit while Carl said on the radio, "Get me to deep water. I need two fathoms."

With the towline secured, the port door winched up and tied off again, and Dave back in the wheelhouse, the rescue boat pointed into the waves. The coxswain drove it straight out into the channel, beam to the *Honi-Do,* trying to turn the bow off the bar. At the same time Carl put the *Honi-Do* in gear and throttled up the engine, adding forward thrust to the rescue boat's swiveling pull. Dave swung the spotlight out into the channel, and we watched in silence as surf broke in white snow over the inflatable. The boat was stunned and turned and occasionally took on quartering seas. Once it swung perilously broadside to the surf. With a terrier's relentlessness, the boat kept turning back into the waves and digging for the barrier beach, which was invisible in the darkness. Finally, improbably, the bow of the *Honi-Do* described a small arc, its forestay moving in slow motion from the left of the fourth flag to its right. Then the pitching of the dragger evened and moderated as gradually it followed its bow and lifted free of the bar.

"We're off," Carl said on the radio.

"Roger," said the coxswain. "But if we get hung up again, we're not going to have enough power to get you turned fast enough."

Carl acknowledged that and signed off.

"I'll bet those guys were shitting in their pants out in those waves," Dave remarked.

"I was nervous myself, feeling us roll," Carl said.

"What time is it now? Nine-thirty? Isn't that around when we left this morning? Then the tide should be about right for that forty-four-footer they got."

"I don't know."

Dave stared out the windshield, craning his neck. The spotlight ran down the length of the towline. Ahead of us the stern of the rescue boat was just visible, showing up like a piece of flotsam amid the whitecaps. "Those guys looked awful small when I was looking down on 'em from the bow," he said.

The clouds were breaking up overhead, and in the shadow of the new moon the stars were as white and lucid as the lights of Chatham in the widening channels of the sky. By then the inflatable had carried us past the end of North Beach and into the long neck of the harbor, which itself was braided with sandbars, boiling with chop. We moved deliberately from flag to flag, hugging each one like a golf ball rimming out of a cup. Soon the grind of the surf on the barrier beach was audible to starboard, the slapping of waves on North Beach to port, and then the winking lights of the town floated in the distance ahead of the bar. Carl murmured almost to himself that the coxswain had missed a turn, that we should have backtracked before cutting to the next buoy.

Less than half a mile ahead in the darkness lay the hump of Tern Island and the outer grounds where a number of other boats now pulled at their moorings. The coxswain called on the radio to say that he didn't want to tow the *Honi-Do* into that area, where it might strike other vessels, but wanted to tie his own boat alongside and raft us in.

"Should I drop a door?" asked Carl.

"Negative," said the coxswain. "We can tie up and get her moving fast enough."

Carl mentioned to Dave that he wasn't so sure about that. He was proven right when the rescue boat pulled alongside and its crew botched its first attempt at a tie-up. The coxswain swore like a sailor, then a trooper, then like one of hell's drill sergeants — what Carl described as going banshee — as his crewmen fumbled with the lines on the pitching boat and as the tide pushed the *Honi-Do* hard against the bar. Finally, by the time the rescue boat was secured, there was nowhere for it to go. Carl throttled the engine, punched the gears back and forth between forward and reverse, and threw up his hands. We were grounded again.

The coxswain came aboard, thin-lipped, rock-jawed, throwing off steam. Carl took him into the wheelhouse and showed him on the video plotter the tuck in the channel that required a boat to backtrack after rounding the last buoy. The coxswain said that the buoy wasn't where the video plotter showed it; his chart showed a different location. Whatever, said Carl. He then suggested that the rescue boat pull a towline off the *Honi-Do*'s stern and try to back it off the bar. But the coxswain insisted that the channel was open from that buoy for a straight shot to the next and they should try to pull us free as they had done before. Carl dropped the port door on a short line, lightening the boat and also providing a fulcrum to ease the movement of the bow. That done, the coxswain harried his crew out into the channel again, but this time to no avail.

When the rescue boat abandoned the effort and veered back toward the *Honi-Do,* Carl rang up Chris Armstrong, who was still aboard the *Overdraft* in the harbor. "I'm curious," Carl said. "Is that truck still there?"

"It's still in the parking lot," Chris replied.

Carl called up the Coast Guard station. Could he meet with the station chief if he came ashore? Affirmative.

Carl had little more than ten totes of shrimp aboard, but those shrimp very likely had cost this boat its rudder, or at least the expense of repairing it. That made Carl all the more anxious to be able to give Mark a check, however modest, for the shrimp he had caught that day. He was also sure that the bars had flooded enough now for the Coast Guard's bigger vessel, the forty-four-foot steel surf boat, to be able to operate out there. He didn't want to debate that with the coxswain, who plainly had his own way of doing things; instead he wanted to talk personally with the station commander, whom he knew. And then he wanted to put his face in front of the driver of that Chris King truck, to throw himself across the ramp of the parking lot if necessary. When the coxswain pulled alongside again, Carl asked to be taken ashore.

He hauled up the port door and dropped the starboard as an anchor, paying out about thirteen fathoms of cable, and told Dave to call him on the VHF if the cable tightened enough to suggest that the *Honi-Do* might be drifting and dragging the door behind it. Then he dropped over the side into the rescue boat, strapping on the life preserver that was handed to him as he clambered into its cockpit. Its jet unit muttering, the vessel bobbed and slapped against the *Honi-Do*'s side and then nosed away into the channel, its bow lifting, its crew and passengers hunkering down as the surf rolled beneath it. It took only a few seconds for the boat to disappear beyond the range of the *Honi-Do*'s spotlight.

The dragger was left to the enveloping darkness, the monochromatic glow of its instruments, the moan of the wind through the rigging and paravanes, the slap of the chop against the hull. "I've got a bad feeling we're going to eat our catch," Dave said, his words thick now with the swelling of his cheek. I couldn't imagine poor Dave eating anything. What's worse, I considered, than being hung on a bar in rough weather outside the cove at the end of the day, knowing that your paycheck is about to drive out of the parking lot, knowing as well that things could possibly get a lot worse? Being there with a toothache, I supposed.

Three survival suits lay stacked like corpses in front of the companionway. The paperbacks were scattered across the console, one of them open on its spine, its pages flipping languidly with the rocking of the boat. Above us the moon sailed invisibly in a patchy sky. Dave checked the cable on the starboard door, gave it a little more length, checked it again half an hour later when the video plotter indicated we were moving. By then it was as tight as piano wire. "Guess we better try to get hold of Carl," he said.

Mark Leach, the president of the Cape Cod Commercial Hook Fisherman's Association, isn't quite ready to give up. Rebuffed in his attempts to persuade the New England Fisheries Management Council that long-liners and jig fishermen should be excluded from the provisions of Amendment Seven, he has published an editorial in the February issue of *Commercial Fisheries News* charging the council with violation of the Magnuson Fishery Conservation and Management Act.

"The act," writes Leach, "requires that fishery management plans 'include readily available information regarding the significance of habitat to the fishery and assessment as to the effects which changes to that habitat may have upon the fishery.' See 16 U.S.C.{}1853(a)(7). 'If man-made environmental changes are contributing to the [decline of spawning stocks or average annual recruitment] . . . councils should recommend restoration of habitat . . .' See 50 C.F.R.{}602.11(c)(7)(iii)."

Leach argues that the towed gear of draggers is causing just such "man-made environmental changes," charging that the sort of rock-hopper gear that ruined Fred Bennett's flower vase in Hyannis — gear that allows draggermen to work rough bottom that previously would have rimracked their nets — is having an effect analogous to "strip-mining old-growth forest." He says that the council has refused to recognize the effects of towed gear on habitat and fishery recruitment because of its domination by dragger interests. The CCHFA, however, "plans to continue the fight against the council's failure to properly address the habitat requirements of the Magnuson Act, including litigation if necessary."

Of course, not only the New England council but every fishing fleet in the world is dominated by dragger interests. This is because for millennia — at least since 2000 B.C., when Mehenkwetre, a prince of Egypt's Theban dynasty, was buried with models of a pair of reed boats towing a weighted fishing net — trawl nets have proven to be the most efficient means of gathering large quantities of fish. Nor is the question of such nets' impact on habitat and species recruitment unprecedented. The text of a petition brought before England's House of Commons in 1366 has an eerily familiar ring to it:

> Some fishermen who have during the past seven years by a new craftily contrived kind of instrument, which among themselves is called Wondrychoun [probably from the Dutch *wonderkuil*, "marvelous fishing trawl"], made in the form of a drag for oysters, which is of unusual length: to which instrument is attached a net so small a mesh that no kind of fish, however small, that enters it can pass out, but is forced to remain within it and be taken. And besides this, the great and long iron of the Wondrychoun presses so hard on the ground when fishing that it destroys the living slime and the plants growing on the bottom under the water, and also the spat of oysters, mussels, and of other fish, by which the large fish are accustomed to live and be nourished.

This craftily contrived instrument was always controversial. In 1583 the Dutch banned trawling for shrimp in their coastal estuaries, and in 1584 the use of a trawl in France was made a capital offense. In 1631 a trawl was defined to the Privy Council of England's Charles I as "a net . . . so as it sweepeth at the fish which lieth upon the ground, great and small, and bringeth the verie sand up with the net: there is nothing can 'scape." The dragger interests of that time responded that "the trawl did good by raking up the buried food on the bottom to bring it up to the fishes, and the fishes otherwise could not get the food." Unconvinced, Charles banned the use of trawls, and the Royal Navy (until distracted by war and piracy) burned any such nets it found aboard fishing vessels.

But there were always more such nets to be found, because they worked wondrously well. Wooden timbers fixed across the mouths of trawl nets for groundfish served to hold them open underwater; these came to be known as beam trawls. Towed slowly behind sailing vessels, both beam trawls and shellfish dredges were small, light, and confined to shallow waters until the middle of the nineteenth century, when the coincidence of two inventions allowed fishermen to expand the scale and effective depth of their gear. One was the steam engine; the other was the otterboard (what is called simply a door on the *Honi-Do*). This is held by some fishing historians to be derived from an ancient device for moving a fishing line or seine net out farther into a river, but others credit it to English sport fishermen who used small boards as depressors to keep their trout lines flush to the sides of their skiffs when poaching from closed streams. Whatever its origins, the first steam-powered otter trawl was built in Scotland in 1892, and within three years, the otter trawl was the standard fishing rig of the British North Sea fleet.

In America the otter trawl debuted on Cape Cod, on a vessel loaned on an experimental basis by the U.S. Fisheries Commission to several Cape fishermen in 1893. The rig caught on more slowly on this side of the Atlantic. The first American steam-driven otter trawler, the *Spray*, was launched by the Bay State Fishing Company of Boston in 1905, and by 1912 there were 11 such vessels — as compared to 1341 out of England and Wales — working the offshore waters of the northwest Atlantic groundfishery. The advent of gas and diesel engines further increased the power of the otter trawlers, and then Clarence Birdseye's quick-freeze technology greatly enlarged the market for New England groundfish. The harvest of 42,500 metric tons of haddock on Georges Bank in 1924 climbed to 116,000 metric tons, taken by 321 vessels, by 1929. Then came the crash in Georges haddock and renewed criticism of the capabilities and unintended outcomes of this modern Wondrychoun.

At the same time, Harald Salvesen, the chairman of Christian Salvesen Ltd., a Scottish general shipping and whaling firm that was a world leader in factory-ship whaling, foresaw the day when the quarry of his own fleet of ships would be scarcer than Georges haddock. Salvesen argued as early as the 1930s for the whaling industry to adopt self-imposed quotas on certain species, but his competitors ignored his warnings. By the time the International Whaling Commission was established in 1946, Salvesen had already despaired. Whale stocks by then were in deep decline, ferocious new competitors had appeared in the fleets of the Soviet Union and Japan, and Salvesen predicted — correctly — that the IWC's control systems would be unenforceable and ineffective. Finally he decided to get out of whaling and get into fishing instead, but to do so with the technologies that had made his factory whale ships such deadly and efficient hunters: stern ramp loading, on-board processing capabilities, and the ability to make long distant-water voyages.

The large otter trawlers, or draggers, then working the North Atlantic were limited to the range dictated by their ability to deliver their fish fresh, and were all side-trawlers, or eastern rigs. In order to haul back, such ships must stop dead in the water, broadside to the seas, and wait for the net to drift clear of the propeller. Then they bring their nets aboard over the rail, a difficult and dangerous operation in heavy seas. The advantage of such a rig, however, is that the entire net doesn't have to be emptied at once; instead, discrete portions of a swollen cod end may be cinched off, brought aboard, emptied, and returned to the water. This avoids the danger of having the net burst under heavy loads, and of perilous stern pitching in giant ocean swells.

The advent of synthetic twine after World War II, however, increased the carrying capacity of trawl nets, and Salvesen's experiments with a converted English minesweeper (overseen by Sir Charles Dennistoun Burney, the inventor of the stabilizing paravanes sported by the *Honi-Do*) convinced him that the technical problems of stern-ramp haulbacks were at least solvable. On-board processing capabilities were provided by a pair of crucial inventions, one American, the other German. The first was Clarence Birdseye's multiplate freezer, a stack of metal plates through which ammonia refrigerant circulated in small tubes, which could be opened to receive packaged foods and then hydraulically closed to variable pressure settings. Designed in 1933 to be used in vegetable fields and orchards for "field-fresh" freezing, this freezer had the great virtue of small size and economy. The second was the Model 99 Schwanzlaufer, an ingenious fish-filleting machine developed by a small West German firm, Nordischer Maschinenbau Baader, which was willing to make the wholesale design adjustments necessary for its device to work at sea. The Baader Schwanzlaufer produced fillets at the rate of forty-eight per minute, and the racks that the machine discarded could be converted on board to fish meal, which since the 1920s had been used in Europe as a growth agent in chicken and livestock raising.

In early 1953, Salvesen took the plans for his new ship to an Aberdeen shipyard. The following year saw the launching of the *Fairtry*, an otter trawler the like of which the world had never seen before: 2600 gross tons and 280 feet long, with the sort of above-water crew accommodations that swiftly earned it the sobriquet "the floating Ritz." Salvesen had learned the importance of good living quarters for his crews during the eight-month voyages his whaling ships had made, and a photograph in London's *Fishing News International* of crewmen lounging in a cabin was captioned: "Individual lockers are given each man and there is an air of attractive roominess about each berth. Selection of mates is left to the men themselves. Officers have single berths, petty officers are in double berths, and other hands in four-berth cabins. Ample showers (hot and cold) are provided throughout the ship, and cinema is regularly shown in the crew's mess."

On its maiden voyage, the *Fairtry* steamed to the Grand Banks off Newfoundland, which was well beyond the range of England's fleet of side-trawlers. A number of unanticipated problems arose, the most serious of which was that the ship simply caught too many fish. Prodigious hauls of twenty tons or more choked the trawler's stern ramp, crushed the trapped fish, and burst cod ends made out of even the stoutest synthetic twine. But Salvesen was right in believing it could be made to work. Tows were reduced from an hour to twenty or thirty minutes, winches were fitted with springs and the net with enough slack wire to reduce the pitch and surge of the cod end in heavy weather, and the cod end itself was made double-tailed, like a pair of pants, to protect the fish from being crushed. Other wrinkles were smoothed out, and the *Fairtry* proved a great and immediate success, returning from one 37-day trip to the banks with as much as 650 tons of cod fillets — catches the like of which the world had never seen before.

While fishing on the banks one morning in the summer of 1956, however, the captain of the *Fairtry* discerned a ship on the horizon that was at once both strange and oddly familiar. As the vessel approached, he began to understand why. "I couldn't believe my eyes," he later recalled. "The ship was the *Fairtry* exactly! Only the name was different. She was called the *Pushkin.*"

In 1953, before the *Fairtry* had even been completed, the shipbuilding firm constructing it in Aberdeen had received a tender from the Soviet Union for the building of twenty-four factory trawlers of the *Fairtry* design. Along with the tender came a request for a copy of the ship's plans, for what was described as preliminary study. The shipbuilding firm, J. Lewis & Son, requested a substantial cash advance, which the Soviets refused, claiming it was against their business regulations. Britain's Conservative government at that time was quite interested in expanded trade with the Soviet Union, and at the government's urging, the plans were finally sent to Moscow. And the contract for the construction of the trawlers? Well, that would have to be written according to Soviet law, said the Russians, and the terms of the contract proved to be so one-sided that negotiations with J. Lewis & Son ground to a halt. Instead, the twenty-four trawlers, exact replicas of the revolutionary *Fairtry,* were built at a shipyard in West Germany. "Thus occurred what was probably one of the most important and certainly the fastest transfers of technology in the history of commercial fishing," observes the historian William W. Warner. "With virtually no prior tradition of high-seas fishing, the Soviet distant water fleet was off and steaming."

By 1959 there were thirty-five Soviet factory trawlers working the Grand Banks, twelve of them of a newer and larger class than the *Pushkin.* By 1965 the Soviet fleet had increased to 106 factory trawlers — along with many mother ships, side-trawlers, refrigerated fish carriers, oceangoing tugs, research and scouting ships, food and freshwater supply ships, and tankers — and they ranged the northwest Atlantic from Greenland to Georges Bank. Also off and steaming were the distant-water fleets of other nations, as strong demand, rising prices, and declining fish stocks at home ensured that the great ships would be big moneymakers: West Germany first, of course, and then Japan, Poland, East Germany, Romania, Spain, Portugal, France, Norway, Italy, the Netherlands, Belgium, Iceland, and Denmark. Surprisingly, Britain lagged behind in this race; British consumers had no interest in frozen fish, thanks to poorly prepared supplies foisted on the British armed forces during World War II. It was a decade from the launching of the *Fairtry* before this aversion began to soften and other British firms began to build factory trawlers.

In the United States, the conditions that made the descendants of the *Fairtry* so attractive in European and Soviet-bloc countries — a strong appetite for fish and a scarcity of fish in local coastal waters — did not then prevail. American politicians, however, were loath to lag behind in any international race, particularly one that was being won so handily by the Soviet Union. At the same time, the owners of America's suddenly quaint eastern-rig trawlers were having a hard time, owing to weak demand, low prices, rising construction and insurance costs, cheap foreign imports, and now the appearance of these craftily contrived leviathans, these Wondrychouns past imagining, in their own back forty.

Advocates of federal support for the fishing industry, such as

Senator Warren Magnuson of Washington, chairman of the Senate Committee on Interstate and Foreign Commerce, phrased their appeals as both pep-rally cheers and a call to arms. The fisheries, Magnuson said, had "contributed so much in making the United States the most powerful nation in the world." Others emphasized the importance of independence from foreign sources of protein and the usefulness of fishing vessels in time of war. Even without war, fishery programs constituted "a vital part of our national defense and security structure," Magnuson argued. "Our intelligence sources tell us that as a part of the strategy employed by the Russians in this Cold War, they . . . are presently implementing a long-range commercial fishery program designed to make Russia the leading fish-producing nation in the world by 1965."

In 1960 Congress passed legislation providing federal construction subsidies to cover the difference, up to a third of a vessel's total cost, between the costs of boat construction in the United States and abroad. Four years later, and a year after nearly three hundred Soviet ships were counted on Georges Bank and other New England fishing grounds, this law was beefed up by the Fishing Fleet Improvement Act, which expanded the subsidy to half of a vessel's cost and required subsidized vessels to be of advanced design and equipped with gear that would allow them to operate in "expanded areas." Almost imperceptibly, the original goal of federal fisheries legislation — helping offshore groundfishermen turn a modest profit — had been superseded by a sort of nautical version of the space race. With this, the stage was set for the construction of the *Seafreeze Atlantic,* the first American contestant in the factory trawler wars and the first bad sign that America would set about saving its fisheries in entirely the wrong way.

In 1966 American Stern Trawlers, a New York–based subsidiary of the American Export–Isbrandtsen Line, applied for subsidies for two factory trawlers: one to pursue groundfish in the North

Pacific (a vessel later known as the *Seafreeze Pacific*) and the other to work the North Atlantic out of Gloucester.

The subsidies had the support of the National Marine Fisheries Service and were approved over a number of objections. Other boat builders were unhappy that the entire federal subsidy that year would be absorbed by these two vessels. New England fishermen were not cheered by the prospect of another factory trawler off the coast, no matter what flag it flew, and fishermen's unions suspected that the owners would hire foreign crews. Dissenting voices within NMFS pointed out that America's strength was in its coastal fisheries, that dispatching expensive ships to "expanded areas" made no sense, that even Britain's famous *Fairtry* was by then losing money. NMFS's director, Harold Crowther, thought otherwise. "We hear many claims," he said, "of the need for the U.S. fishing industry to get out on the high seas and compete with foreign fleets such as those of the Soviet Union, Japan, Canada, and others." These ships, he believed, would show the world that the Americans were still in the game.

The *Seafreeze Atlantic* was launched in the fall of 1968. Built at a cost of $5.3 million, half of which was subsidized, the ship was 296 feet long and displaced 1593 gross tons. It boasted all of the technological capabilities of the *Fairtry* as well as the ability to adjust its nets for midwater trawling, a technology only just perfected by the West German trawlers. The ship was skippered by James Ackert, then head of the Atlantic Fishermen's Union, who commanded seventeen officers, many of them West German. Of the fifty-six crewmen, only twelve were fishermen. The rest were defined as factory workers, and hired as such. Moreover, instead of profit-sharing, via which the crewmen of New England vessels had been paid since the first salted cod crossed the Atlantic, the *Seafreeze*'s crewmen were paid flat salaries: $8000 per year for fishermen, $5200 for processing workers. At those rates, both fishermen and factory workers could do better ashore. "I tried to get them to offer more," said Ackert, "but the accountants said that was the amount of money the company could pay."

Ackert also recommended that the *Seafreeze* stay on Georges Bank and harvest herring there for that fish's substantial European market. American fishermen had never done so before because of the herring's poor quality by the time their boats arrived back in port. But the *Seafreeze Atlantic*'s freezing capacity, said Ackert, would solve that problem, and the crew would be closer to home. Instead, company management insisted that Ackert follow the foreign fleets looking for cod off Iceland, Greenland, and Labrador. By this time, Ackert was feeling forebodings. The night before the ship's maiden voyage, he recalls, he visited a friend who worked for NMFS. "It's going to fail, isn't it?" he asked. His friend assured him that it would.

The problems that arose with gear and equipment on the *Seafreeze*'s first cruises to the Arctic were not unanticipated by Ackert, and were typical of shakedown cruises on any new vessel, particularly when working among ice floes. What was unusual was the inability of the crew, many of whom had never been to sea before, to adjust the gear, recommend engineering changes, or even understand who was responsible for which tasks on deck. Most were recent immigrants who had trouble understanding the German officers' accented English. All lacked the personal profit motive that goaded Brian and his mates on the *Cape Star* and the *Bell* through the alternating periods of tedium and frenzy that typified offshore dragging even in those old side-trawlers on Georges Bank.

And all the crewmen, inevitably, hated the *Seafreeze*'s lengthy cruises north, wondering along with the rest of the world what the Yanks were doing in the Arctic when Georges Bank was so rich and close to home. Fish were left to spoil ungutted in pens. Injuries were frequent. Most of the crew resigned whenever and wherever the *Seafreeze* went into port, so that Ackert had to hire a new and equally inexperienced set of "factory workers." Brian knew a fisherman who had shipped on the *Seafreeze,* who told him that the crew used to stand at the rail and throw food overboard, emptying the stores, in that way hastening the next visit to port.

Ackert was replaced in March 1969, and the next month a crewman from Gloucester died when a falling block hit him in the head. The processing workers and some fishermen refused to work after the death, and when the *Seafreeze* docked at the island of St. Pierre, south of Newfoundland, in order to put the body ashore, the crew went on a rampage of vandalism throughout the ship. "I've been twenty-two years at sea," said Captain Cecil Benson. "What happened on this ship, as far as I'm concerned, just doesn't happen on any ship." At that point Ackert resumed command, arriving in St. Pierre with another replacement crew and steaming north again, only to have a boiler break down at sea. Ackert caught pneumonia, as did eight crew members. The *Seafreeze* returned to Canada, where the demoralized Ackert relinquished its bridge for the last time.

The ship blundered about the North Atlantic for two more years. In 1971 it was tied up permanently in Norfolk, Virginia. Its sister ship, the *Seafreeze Pacific,* fared even worse, going into mothballs after less than a year's service. By the time American Stern Trawlers sold out to new owners in 1974, the company had lost $11 million.

That was the year that 1076 European and Soviet-bloc fishing vessels swarmed across the Atlantic to work North American waters. Their harvest of 2,176,000 tons of fish was ten times what New England fishermen were able to land, and three times the landings of Canadian Atlantic fishermen. But in fact the heyday of the *Fairtry* and other factory trawlers had already passed. William W. Warner reviews the fishery statistics of the late 1960s and early 1970s in terms that sound . . . well, like déjà vu all over again:

> Huge as the total catch [for 1974] might seem, the catch per vessel was down and the fish were running generally smaller than before, even though the foreign vessels fished longer hours with improved methods over a larger area for a greater part of the year. In fact, the foreign catch had been better for five of the preceding six years — slowly declining from a peak of 2,400,000 tons in 1968 — with

fleets of equal or lesser size. When this happens or when the yield per unit of effort starts to fall off, as management experts prefer to say, it is nearly always an early and sure warning of the general decay of major fishing areas. First observed in the Old World with the herring and sole of the North Sea and the cod of Norway's Arctic grounds, the phenomenon now seemed to be repeating itself in the Northwest Atlantic. Everywhere the factory trawlers went, in other words, more were fishing for less.

With the passage of the Magnuson Act in 1976, with that law's two-hundred-mile exclusive economic zone and the raising of similar barriers elsewhere in the world, nearly every major fishing ground became subject to some form of national sovereignty. The great leviathans were either banned entirely from these grounds or subject to such low quotas as to make them too costly to operate. Hundreds followed the *Seafreeze Atlantic* into mothballs, though not before they had removed, in roughly twenty years of operation, more than 72 billion pounds of fish from North American waters. Harald Salvesen, who had lamented his inability to stave off the commercial extinction of the whale, had abandoned whaling and made a machine that beguiled us once again into that age-old dream of defying the limits of nature through some miraculous tool — a machine that nearly, but not entirely, brought about the commercial extinction of the Atlantic cod and haddock.

The *Honi-Do* is that old-fashioned sort of fishing vessel that has no on-board processing (except the galley in which Dave Reed boiled up fresh shrimp for dinner that night) or distant-water capabilities. Nor can it accomplish midwater trawls. But its electronics are of the sort the West Germans refined to space-age perfection, its paravanes are Sir Charles Dennistoun Burney's, and its stern-ramp haulbacks are straight out of Salvesen's dream. The net, spun tightly around the reel, its corks and ground gear and cables and varieties of twine crisscrossing and overlapping each other, looks like something stitched together out of the odds and ends of a scavenger hunt. But under water, flared by its doors and at sail under its corks, the net is an object of terrible beauty, "a harmonious and efficient unit the shape of which is determined by minutely calculated gradations in both mesh size and twine diameter and the careful testing of stress points," observes Warner.

The nature of the Wondrychoun's impact on the ocean floor, whether benign, catastrophic, or something in between, is an issue as hotly debated at council hearings in Hyannis in 1995 as it was before the Privy Council in Britain in 1631. On this question today the scientific community is firmly, unequivocally, without a shadow of a doubt . . . undecided. Balancing the disintegration of Fred Bennett's flower vase is the example of several rich fishing grounds, such as the North Sea, that have been trawled for decades and are still productive. Richard Langton, of Maine's Department of Marine Resources, the author of one of the few recent studies investigating the impact of draggers in the northwest Atlantic, admitted to *National Fisherman* that he and other scientists still can't draw a clear conclusion in the matter, "but there should be a level of concern about the potential long-term impact."

Not really so old-fashioned, the *Honi-Do,* the *Overdraft,* the *Joanne A,* and even the old eastern-rig trawlers are more akin to the small and opportunistic mammals that succeeded the dinosaurs, with 1976, the year of the Magnuson Act, being the year that the comet struck and the dinosaurs died (or, in this case, migrated elsewhere). The haddock were mostly gone, but there were still cod on Georges Bank, along with viable numbers of flounder, plaice, pollack, hake, redfish, squid, shrimp, and sea scallops. There were to be no more *Seafreeze Atlantic*s. Instead, America, in sole possession now of most of Georges Bank, would follow a different line of attack in hopes of restoring prosperity to its fisheries and reclaiming its preeminence among fishing nations.

Dave Farnham, Mark's father, is trying to interest a yearling ring-billed gull in a piece of dried butterfish plucked from the twine of the *Honi-Do*'s net. The gull stands on the gallows frame with its maculate breast puffed out, its weight shifting from foot to foot, first one eye and then the other darting from Dave to the fish and back to Dave again. The old man smiles, clucks, talks to the bird, tries to soothe it into taking the fish from his fingers. Finally he gives up and tosses the butterfish into the harbor. The gull rises from the gallows frame and floats over the gunnel just as a bigger herring gull swoops down to scoop up the fish. The yearling gull hovers in confusion for just an instant, then rises to settle on a piling, staring at Dave in an accusatory manner.

To the old man, this is only another sign that the world ashore is a fallen one and the only true paradise lies out to sea. "The flycatchers used to come right into the wheelhouses of the boats at sea," he says. "They'd clean out all the flies and then perch right on your shoulder. They don't do that ashore."

One uses the term "old man" advisedly with Dave Farnham, who still has the backbone and staying power to dig his limit of steamers on a single turn of the tide. He wears a sheepskin jacket of the sort popular in the 1970s, when he came to the Cape, and suggests in his manner the same sense of momentum as his son, though Dave does this through a wiry sort of agility and athleticism rather than Mark's raw physical power. Often smiling, or with his jaw lined into the pleasing verge of a smile, he does not exude Mark's pugnacity. "He gets that from his mother's side of the family," Dave assures me. "He and the other kids would fight on the way home from church."

He rests his elbows on the *Honi-Do*'s port rail. Above his head the sky is an icy blue, with clouds like chips of frost flying slow and high. Beneath his knotted hands a red-breasted merganser, a male, breaks the surface of the water, materializing out of nothing, like a conjuror's prop. Dave looks at the bird, which carries some sort of bait fish in its beak, and remarks that things have changed too fast, even for a man who didn't come to fishing until middle age. "Fishing now is nothing like it was when I started," he says. "You used to be able to catch big cod — fifty-pounders, sixty-pounders — right outside the harbor here. Instruments?

Well, a fifteen-dollar compass was all we had. Fishing didn't pay much then, but it was fun. It was a lot better than working in a tool shop."

A scuba diver surfaces near the boat's stern as abruptly as the merganser did, and he startles the bird into an eye-blink dive. The diver spits out the mouthpiece of his regulator, lifts his face mask. "Jesus," he says, "I can't believe the size of that rudder down there."

Carl rises like a stowaway from the *Honi-Do*'s lazarette. He's been working there since seven this morning, trying to pull a six-inch block bearing away from the internal rudder shaft. He peers over the pen board and tells the diver that it takes three men to lift the rudder whenever the boat is hauled out. "Can you see what happened?" he asks.

The diver's face, framed by his mask and wetsuit, is white with cold. Saltwater flicks like beads of ice from his mustache as he talks. "You know that steel plate that connects the tiller arm to your rudder post? All the bolts are sheared off that plate except one. That one bolt is all that's holding it on."

Carl nods, grateful at least that the rudder isn't lying somewhere on the bottom out there on the bar. He has some new two-inch bolts handy just in case that was what happened, and he drops them down on a rope to the diver. Then he goes forward to talk to Dave. A fierce tidal rip is pouring through the cove, and under its pressure the *Honi-Do,* moored to the pier in front of Chatham Fish & Lobster, is grinding against the pilings. Dave looks contemplatively at a man in a dinghy rowing out to another boat in the cove, pulling as hard as he can just to stay in place against the tide. Carl says, "I haven't heard anything about the auction results on that shrimp yet. I know Billy Martin is paying a dollar a pound for shrimp now. I think we had about a thousand pounds."

Mark's father was here when the shrimp finally got unloaded. When Dave Reed called Carl at the Coast Guard station last night to tell him the *Honi-Do* was drifting, Carl told Dave to drop the second door straight down, tie it off tight, and then let out about 150 feet more cable on the first door. Then Carl telephoned Dave Farnham to tell him that the boat was in trouble.

Carl's instructions to Dave Reed succeeded in stopping the boat's drift. Half an hour after that, the Coast Guard's big steel surf boat came steaming down the channel with Carl aboard. In the spumy darkness, the first glow of its lights looked like phosphorescence at the bottom of the sea. Then we heard the rumble of its twin diesel screws and saw the glow resolve and harden into the beams of its twin spotlights. The surf boat seemed to skate like a hovercraft over the chop, and to the crew of a disabled dragger in bad weather and cold water, there was indeed something transcendentally heartening in its noise, its stabbing lights, its flying bridge, its orange-jacketed crew, even the same drill-sergeant coxswain who had skippered the smaller boat. The surf boat nudged its bow into the port side of the *Honi-Do*'s bow, was washed back by a wave, then jabbed itself into the bow again long enough for Carl to climb aboard with a towline. Dave Reed tied the line to the mooring bit while Carl went to the radio and called up the coxswain, saying, "We went out of Aunt Lydia's Cove at nine this morning without bumping. It'll be close, but we should be able to tie up inside by now."

By this time the coxswain either had had his fill of grounded draggers or was proceeding on the instructions of the station commander not to raft the boats together, just to tow the *Honi-Do* in. Dave and I went aft to secure the doors once Carl had winched them in, and then Carl got on the radio to the *Overdraft*. "Is that truck still there in the parking lot?"

"Still waiting," Chris Armstrong said. "He says he's got to be in New York by five A.M. — that's all that matters."

Carl clicked off the microphone. "I wish to hell they'd told me that before."

By then the *Honi-Do* was under way, moving out into the channel, the water filling in underneath it, the flags moving past again, the towline stretching beyond the reach of the spotlight to the dark shadow and bright running lights of the surf boat's stern. In Aunt Lydia's Cove the lights were burning in the parking lot, on the pier, at Chatham Fish & Lobster. The other boats were black hulks illuminated in sections by the surf boat's lights, and the cove itself was rolling with chop as the tide poured in. In close quarters, with the other boats ringed about us, the coxswain was unable to keep the towline taut, and the moment it slackened, the rudderless *Honi-Do* began turning like a pinwheel, its stern cracking hard into the wheelhouse rail of the *Wendy-Jean,* a thirty-six-foot long-liner.

The coxswain, his cussing raised to new heights of metaphor and onomatopoeia, brought his boat alongside the *Honi-Do,* causing its spinning to slow and stop, and then ordered his own vessel tied lengthwise to the *Honi-Do*'s starboard rail. Rafted like that, swiftly and successfully this time, the surf boat steered the dragger without incident into the pier. There Dave Farnham leaped aboard to help tie it off, then to help in lifting ten totes of shrimp up to the driver of the Chris King truck, who hastily carted the totes out to the refrigerated truck in the parking lot. Above us the sky opened up, the shoulders and raised club of Orion, the horns of Taurus, the soft blur of the Pleiades all brightening in the south.

Now Mark's father asks Carl if there was any damage to the *Wendy-Jean* last night. Carl says he called Terry Picard, the owner of the *Wendy-Jean,* and told him what happened. Picard, whose boat occupies a position right at the entrance of the mooring area, where it tends to get dinged right and left by incoming vessels, took a look and said all that got broken was an antenna fitting. "Terry told me that us and the *Overdraft* are the only two boats who ever tell him when they hit him," Carl says.

The diver is having trouble getting the new bolts through the rudder shaft plate. He says the bolts are too short, and Carl suggests that the rudder is probably a little bit out of alignment. Maybe the wheel puller he was using on the bearing in the lazarette has got it pulled up too high, or else it's dropped down too low. Maybe a crowbar between the bottom of the rudder and the boot could pry it up enough for the bolts to slip through.

Nothing works. The diver finally says that he can't take the cold anymore and climbs stiffly up a ladder at the end of the pier. He drops his tank, strips off his gloves. His hands are as pale and waxy as the broken lengths of a sponge called dead man's fingers drying in the mesh of a net heaped behind him in the parking lot. A second diver, this one with a drysuit, comes in the middle of the afternoon, after the tide has turned and the cove has gone as placid as a pond. He works his fins carefully down the rungs of the ladder off the pier and falls backward into the water. The bubbles from his respirator capture the interest of an eider duck, which veers toward their effervescence just as the train of bubbles makes a sharp turn and engulfs the duck. The eider scoots from their midst, lifting its wings, and goes back to prospecting for mussels along the pilings.

Bill Amaru is on his way out to his boat, and he stops on the pier at Carl's side, his graying curls lifted by the afternoon breeze. "So what did you do, Carl?" Carl tells him about the bolts sheared off the rudder shaft. Amaru nods. "So you had a hole in your stern and almost no rudder. I thought I was going to have to come pick you up off the boat."

Carl nods as well, says he's glad it didn't come to that. Once Amaru leaves, Carl tells me that really we were never in much danger of taking on water, since the connecting plate between the tiller area and the rudder shaft is outside the hull. "It's an unusual design, and I've been trying to explain it to people all day long," he says. "I didn't feel like going through it one more time. Let it pass."

The diver is up again, saying he might be able to get the new bolts all the way through the plate once he gets the remaining original bolt out of there. Half an hour later he surfaces with that bolt and hands it to Carl. The three-eighths-inch steel shaft of the bolt is curved into an *S* like a piece of spaghetti. Carl looks at it the same way he looked at that connecting rod last summer, his mind filling with the way one circumstance penetrates another and energy builds up along the way, until a butterfly creates a hurricane, until a trivial misstep — a weak link, a submerged object, a change in wind direction, an unspoken word, an inconclusive meeting, an ill-considered subsidy — bears explosive and irreversible consequences.

He says only a week ago the *Honi-Do* ran aground out here with Mark at the wheel: "We were backing up, trying to get over a bar, and we ran into an old mooring. That pushed the rudder over hard and we tore off some mounting brackets in the steering. Had to get towed in." He turns the bolt over in his hands and speculates that maybe this and the other bolts on that plate started to go then. He gives a short laugh, clearing his mind, parrying with fate. "Well, that's what we'll tell Mark, anyway," he says.

Amaru is already back in from the *Joanne A*. He stops at Carl's side and faces into a breeze that has backed to the southeast, the first meteorological misstep on the path to more bad weather. "Guess I'm sitting on my ass again tomorrow," he says. "I don't want to try coming in here in the dark with a southeast swell developing. That's the story of this whole winter. You get a little tiny crack of good weather, and then you're shut down again."

It has been a hard winter. Mike Russo got in only five days of fishing last month, and I heard today that Dan Howes has been spending a lot of time ashore in Maryland. Brian has been jigging for cod whenever he can get out in the *Cap'n Toby*, but last month his steering came uncoupled while he was crossing the bar at Nauset Harbor, and in the eight-foot swells there that day, he thought he was going to get rolled and beaten to pieces. Fortunately the wind was out of the southwest; it blew the boat off the beach and also out of the breakers, where Brian was able to make repairs. "It was weird," he told me. "The whole thing took maybe fifteen minutes. It wasn't anything I did good or bad. It just happened, and then it was over, and it was time to go back to business — the archons toying with me again."

There is a photo in one of the Cape newspapers today of that shorefront house below Holloway Street in Chatham: a house built on sand, now reaping the whirlwind. I'm reminded that it was in this month in 1928 — on these same two days, in fact — that Henry Beston's Outermost House was lashed by what he describes as "the great northeast storm of February 19th and 20th. They say here that it was the worst gale known on the outer Cape since the *Portland* went down with all hands on that terrible November night in '98." On the second day of the gale, Beston abandoned his house for a walk on the dunes:

> A third or so of a mile to the north I chanced to see a rather strange thing. The dune bank there was washing away and caving in under the onslaught of the seas, and presently there crumbled out the blackened skeleton of an ancient wreck which the dunes had buried long ago. As the tide rose this ghost floated and lifted itself free, and then washed south close along the dunes. There was something inconceivably spectral in the sight of this dead hulk thus stirring from its grave and yielding its bones again to the fury of the gale.

Exactly sixty years later, during the February rampage of the Great Blizzard of '78, the last remains of Beston's little house were pounded apart and swept out to sea through Nauset Harbor, out into the mansions of the true Atlantic house.

≈ 13

In Cod We Trust

THE DAWN BREAKS open like an angry wound, the northeast wind slicing a V-shaped cut, rimmed in iodine red, between the ocean horizon and the dome of the cloud cover. Then, minute by minute, as the clouds fold back and the sky opens up, the inflammation drains away. The sky turns healthy fluorescent flesh tones of peach and pink. The blue-water swells advancing from the horizon are limned in light and rise with a bronze burnish at their crests. My camera is jammed, but I only have black-and-white film in it anyway. Mike says it reminds him of sunrises on days when cod seemed to be everywhere, on days when everything, from beginning to end, just fell into place. He says sometimes he brings a camera out here himself, just to capture for Susan some of the things he sees, such as dawns as clean and pastel and full of promise as this one.

The *Susan Lee* is bobbing, corklike, above a hummocky stretch of bottom southeast of Chatham, long out of sight of land and twenty miles into the ocean's wilderness. Ashore, swaying clouds of blackbirds are now blowing through the marshes, and palm warblers are picking for nesting materials along the bare wires of denuded vines: poison ivy, Virginia creeper, catbrier. The dunes are softening on the beaches, shedding the granitic compression stamped on them through recent months by the wind, the cold, the brittle winter light. Their slopes take on an olive tint compounded of blanched sand and the April effulgence of the poverty grass, a plant that Thoreau said deserved a better name, whose first hue in early spring struck Henry Beston as "one of the rarest and loveliest greens in nature."

Where once William Bradford saw nothing but "hideous wilderness, full of wild men and wild beasts," where once Thoreau saw famished dogs ranging in packs, the carcasses of men and beasts together rotting in the sun and waves, now the zoned and regulated environs of the golden-arches and car-payment culture observe their own seasonal rhythms: shopowners review their inventories, restaurateurs their summer marketing plans, motelkeepers their staffing needs. In the bays, oystercatchers, newly returned from the South Atlantic and the Gulf Coast, thrust their orange bills with eye-blurring speed into the open shells of oysters and mussels, seeking to cut the bivalves' adductor muscles before their shells snap shut. If the oystercatcher misses and an oyster clamps down on its bill, then the catcher itself is caught, and will drown on the rising tide. Such small and savage dramas, like the fall of Icarus in the Auden poem, are played out largely unwitnessed.

The wilderness that Bradford's men hacked away from Plymouth, rolling it back, steadily westward, occupies precisely its original dimensions on this eastern side of the shoreline. Thoreau once regarded a number of fishing craft at work and found himself struck by the empty distances that separated them: "Though there were numerous vessels at this great distance in the horizon on every side, yet the vast spaces between them, like the spaces between the stars . . . impressed us with a sense of the immensity of the ocean, the 'unfruitful ocean,' as it has been called, and we could see what proportion man and his works bear to the globe."

Farther to the east, a hundred miles deeper into the true Atlantic house, lies Georges Bank, where that turbulent mix of cold air from the continent, warm water from the Gulf Stream, and shoal-water depths breeds the sort of monster that devoured fifteen schooners at once in 1879, and that in 1841 left Nauset Beach strewn with fifty corpses. Even at rest, even here over Joe

Bragg's Ridge, just twenty miles off the beach, under a mild fair-weather breeze and beneath the sort of dawn that laps at the gates of heaven, the soul is struck dumb by the ocean's breadth and void, by a thirty-two-foot boat's dust-mote proportions, by the lineaments of the swells' indifference. It's a dumbness that, at least under these sweeping and propitious conditions, exhilarates as much as it chills, that makes you gloat on the stump of your own loneliness, that makes you pity the poor landsman, that makes you fumble for your camera and know that the snapshot is going to be as speechless as you are in saying just what it was you saw.

Mike lays down his first set of gear at 6:30, twenty minutes after sunrise and precisely at slack tide. His long lines rest in fish totes in the cockpit of the *Susan Lee*. The lines are divided into bundles of three hundred hooks each, fifteen hundred feet in length, with each hook attached by a six-inch gangion to a main line of quarter-inch nylon. The bundles lie loosely coiled each in its own fish tote, their hooked gyres separated by layers of wet newspaper, their ends marked by green streamers. The hooks have already been baited with slices of squid. The pieces are white or gray, doughnut-shaped or nozzled. They look like plastic plumbing components.

After Mike has tied two bundles into a single long line for his first set; after he's tied a twenty-five-pound anchor, a thirty-inch plastic buoy, and a nine-foot-long radar-reflecting high-flier to each end; after he's checked his loran coordinates and notebooks on previous sets in this area; after he's pitched one end of the bundle over the side with its anchor, buoy, and high-flier — it's then that the *Susan Lee* runs up one side of the ridge, and the lines and hooks slide whirring over a transom rounded off by layers of smooth fiberglass. Some of the hooks are thrown out to the full length of their gangions like spokes as they run over the transom. Sometimes pieces of bait are thrown free, or scraps of paper fly loose. With the boat on autopilot, Mike stands in the stern and hoses the bundles down as they empty from the totes.

Mike says that this area can't take too much fishing pressure, so if he knows another long-liner has been working here within the previous week, he won't go. "Not many guys are going out here now. The fishing's been slow all over," he says. It's also deepwater fishing, which is very difficult. Joe Bragg's Ridge is one of a series of egg-shaped hummocks rising about thirty fathoms above the floor of the wide plain underlying the Gulf of Maine, which in this area is about ninety fathoms down. Trying to put your gear exactly where you want it at this depth, Mike says, is "like trying to drop a golf ball into a Dixie Cup from a hundred yards up. Usually you can predict from your first set how your gear is going to sag, once it's on the bottom. But if the tide is swinging around and if your anchor's not in just the right place, your gear is going to get all fucked up. A lot of times gear falls right off the edge of this ridge, especially if you set it too long. If you're just working long stretches of hard, flat bottom, you can set your gear as long as you want. It's all basically trial and error wherever you go, but especially here."

Mike sets five lines of gear, all running on a parallel north–south bearing and two bundles long. This takes only thirty minutes and brings the *Susan Lee* to the southeast corner of the ridge. The high-fliers, which carry red flags on top, rise and fall with the swells; in the stop-frame of the imagination, they mark the holes on a golf course mapped out on a rolling blue-gray tundra. Mike guns the boat and wheels her northwest, running for the upper end of that first set with the horizons yawning on either side and the last of the clouds breaking up into the disjunct and weightless notes of a Thelonious Monk piano solo.

On January 26, by a margin of eleven to three, the New England Fisheries Management Council voted to approve Amendment Seven to the Northeast Multispecies Fisheries Management Plan. Pending its final approval by NMFS, the new set of restrictions goes into effect in just a little more than ten weeks, on July 1: area closures of 4700 square miles of ocean, reductions in allowable days at sea and a mandatory twenty-day spring layover, the elimination of the last open-access groundfish permit categories, a five-hundred-pound possession limit on haddock, and total allowable catches on cod, haddock, and yellowtail flounder.

To spokespersons for organized groundfishermen, this is a countdown to Doomsday. According to Maggie Raymond of the Groundfish Group of Associated Fisheries of Maine, "This plan will have devastating financial consequences for the New England groundfish industry and those communities dependent on that industry. Our infrastructure will be lost, and this consequence is irreversible." Ed MacLeod, representing the Gloucester Fishermen's Wives Association and several Gloucester-based draggers, said, "We think that everyone involved has to ask themselves if the financial pain and resulting consequences . . . are worth the possible gain that may or may not be a reality."

But Mike isn't worried much, at least not at this point. The roller-coaster ride of the past summer — Are they going to shut everything down or not? Should he buy another load of bait or not? — is over, and he has a pretty good guess now as to what's over the horizon. As a long-liner, a hook fisherman, he still doesn't like being lumped in with draggers, particularly those big draggers out of New Bedford and Gloucester, but he appreciates the implicit distinctions made in the provisions of Amendment Seven, which on the whole will be harder on those draggers than on such small vessels as the *Susan Lee*. Days at sea will be limited to 139 for both small boats and large on June 1, and 88 next year. But Mike gets no more than 100 to 130 days of fishing for cod per year anyway, and days may be split into twelve-hour segments; so long as he's making short day trips, as he usually does, those 88 days of the amendment's second year are as good as 196 days for the *Susan Lee*. That twenty-day layover in the spring will hurt, but this spring, at least, has been slow for cod anyway. A five-hundred-pound limit on haddock may be painful to a big boat making ten-day trips, but not to small boats like his. Likewise, the area closures are mostly in the offshore waters worked by the big draggers and scallopers. He considers the change in his permit from open access to limited access an upgrade, since that caps the number of his competitors and means that down the line, one way or another, his groundfish permit is going to be worth money. And the limits on hook numbers that were an early part of the package do not appear in the final version of the amendment. "As far as the hook fishery goes, we come out smelling like a rose," Mike says. "Maybe it's because those guys realize we don't have the impact on fish populations that those other fisheries do."

Some other long-liners are less sanguine. With their small profit margins, they're uneasy about the various ways the amendment's screws may be tightened in subsequent years if its goals in the reduction of fish mortality aren't met. They're nervous about the total allowable catches, which may be gobbled up early in the year by the big boats while their own vessels are weatherbound. And like Jim O'Malley at the lobster hearings, they're apprehensive about the ease with which TACs can become quotas, and quotas (or even days at sea) can become individual and transferable. The preferred name for that process now — a euphemism that also connotes boat buyouts, bankruptcies, limited access, and anything that reduces the overall number of fishermen — is "consolidation." Mike says, "That's years down the road, that ITQ stuff, if it happens. They've got to get to the other side of Amendment Seven first."

I point out that no one ever got to the other side of Amendment Five — that it was determined that the regulation's goals were not being met and then the rules changed. Mike shrugs. He sees himself adapting to the exigencies of that environment as he has to this one, imagining that if he finds himself short on days at sea, for example, he will be able to figure out a way to buy someone else's ("Tyson Foods owns draggers. If they came in here, they'd be buying up the quotas or days at sea of draggers, not long-liners. I wouldn't have to compete with them").

Nothing has ever been guaranteed for a fisherman anyway. Thoreau, contemplating the Cape's stormy shores in October, said that "he who waits for fair weather and a calm sea may never see the glancing skin of a mackerel, and get no nearer to a cod than the wooden emblem in the State House." Mike says it in his own way when he compares working the inshore waters to going thirty-five miles out to areas down near Nantucket Shoals, as he does sometimes now without a crewman aboard: "No fear — coconuts for balls. But hell, you can fall off the end of the pier and drown. The more risks you take — weather, distance, whatever — the more benefits." This volatile sort of regulatory environment, with its own weather systems and navigational problems, is just another category of risk, and so also another means, if Mike plays his cards right, of maximizing his benefits.

He wishes he hadn't lost Ty Vecchione, who didn't last long at the Massachusetts Maritime Academy and is now a sternman for a Chatham lobsterman. Ben Bergquist gave him good help this winter, and he was supposed to stay on until May 1, but he's gone for an early start on taking over his father's lobster operation. "I've got mixed feelings about that," Mike says. "March was slow, and things do tend to go sour when the checks stop coming. But usually May first means May first. Ben's got a lot of stuff going on, though. I shouldn't be upset. Funny thing, though — the best weeks I ever had were the weeks right after Ty and Ben quit. I got three thousand pounds the day right after Ben quit. Called him up, told him he lost a month's pay. My ass was dragging, though."

Mike is on the lookout for another experienced crewman, and he is plotting his next set of moves. He has already pulled the trigger on an autopilot for the *Susan Lee,* a Robertson AP 300X, a $4000 item that is top of the line for a boat of that size. On a trip I took with Mike and Ben last February, I saw Mike ceremoniously punch heading, speed, and steam time to the fishing grounds into the new device after we'd cleared the harbor. "Now what are you going to do with yourself?" Ben asked. Mike went down into the cabin, wrapped himself in a blanket, and slept for an hour and a half. Now that he is fishing alone again, the autopilot is the next best thing to a good crewman. "I don't have to worry about the wheel, and I can set my gear a lot faster," he says. "Especially in July out there in the channel, if it's foggy or sloppy with a lot of other boats around, it's very competitive. It's a gong show out there. If you don't get your gear in quick enough, which is hard to do if you're alone, you're going to get squeezed out of a good spot after one set."

The next move — possibly — will be a much bigger gamble than the autopilot. He still doesn't like the freezer setup in that rented bait shanty in the industrial park, and he's heard that another shanty there, with better freezers, is about to come up for sale. Aside from the improved bait quality, Mike could also use the tax write-off that such a purchase would provide against his next year's earnings. He likes the idea of putting his money into equity on a property instead of rent, and even of collecting some rent himself from the tenants he would probably be able to find.

Buying the shanty would set him back quite a lot, more than twice what he paid on the *Susan Lee,* but he knows he has the track record now — the earnings statements, the business plan, and so forth — to swing the bank loan. And the man whom Thoreau saw as marching to the beat of his own drummer is, in Mike's mind, the man singled out for success in this business. As he puts it, "I'd be sticking my neck way out, I know it, but one thing about Murphy's Law — the quickest way to succeed is to be out of step with the public. I'm not going to be thirty forever, and I don't want to be doing exactly the same thing I'm doing now when I'm forty. And I wouldn't mind being able to slack off some when I'm fifty. So while everybody else is retreating, I figure this is a pretty good time to advance. That's what I've been doing all along, really, and it's been working out. I'm not being blasé. I've made some mistakes. But all in all, so far, so good."

"It's the wrong fucking fish."

Mike says that it's spotty wherever you go in the spring. Usually in March the cod follow the intrusions of warmer water into the channel and the Gulf of Maine, but these intrusions are hard to find. Satellite readings will reveal surface temperatures but nothing of what's happening underneath, and are useful only as general indicators. Mike characterizes his strategy at this time of year as "take the dog and throw it at the chart." He says that generally the cod end up finding him rather than vice versa. A week ago he considered himself found: three thousand pounds on that day after Ben quit, and then loads of two thousand pounds each on the two trips after that.

But today, spiraling up from the black depths of Joe Bragg's, hook after hook bears the writhing form of the wrong fucking fish, the cod's alter ego, its evil (and low-priced) twin. The line is hauled in over a crucifier, a twelve-inch roller set behind a pair of short upright posts spaced only a few inches apart. Fish that are undersized or unwanted bycatch (such as skates) are bumped off the hook by the crucifier to drop back into the water. The fish that Mike wants are gaffed as they come up to the rail and then tossed into a bin behind the wheelhouse bulkhead. There are very few skates in this set, and only a few vagrant cod, cusk, and redfish. Nearly everything that comes aboard is a spiny dogfish, *Squalus acanthias*.

William W. Warner calls the dogfish "the curse of Georges Bank." If the occasional great white shark seen off the Cape or in the Gulf of Maine is the *Tyrannosaurus rex* of these coastal waters, then these small five- to eight-pound sharks, packs of which can number in the thousands and blot out the whole screen of a sonar fish-finder, are the more worrisome velociraptors, at least from the perspective of other fish. Henry Beston came upon a school of sand lance that had been driven into the beach one evening by a pack of dogs: "The surf was alive with dogfish, aswarm with them, with the rush, the cold bellies, the twist and tear of their wolfish violence of life." Henry C. Bigelow and William W. Welsh, authors of the classic *Fishes of the Gulf of Maine,* also describe that violence, and its apparent costs to fishermen:

> Strong, swift-swimming, voracious almost beyond belief, the dogfish entirely deserves its bad reputation. Not only does it harry and drive off mackerel, herring, and even fish as large as cod and haddock, but it destroys vast numbers of them. Again and again, fishermen have described packs of dogs dashing among schools of mackerel, and even attacking them within the purse seines, biting through the net, and releasing such of the catch as escapes them. Often, too, they bite groundfish from the hooks of long-lines, take the baits and make it vain to fish where they abound.

These days it is no longer vain to fish where dogs abound. "The dog is a far better food fish when fresh than is generally appreciated," concede Bigelow and Welsh, and a substantial market exists in Europe, where the spiny dog is labeled "rock salmon" and used commonly in fish and chips. A domestic market is building as well. On this side of the Atlantic, rock salmon is called "Cape shark," and harvest volumes are five times what they were in the late 1980s. But knowing that the cod are back, Mike grimaces in disappointment as he works his way down his first set of gear and as his bin fills with rock salmon. "As long as they're worth two to three dollars each, I'll kill 'em," he says in the manner of someone who would rather be spared the trouble.

Yet in their own wolfish way, the dogs are impressive, even beautiful fish. Two to three feet long, they are scaled-down prototypes of the sharks with which we are familiar, and their miniaturization makes all the more arresting a shark's fearful symmetry, its disturbing marriage of streamlined, rocket-age design and predatory malice. Slate-colored or a sandy brown above, flecked with orange to mauve spots along its flanks, a dog has an underside that looks like molded plastic, its color white enough to trouble Ahab, the skin wrinkling like a rug when the fish turns or twists. Its fins are slabbed like airfoils, notched at their bases, and the twin dorsal fins are fronted with venomous spines. The teeth are small but crisp and keen and splayed toward the corners of a mouth slashed as though with a scroll saw, like an afterthought, out of its jaw. The skin is dry and gritty, the fish as easy to pick up and handle as a billy club.

The dog is of a different empire from the cod or haddock. Its cartilaginous skeleton aligns it with the skate as a member of the class of fishes called Chondrichthyes, as opposed to the bony and more numerous Osteichthyes. The dog is ovoviviparous as well, bearing its young live and in small numbers, in litters of two to fifteen, after a gestation period of eighteen to twenty months, longer than an elephant's. In the bin behind the bulkhead some pregnant females, dying, have dropped their young. The pups, miniatures of miniatures but entirely sharks, lie with yellow yolk sacs attached on stalks to their breasts like auxiliary fuel tanks. They rest stillborn in a pseudo-mammalian gurry of blood and bile and amniotic fluid. The gills of the adults flare, their ear holes open and close. Occasionally a tail thrashes.

There is nothing mammalian, nothing poignant, about a dog's eye. The organ is crocodilian, a hooded gemstone, its pupil a slice of emerald driven like a rivet into a disk of burnished bronze. It's the sort of eye inked into the brows of Saturday morning cartoon monsters and fiends. If it catches the sun just right, the eye glows like that of a fox caught staring into the beam of a headlight. To be caught yourself in the glint of this green inner fire, here in the light of broad day, is a little unsettling. There is no conversation in it, no blush of circumspection, recognition, or retreat. The glint fades as the fish dies. Then no trick of perspective or positioning will recall it. Then the dog is just one of a growing pile of low-priced fish flesh — tasty enough, but not quite what the chef had in mind. A lot better than nothing, if you're a fisherman, but still the wrong fucking fish.

Though European markets exist even for skates, dogs and skates occupy the bottom niches of a market report and are lumped together as trash fish in general discussions of what fishermen are liable to come home with. In 1968, a year before the launch of the *Seafreeze Atlantic,* NMFS conducted scores of trawl surveys in New England waters, measuring by that means the relative abundance of various species, and found that their samples contained, on the whole, 43 percent groundfish and 40 percent trash fish. In 1993 those same surveys reported only 9 percent groundfish and a whopping 78 percent trash. This is roughly the same proportion indicated by Mike's survey today. He gaffs yet another dog and sends it thumping to the floor of the cockpit. These great packs of Cape sharks now roaming Georges Bank and the Great South Channel are the habituated villains of a fallen world, the jackals let loose by some sort of original sin.

Something went wrong.

Consider the fate of the *Patricia Marie*. In the summer of 1976, only a few months after Gerald Ford signed the Magnuson Act into law, cod landings were down for American boats on Georges Bank and off Cape Cod. The foreign factory trawlers were still plying Georges, and had until the following March to pack up their operations and move elsewhere. Cape fishermen were looking ahead to winter, when catches would be even lower, expenses higher, and they were generally nervous. Their boats were scouring Stellwagen Bank, up and down the channel, and all over the Nantucket Shoals for any schools of cod. Then one day a small dragger working the Pollock Rip, a shallow ground a few miles east of Monomoy, came up with the wrong fish in its net: not cod, but sea scallops, *Placopecten magellanicus*. A lot of them.

In fact the dragger had stumbled upon a motherlode of scallops, a cousin to that extraordinary vein of bay scallops found in Pleasant Bay in 1982 and 1983. "You could throw a bushel basket in the water and catch them," recalled one fisherman. Another said that "you could get solid scallops for as long as your body — and the boat — could hold out." With shucked scallops selling for $2.50 per pound, a good price in those days, fishermen were not inclined to quit early.

The *Patricia Marie* was one of the first of the many boats to flock to the rip. She was a 1940s-vintage eastern-rigged dragger out of Provincetown, a little more than fifty feet long, and with her bow and decks made out of the same wood used on the flight decks of aircraft carriers, she was considered the sturdiest boat in the Provincetown fleet. Her skipper, Billy King, was an experienced fisherman in his mid-forties with a streak of the flamboyant in him. He painted the boat's bow a Chinese red, used a Playboy bunny head as its insignia, and dressed his mast with the most colorful flyers and banners in the harbor during Provincetown's annual blessing of the fleet. During the blessing one year, King happened to ram the dock on which the bishop stood, and for that one misstep he was known thereafter, affectionately, as Cap'n Crunch.

King had given up on cod and had been dragging for yellowtail flounder when he heard about the boom at Pollock Rip. He immediately geared his boat for scallops instead. "We'd fish for four days at a time," recalls Chris King, Billy's son (who is the Provincetown buyer who bought Carl's shrimp last February). "We'd drag for ten minutes, and I mean big drags, and we'd fill them with scallops. After three hours of towing, the deck was completely filled. Then we'd steam off two or three miles away and cut for twenty-two, twenty-three hours. We'd do that for four days, and on the fifth day we'd come home. We'd have six thousand, seven thousand pounds of scallops on board."

Billy King grossed $125,000 in five months. Often there were fifty to sixty boats pounding the scallop beds there on a given day, everything from thirty-foot Broncos out of Harwich to draggers from New Bedford twice the size of the *Patricia Marie*. Boats were close enough for playful deckhands to throw eggs at each other during tows. Meanwhile, dredge after dredge came up bursting with scallops. There seemed to be no end to them, though the boat captains knew better. So they worked like madmen around the clock, grasping for their share of the bonanza, taking their shot at whatever debts the money might pay off, whatever dreams it might help realize, before the tows inevitably started coming up empty. The *Patricia Marie*, like a lot of other boats, filled her hold to overflowing, and then the crew started piling the scallops in mounds on the decks. Finally she joined other vessels in steaming to port, as low as a lily pad, the gunnels barely above the waves.

At the end of the summer King plowed some of his money back into the boat. He built a whaleback onto the bow, a turtle-back onto the stern, raising the profile above the water so the crew would have more protection and increasing the work and storage space amidship. The *Patricia Marie* was back at work by October, and back with the rest of the fleet pounding the Pollock Rip. Chris King was still in high school then; for a while it looked as though he would have to miss some school and be part of the crew for a trip during the third week of October. But finally Billy was able to come up with a full complement, and Chris stayed home.

On the night of October 24, with a crew of seven aboard, the *Patricia Marie* was mounded high with scallops and steaming back to Provincetown in following seas that had grown to ten feet in twenty-five-knot winds. It was rough weather, what Cape fishermen call snotty, but nothing terrible. The boat was a few miles off Nauset Beach, and King was chatting on the radio with the skippers of other vessels. He stopped in midsentence to tell a friend that he needed to check on something. Then the *Patricia Marie* disappeared, her blip vanishing as though vaporized from the radar screens of other vessels.

The boat closest to King's was the *GKB*, skippered by Michael McArdle. "We heard men screaming in the water for help," McArdle told a reporter. "But by the time we got there, the men and the boat were gone. She went down in a matter of seconds." A twelve-hour search for survivors proved fruitless. King's body was found late the next day, clinging to a buoy; he had not drowned, but died of exposure in the fifty-degree water. The body of one crewman came up in a dragger's net a few days later. Four other bodies turned up weeks later, one at a time, in fishermen's nets. Each recovery opened new wounds in Provincetown's grieving fishing community. The body of crewman Dicky Oldenquist, a charismatic man with an almost mystical sense of where the fish were, was never found.

After the accident, a team of divers found the *Patricia Marie* in 135 feet of water. She was sitting upright on the bottom, listing slightly to port, undamaged, still loaded with scallops, with even the dredges loaded full. In fact she looked exactly as though she were still steaming to port. The only plausible explanation for the sinking seemed to be that, running low to the water in a following sea, the boat took a freak wave over the transom. Usually waves that break on deck drain quickly back into the sea through a boat's scuppers. But this time probably the scuppers were blocked by all the unshucked scallops on deck, and the water ran forward, adding its weight to the whaleback King had just built and pushing the bow down. Then the diesel engine, instead of powering the vessel up the back of the next wave, drove her down toward the bottom as though she were a submarine. There was no time for life rafts, survival suits, or even a mayday. The boat dropped beneath the surface like a rock.

To the newspapers, the moral of the story of Billy King and the *Patricia Marie* was clear enough. They said King had been made reckless by greed and had earned the wages of recklessness at sea. When the dredges finally came up empty at the Pollock Rip that winter — the scallops had either moved or been fished out, no one could say for sure — the moral took on wider ramifications: greed destroys natural resources as well as endangering men's lives.

Other fishermen were more circumspect. They knew how difficult it was to meet expenses, to turn a profit, especially with the factory trawlers working offshore, and how rarely, under any circumstances, the sea offers opportunity for the sort of money that many people take for granted — people who work less hard, at much less risk, with sick leave and health insurance and paid vacations as well. They knew the solemn and remorseless logic of working an unregulated commons: that what you don't take for yourself now, when you have the opportunity, will be taken only a moment later by your competitor; that if your competitor is taking gambles, such as loading his deck with scallops, even plugging his scuppers with them, and winning, then you either match his gambles, trump them, or be satisfied with what he leaves you. They knew that any trip out of port, no matter how fair the weather, how propitious the forecast, involves a risk of both money and lives, that the line between recklessness and greed — or, more generally, the line between recklessness and meeting this month's mortgage or boat payment — is a mathematical figure of a thin and shady kind. Distinctions between balls and coconuts, in other words, sometimes can't even be settled by an autopsy.

They knew also that everything — their livelihoods, their communities, their culture, their collective soul — depended not only on ample numbers of fish in the sea but also on their skill in finding and removing those fish. This is a paradox that cuts to the heart of any hunter, whether of the terrestrial or marine variety. An element common to the cultures of subsistence hunters throughout the world and down through history has been an elaborate system of taboos meant to propitiate the spirits of the slain animals, to ensure their resurrection and so to resolve that paradox. During the winter of 1976 on Cape Cod, with the scallops gone from the Pollock Rip and the cod still hard to find, fishermen and their families awaited the elaboration of a new and millennial system of taboos, one that promised to bring all the slain animals back again. This would be the task of the Magnuson Act.

In the early 1970s it looked as though New England, where the American commercial fishing industry had been born, was where that same industry had gone to die. Between 1967 and 1972, the number of working boats in Gloucester shrank by nearly half. The Gloucester fleet's defining moment seemed to have come in the late 1960s, when the veteran draggerman Joe Pallazola decided to call it quits. He towed the *Rosie and Grace,* his 110-foot converted World War I submarine-chaser, out past Marblehead, dynamited the vessel, and watched it sink.

But the industry wasn't dead yet; in fact it was running a stronger pulse than many of its observers realized. Though the foreign trawlers were vacuuming up groundfish on Georges Bank, they were not selling those fish to American markets. In 1966, Pope Paul VI and the conference of bishops in the United States had ruled that American Catholics no longer had to abstain from meat on Fridays; while this was a short-term market setback for the fishing industry, it began a long-term change in the image of fish in the United States, from that of a food eaten as a matter of penance to one eaten instead for luxury or health reasons. Cod was hard to find, but demand and prices were up. As the fleets shrank in Boston and Gloucester, the earnings of those fishermen who could stay in business were rising. By 1974, the year of the industry's "Sail on Washington," a crewman on a Gloucester offshore vessel was earning between $15,000 and $20,000 per year, more than such crewmen had ever earned before.

Two years later, when the Magnuson Act expanded the breadth of America's fisheries by some two million square miles, the response of the fishing industry was an optimism of the most guarded sort. Part of that had to do with the folk wisdom that counsels you to be careful what you say. Those who were making a living then in the big fishing ports were doing so because the New England fleet had shrunk. There was no sense inviting extra competition by cheering too loudly now that the other variable, the acreage of the waters in which the fleet could work, had so vastly increased.

And another part had to do with the wisdom that warns you to be careful what you wish for. The industry had lobbied for years for the banishment of the factory trawlers from Georges, and after long and strenuous objections from the State Department, Congress finally had managed that. Fishermen had prevailed, however, only by emphasizing the trawlers' potential for ecological devastation and thus enlisting the support of environmental groups and congressmen sensitive to such issues. Just getting rid of the trawlers was not enough: not only environmentalists, but also NMFS, the National Oceanic and Atmospheric Administration (NMFS's parent agency), the academic community, and state marine fisheries agencies insisted that conservation-oriented management provisions be attached to any extension of fisheries jurisdiction. Fishermen had scant enthusiasm for federal management but could hardly object to its goals. So they got what they wished for and more: instead of simply a one-sentence ban on foreign fishing, a complex eighty-page law that sought more or less to invent the science of fisheries management.

They also got substantial industry representation on the eight regional fishery management councils established by the act. Each council's members would include the director of each state's marine fisheries bureau, the regional director of NMFS, and eleven others "knowledgeable or experienced with regard to the management, conservation, or recreational or commercial harvest of the fishery resources" of that region. Certain of these seats would go to environmentalists, academics, and consumer advocates, but there were enough seats to make industry control of the regional councils a probability. This was grounds for optimism, as was the law's momentous definition of the "optimum yield" of a fish resource: the amount of fish providing the greatest overall benefit to the nation, "which is prescribed as such on the basis of the maximum sustainable yield from such fishery, as modified by any relevant economic, social, or ecological factor."

That last clause was important. Thanks to having so long endured the embarrassing plunder of Georges Bank by Communist-bloc nations, among others, and thanks to the growing political sophistication of fishermen's organizations, which had learned to play the game more like Jimmy Hoffa and less like the flower children, the Magnuson Act was written not only to protect fish but also to protect fishermen and their communities.

Since the well-being of the latter depended on the strength of the former, there seemed a workable logic to this.

The law then, if everything fell into place, was to be the key to a new sort of Eden, a place not where the commercial fishing industry would die, but where America's abundant natural resources, the Yankee know-how of its science and technology, and the enlightened rule of its people would strike a historic balance between appetite and consumption. In America, observed Tocqueville as he praised Americans' readiness to challenge natural boundaries, "the idea of novelty is indissolubly connected with the idea of amelioration." The novelty here would be that the efforts of man would not supersede but define and respect a natural boundary. This, finally, would be to the amelioration and greater wealth of both its fisheries and its fishermen.

Not only were the New England fisheries America's original natural treasure chest, they were still a treasure chest, still Siegfried's fabulous horde of gold, supplying 23 percent of the fish to world markets. For all the efforts of the factory trawlers, stocks of the prolific cod were still strong, though depleted, said the biologists, and in need of some rebuilding. At least Soviet factory trawlers were not about to be replaced by American ones, thanks to the failure of the *Seafreeze Atlantic.* The only threat to New England's treasured groundfish was a fleet of reduced fishing capacity working on what was designed to be a short leash. From the Pollock Rip on out to Georges' Northeast Peak, the hunter was about to lie down with the hunted.

The first serious jerk of the leash came as early as December 1977. Of the eleven voting members of the New England Fisheries Management Council appointed by the secretary of commerce in 1976, seven had ties to the groundfish industry. But the council was serious about its stewardship of the region's fisheries, particularly its groundfisheries, and accepted NMFS's recommendation that a management plan be implemented immediately — through emergency regulation, with little debate or public review — following the United States' 1976 resignation from the International Commission for Northwest Atlantic Fisheries. Central to the plan were regulations that came directly from the ICNAF: annual regional quotas on cod, haddock, and yellowtail flounder and tight restrictions on the amounts of these species that might be taken as bycatch while fishing for other species. Once a quota was met, the secretary of commerce or NMFS had the power to close that fishery.

Then, only four months into the plan, in July 1977, fishermen in the Gulf of Maine found that they had already harvested enough cod to close that regional fishery for the rest of the year. Georges Bank fishermen found themselves in a similar fix by the third week in August, though boats in both regions could still bring in small amounts of cod as bycatch. These conservative regional quotas, however, were not what fishermen had anticipated from the open waters of the Magnuson Act. The quotas' existence changed fishermen's behavior, forcing them to fish as hard and fast as possible while they still could. This "derby fishing" favored the bigger boats, which could fish more days across wider areas and for larger volumes. Bycatch restrictions also meant that thousands, and then millions, of pounds of groundfish brought up already dead in gear targeting other species had to be shoveled over the side to rot in the sea. It was the sort of waste that would have appalled any subsistence hunter or fisherman, that would have run counter to a whole host of taboos.

Then again, the bycatch wasn't always discarded. Fishermen's longstanding distrust of NMFS, their disappointment with this sort of debut by the council, and the impossibility of NMFS's small number of officers really enforcing council regulations all gave rise to an outlaw mentality aboard more and more vessels as skippers saw their competitors cheat and get away with it. Some found loopholes in the regulations, while others simply steamed right through them, unloading at night, or unloading partly in one port and then in another, or unloading at the dock of a processor willing to define their illegal cod or haddock as pollack. On the rare occasions when a fisherman or processor was caught and cited, fines were never levied, or were so small as to be chalked up as another cost of doing business. A 1987 council report noted that the typical violator of regulations on Georges Bank grossed nearly $225,000 per year in extra revenue, while the average fine for fisheries violations was less than $2000. Fishermen who respected the regulations or feared for the cod, or both, were enraged as the years went by and the groundfish stocks dropped through the bottom and the violations only increased. But their complaints went unheeded by an enforcement agency that was undermanned and a judiciary that generally viewed such violations as trivial.

In late December 1977, NMFS shut down all groundfishing in New England for the rest of the year. That meant only for a week, in fact, but the wholesale closure hit the fishing industry like a punch in the face. This was a new experience for fishermen, and a frightening one for a workforce neither eligible for unemployment insurance nor provided with the sort of subsidies and supports available to farmers for not farming. A government order telling them that they could not go about their jobs, could not bring in any income, was shocking also for its revelation of the power that the federal government and NMFS now held over them — a power unique in the American economy.

In the firestorm of protest that followed, the council divided NMFS's annual quotas into quarterly increments. When the groundfishery closed again in March 1978, after months in which many small boats could rarely get out of the harbor, busloads of Chatham fishermen descended on the council's March meeting waving such signs as "In Cod We Trust" and "The Fishing Industry, 1620–1978, May It Rest in Peace." Another closure followed in May. When it became apparent, finally, that quota overruns would force a murderous three-month closure that year, from September to December, the council appealed to Secretary of Commerce Brock Adams to approve an early start to the next fishing year, and therefore a new set of quotas, as soon as possible. Adams agreed, and along the New England coast, in what could be celebrated as a landmark in political gimmickry, New Year's Day came on October 1, 1978.

And so it went, with the council at first a shell-pocked no man's land between the fishing industry and NMFS and then more or less an occupied territory, siding with the fishing industry as NMFS was forced to admit that its stock assessments could not be 100 percent precise — that it was at least possible that there were more cod out there than the scientists thought — and as the council remembered its mandate to consider social and economic factors in its deliberations. By 1982, when Brian was working offshore on the big New Bedford scallopers, the council had discarded NMFS's quotas, rejected the agency's recommendations on limited entry, which the industry opposed, and implemented instead a set of minimum-size regulations on groundfish and mesh-size regulations on otter trawls and gill nets. These did not do the trick.

One of the reasons that the quotas were always so quickly exceeded, of course, had to do with the steadily increasing numbers of hunters appearing on the commons. In the first three years of the Magnuson Act, the total number of New England fishing vessels leapt from 825 to 1423. The federal Fisheries Obligation Guarantee program, which provided government-guaranteed loans to build or upgrade boats, was still in effect, as was the Capital Construction Fund, which allowed fishermen to defer paying taxes on their boats. Both programs ended in 1979, but not to worry: the Reagan administration's 10 percent tax credit of the 1980s made it too expensive *not* to buy a boat. With this, if a fisherman (or an investor) bought a $1 million vessel, regardless of how much of that money came out of his own pocket, he was allowed to deduct 10 percent — $100,000 — from his tax liabilities. Then the Reagan tax code's accelerated depreciation credits sweetened the deal. Doctors, lawyers, stockbrokers, and real estate developers flocked to New England ports. Banks sent recruiters down to the docks. Buyers came into boat dealerships waving fistfuls of money.

They were not buying boats like the *Susan Lee*. Draggers were the weapon of choice — the bigger the better. The 590 trawlers that fished out of New England in 1976 had grown to 1010 by 1982; by 1986 the number of large fishing vessels (at least 80 feet long and 105 tons) had tripled. Vessels of more than 150 tons increased fivefold. Some were built to hold up to 200 tons of fish and routinely caught more than half that much each trip out. While these hauls seemed modest compared to those of the factory trawlers, this new generation of mid-sized hunters made up the difference in their numbers and state-of-the-art technologies. "In the old days, fishermen might catch ten to twenty percent of the stocks, a rate that fish can match at normal reproductive rates," the NMFS scientist Vaughn C. Anthony told the *Boston Globe*. "Now, with all the gizmos, the fishermen can hit a productive ground and haul out seventy percent of the stock in a year. . . . This is the first generation of fishermen with the technological capacity to go out there and pretty much catch all the fish there are to be caught."

Meanwhile, NMFS was in a retreat that in the 1980s turned into a rout. The Reagan administration gutted its annual budget at the same time that Congress greatly expanded its regulatory responsibilities. "As a result," observed the industry analyst Charles H. Collins, "NMFS incurred the largest percentage of budget reductions of any of the federal natural resource agencies during the Reagan years. Isolated within Commerce, NMFS had little support from the secretary's office or NOAA. Its management and enforcement budgets crumbled, and as its financial resources declined, fishing industry criticism of its science increased." Nor did NMFS get any support from the environmental community, which was fighting the Reagan administration's determination to lease tracts of Georges Bank for oil exploration and courting the support of the fishing industry as it sought to head that off.

The lions were loose — lions made many and large and ravenous by money — and the slaughter was on. After hovering between 45 and 55 million tons through the first part of the decade, the annual New England catch of cod (at least as it could be officially measured) rose to 76 million pounds in 1977, 87 million in 1978, 99 million in 1979, and 117 million in 1980. The number of groundfish licenses issued by NMFS increased by 83 percent during that period, while overall landings of all groundfish species rose by 50 percent and the real value of cod catches climbed by 37 percent. As the fleet sailed into the 1980s, Gloucester deckhands who had considered themselves blessed a few years before to be earning $15,000 were now making $30,000 to $50,000. Captains of those same vessels were attending to their expenses, meeting all their payments, and then pulling down six-figure incomes.

These were the years of Brian's mighty adventures and steady paychecks on the *Bell* and the *Cape Star* out on Georges Bank. Skilled offshore captains such as Olav Tveit were heroes, feeding the rising U.S. demand for seafood, heading up an American fishing fleet that boasted not the world's biggest boats but arguably its smartest, and doing so during the Decade of the Entrepreneur, proving that with enough piss and vinegar and laissez-faire get-the-government-off-our-backs capitalism, even cowboys who had always scraped for nickels and dimes could be high rollers. It was precisely the sort of happy ending that had been imagined for the Magnuson Act, one that might have been penned by that one-time Unitarian minister from Brewster, Horatio Alger.

But that was not the ending, though some dreamed that it was, believing that the cod would always be there, no matter what. If the fish could survive fifteen years of pounding by the factory trawlers, then the fish were immortal as well as sacred — just as the buffalo had once been sacred, had once seemed immortal.

Landings of cod dropped to 102 million pounds in 1981, and

fell sporadically throughout the 1980s despite steady increases in fishing effort. Demand continued to climb, and prices remained generally high, but suddenly profit margins were thin and getting thinner. Catches might be dropping, but the costs of fuel, insurance, and boat payments were not. Skippers had to get smarter, more efficient, maybe invest in new gear and electronics as the competition got smarter and became more numerous. Murmurings by scientists such as Vaughn Anthony about more drops to come and the need to reduce fishing effort were dismissed by skippers, who said they were still seeing plenty of cod out there. But the skippers had no choice but to see plenty of cod, to resist new regulations and cutbacks, to stay out on the water and keep working: they were still heavily in debt for their bigger and smarter boats, and they were obliged to service those debts. For some Gloucester boats, debt service ran as high as $10,000 a month.

When the council abandoned quotas in 1982, it was like the cap blowing off a pressure-relief valve. Then the cap was screwed back on in a different way with the awarding of Georges' Northeast Peak to Canada in 1984. The drawing of the Hague Line removed the New England fleet from the most distant and productive portion of the bank — in particular, removed its biggest and most productive draggers and scallopers, forcing them to squeeze into the rest of Georges Bank, the channel, and the Gulf of Maine and compete with smaller boats on a reduced field of play. Brian was correct in guessing that the deckloads of scallops and his seasons of financial stability were at an end. He went back inshore, helped revive the Nauset Fishermen's Association, bought the *Cap'n Toby,* and began lobstering in the shadow of the collapse of the offshore fisheries.

Measures were taken to stave it off, but they were half-measures. As catches fell and dealers and processors filled their inventories with imported fish instead (between 1986 and 1992, factory trawlers on the international part of the Grand Banks removed sixteen times their quota of cod, flounder, and redfish), and as prices subsequently fell, the council implemented broader gear restrictions and area closures. "But every time the council makes a change, fishermen became more efficient," observed the Gloucester draggerman Joe Brancaleone. Once the oil industry had been dissuaded from prospecting on Georges Bank, the environmental community weighed in against the council, but whenever council members proposed any of the sterner measures advanced by NMFS and the environmentalists, they had to face scores of desperate fishermen awash in red ink, who succeeded in drowning out the voices in their own ranks calling for sterner measures, or at least real enforcement of those in place. If this was not good enough, then aggrieved fishermen could simply go over the heads of council members to Congress.

As the situation worsened, a confused and politically buffeted council found itself dealing hardly at all with long-term issues of resource management, its mandate under the Magnuson Act, but instead absorbed in short-term issues of resource allocation: big boats versus small, draggers versus fixed-gear fishermen, groundfishermen versus other industry sectors, all of them with their own dilemmas and in serious trouble of one kind or another. "The result has been a paralyzed system of fisheries management," said Gerry Studds, then the chairman of the Congressional Merchant Marine and Fisheries Committee, "that represents everyone's perceived immediate interests faithfully, the fundamental interests not at all." One council chairman, speaking anonymously to the *Boston Globe,* judged that the council's regulations through the 1980s were "about as effective as rearranging the deck chairs on the *Titanic.*"

The Magnuson Act empowers the secretary of commerce to intervene at any time to define and administer his or her own fisheries management plan. But nothing in the law requires the secretary to intervene, and Ronald Reagan's commerce secretary, Malcolm Baldridge, occupied a position between, on the one hand, an industry that had been tempted by government incentives into positions where its members could not financially afford federal regulation and, on the other hand, an administration committed to the wholesale dismantling of federal regulation. In the mid-1980s, NMFS began work on its own management plan, but complaints to the New England congressional delegation swiftly convinced Baldridge to back off. Deck chairs were duly rearranged, and the *Titanic* kept sinking.

What has been described as an abrupt fisheries collapse has not been abrupt at all; rather, it has been a slow and incremental downward spiral, an awful settling of a great ship into the abyss, one porthole at a time. Eight years ago, in 1988, Chatham's Mark Simonitsch rose at a council meeting to suggest that Georges Bank be closed to all fishing indefinitely. He was shouted down. In 1989, after more than sixty years of fishing, the famous Brancaleone family of Gloucester sold their boats, and council member Joe Brancaleone took a job as the assistant night manager of a Burger King in nearby Beverly. It was the sort of demotion that impressed other fishermen once again with the fact that all that counted for skill, achievement, and tradition inside the fishing industry counted for little outside it.

Amendment Five and Amendment Seven have quietly abandoned the old concept of "optimum yield as modified by" any relevant social and economic factors. When a fisherman complains that Amendment Seven will drive him out of business, the council member sitting opposite him thinks silently, with regret, "Exactly." The regulations' implicit goals are precisely a winnowing out of fishermen, whether those who gladly took to sea on the tides of federal loans and tax-rebate money, those who were reluctant to do so but had to compete, or those who stuck to their small boats and took their chances. The Wall Street–minded speculators are long gone. Those who remain are the sons of Toby Vig and Harry Hunt — those who for love or desperation have to fish. Few are ready to go gently. They rage at the dark denouement of a Horatio Alger story gone as wrong as Alger's own story (in 1866 he was drummed out of Brewster for what the town's notice to the American Unitarian Association described as "unnatural familiarity with boys").

Now the New Bedford Seafood Co-op is closing its doors, and Provincetown's Seafood Oceanic is on the verge of ruin. Five offshore draggers were auctioned off in New Bedford over the winter, and also declaring bankruptcy was the Point Judith Fisherman's Co-op, which in the 1960s was the first fishermen's organization to alert Congress to the depredations of the factory trawlers. In Boston, after eighty-one years of business, the New England Fish Exchange is about to be evicted from the Boston Fish Pier, and the Gloucester waterfront is already far down that gentrified path toward a fisherman's version of Desolation Row: bed-and-breakfasts, cozy restaurants, art galleries, and along the docks whale-watching trips and below-retail lobsters. New England's share of the world market in fish has moved its decimal point, shrinking from 23 percent in the mid-1970s to somewhere between 2 and 3 percent now.

The council, at NMFS's urging, has yielded only reluctantly to phasing in limited-entry regulations in its groundfisheries in the 1990s. Fishermen, particularly small-boat fishermen, fought that nearly as bitterly as they fought the quotas that left them tied up at the docks in the 1970s, partly because of the New England tradition of fishing as a communal and often family-based activity, à la the Brancaleones; partly because of the small owner-operator's need to shift seasonally from one fishery to another and target species opportunistically; and partly out of a gut sense of what the statistics bear out, that for every million dollars invested in the industrial-scale fishing seen in most of the world's limited-entry fisheries, one to five people are employed, while each million dollars invested in open-access small-scale fishing employs from sixty to three thousand. But that battle has been lost, along with the great, pestering shoals of cod, and now surviving groundfishermen — or such displaced groundfishermen as Dan Howes — wonder if this first step on the road to privatization of the commons will lead to the sort of changes in their industry that the Dust Bowl wrought in American agriculture, making it the sort of tidy and vertically integrated affair, with most of the money at the top, that governments and politicians have an easier time dealing with, and profiting from.

In New England, the all too manifest accuracy of NMFS's dire predictions during the 1980s has raised the credibility of the agency, and of fisheries science in general, as if that needed another confirmation. It was, after all, a matter of common sense and long experience, an ordinary example of rising effort and declining yield, one more version of Francis Herrick's precollapse stage three in the lobster fishery. NMFS's stock assessments continue to be discouraging, both about the present strength of the cod and about the amount of time necessary for its recovery. Groundfishermen listen more respectfully now, and this is part of a cultural change that parallels the change in the lobster fishery, one that better recognizes the social and economic costs of cheating and abhors the outlaw mentality as much as lobstermen now generally abhor bleaching eggs from eggers. But fishermen cannot forget how cruelly secure the bureaucrats are in their jobs and their employment benefits, how the agency has generally scorned their own observations of fish behavior and numbers and has always promoted limited access, ITQs, and other species of privatization.

Neither, therefore, can fishermen help listening to NMFS's current pronouncements in support of Amendment Seven with at least a shade of cynicism, wondering if they are once again being set up for a fall. "One does not have to believe in conspiracies," said a representative of the Seafarers' International Union during a congressional hearing on Amendment Seven, "to note that the real and actual effect of Amendment Seven is to create an atmosphere in which the only possible salvation is consolidation."

In an editorial column in the *Cape Codder* this winter, Bill Amaru wrote that nobody set out to overfish: "What we did was to compete with each other in an open market. Competition — striving to catch more, more efficiently, and get home before the other guy — that is what fishermen are mainly guilty of." But in the minds of the public at large, fishermen are dismayed to find themselves merely guilty, only slightly more popular than pedophiles, viewed as greedy bastards who are cruel to animals and who loaded their boats so top-heavy with fish that they blocked up the scuppers and now deserve to go to the bottom. They raped America's first and last great public resource, the thinking goes, and it's better now for it to be put in the hands of responsible private interests who will care for it better.

In a sense the waters off Cape Cod, besides providing the amniotic fluid for America's birth, are the nation's last frontier. More than any other regional fishery, they remain a place, albeit diminished now, where anyone without the advantages of a lot of start-up money or a graduate degree can venture into open territory to build an independent livelihood — not as some rootless cowboy, piling up a stake and moving on, but more like a husbandman, staying in one area and bringing the fruits of his labors back to his family and community. And unlike the inhabitants of that world described by Henry Beston as "sick to its thin blood for lack of elemental things," a fisherman lives each day engaged with every elemental nuance of his territory, immersed in both the physics and the metaphysics of life and death, shot through with the adrenaline surge of living so near the razor edge of mortality. He comes to personify, finally, elements of our spiritual commons. If he experiences things that most of us can no longer experience, then merely the assurance of those experiences in some men's lives — like innocence, perhaps, or sainthood — enlarges the boundaries of the commons.

In the *Cape Cod Times,* a letter appeared recently about the two owners of Cape Spray Fisheries, a local fishing and processing company, who were cited for falsifying records and exceeding their five vessels' days-at-sea allocations. NMFS wants to make a point here and is seeking a record $5.8 million in fines against the owners. The author of the letter objected, writing that the men were being punished simply for being good fishermen. Brian disagreed, writing in his own letter to the paper that a "good fisherman" is precisely the one who respects and obeys the regulations that exist for the common good of all: "Remember Laevinus, the Roman consul who said: 'Betray what belongs to the commonwealth, and you will seek in vain to keep safe what is your own.'"

"Shit, I didn't think I set on myself."

A second nylon line has come up tangled in the hooks and dogfish of this line. If Mike runs his lines too close together, the tide can belly one line over another. This results in the sort of expensive mess fishermen call a clusterfuck. That happens easily enough on the slippery slopes of Joe Bragg's Ridge, but Mike thought he had plenty of room between his sets.

As it turns out, this is a piece of abandoned long-line gear with a swatch of gill net wrapped up in it. Mike cuts it away from his own gear, hauls it in to clean it off the bottom and recover what he can. He says later he'll cut the gangions off the long-line gear and use it as a buoy line. The monofilament gill net, fifty feet long, comes up clogged with seaweed, pocked with an occasional rock crab. Mike stows it in the cockpit and goes back to cleaning his long line, wincing once when his hand is jabbed by the spine on the dorsal fin of one of the dogs. "That's the first of thousands this spring," he says.

The dogs come up hooked through the lower jaw. When their snouts bump against the posts in front of the winch, the hooks tear out through their bottom rows of teeth. Many of the hooks are bent, their gangions shredded. "I haven't had woofers like this since December," Mike tells me. "I forgot how much fun it is to have dogs on your gear."

He straightens, stretches, easing his back, his eyes sweeping over the mineral hues of the clouds in the west, the graphite-gray humps of the swells out of the east, the chalky whites of the gulls floating to the *Susan Lee*'s starboard and stern. The birds rest on the waves with the solemn gravity of patrons at an art gallery. Around us the sea and sky are enormous, mobile, planetary. The gulls have found the world's still point, the singular and inconceivable axis of the universe. "This beats staying at home, anyway," Mike notes.

It also beats our day at the Figs last February. The Figs are a piece of ground about seven miles southeast of Joe Bragg's, along a stretch of the slope where the wide undersea shelf of the Cape's outer arm falls down to the deeper plateau of the Gulf of Maine. That day was the maiden run of Mike's new autopilot, and during our ninety-minute steam through the winter darkness to the Figs, while Mike slept below, Ben Bergquist and I talked in the wheelhouse.

Ben is a recent graduate of Vermont's Lyndon College, where he was a natural resources major. For years his father ran a lobster boat and operated a small fish market in Chatham, and last year he bought a larger market. Now he is done with lobstering, and Ben bought out his license, boat, and eight hundred traps. Ben said he was anxious to get going in the spring and figured he had a good future there: "That's a well-managed fishery. There's no bubble about to burst." Whether that was true or not, it was also his only option this side of the bridge, since he didn't have a groundfish permit, nor any prospect of getting one.

The Figs were as empty of other boats that day as Joe Bragg's is today, and in drizzling rain, in seas pitching three to six feet, Ben stood at the transom to set the gear while Mike handled the wheel, studying his notebook and the loran and barking "Yep" to Ben at the precise moments to cast out the anchors, the buoys and high-fliers, the bundles of baited hooks.

At least, the moments were as precise as they could be. A warm front was sliding over the Cape, and the rain was fogging the surface of the water and causing Mike's loran readings to roll back and forth. "We're going for a reasonable proximity," he said. Not that he was at all in the proximity of where he had intended to be working that day. We had gotten a late start because I had overslept — not a capital offense on a fishing boat, but close to it — and it was too late now for Mike to jockey for position on the Cod Fish Grounds, where most of the fleet was working. It was generally thought that the water was still too cold at the Figs for the cod to be feeding there yet, but Mike figured it was worth a gamble, since he had to gamble anyway.

We ran three sets of gear. "There's a puddytat for you!" Mike cried when a ten-pound wolf fish came up early in the first haulback. Known more commonly on the Cape as a catfish, the wolf fish is the color and shape of old lead pipe and comes with a face of studded leather and teeth like the bare ends of a chain-link fence. These are soldered into jaws powerful enough to crush a quahog like a bonbon, which will go on biting even after the animal has been gutted. The chef Julia Child has bitten back, preparing a famous recipe for wolf fish, and the animal is another one-time trash fish now worth money at the Chatham Fish Pier. "But I don't like those things," Mike said. "They've got an attitude."

And then the hooks, one after another after another, came up empty and still baited, the sections of squid dripping in the rain, the steel glinting dully in the leaden light. Mike stared in disbelief as the first set of three bundles of gear, nine hundred hooks, yielded exactly twelve cod, three wolf fish, six redfish (rising out of the depths like glowing coals), and a plastic bucket containing a crescent wrench, a screwdriver, and a load of oily sand. Mike banged his gaff on the rail in fury as the hooks came up. Afterward he gazed at the bare flooring in his fish bins for a long and bitter moment, one that seemed to bring twenty years of unhappy history in these waters to a single crisp loran coordinate in time and space. In the bin behind the wheelhouse, a wolf fish had its jaws sunk deep into the flank of a cod. The wolf fish jerked, released the cod, snapped convulsively at the air, and then seized its own tail, swallowing itself.

The next day I wasn't late. I got up at 2:00 A.M., went down to the harbor, and after a two-and-a-half-hour steam we were on the Cod Fish Grounds at 5:30. Mike said that slack tide was at eight that morning, so we didn't need to start setting gear for another hour or so, but he expected the rest of the fleet to be there as well today, and this way we had our spot staked out. Five other boats were on the *Susan Lee*'s radar screen, though we couldn't hear anybody else near us. A voice on the VHF hailed Mike and wondered if he was alone. "At the moment," Mike said.

"I'm over to the west," the other skipper said. "I might come over there, but not till you're done setting."

"Okay."

The dawn came up with light fog, visibility at less than a mile, and the air in slow motion with a three-knot breeze. A dragger rumbled past the port stern, and then two more went past the bow. As the light filled in to the east, we could make out the profile of another long-liner. Mike said it was the *Seabag,* which belonged to Fred Bennett, the skipper who broke the flower vase at the hearing in Hyannis last fall. Mike took the *Susan Lee* over to within shouting distance of it. "I'm just gonna work around you, okay?" Mike called out. "I haven't worked here the last few days. I've been working all over."

Fred nodded. He stood with a bearded crewman in the cockpit of his trim green-and-white boat, his neat gray hair bare to the breeze. "I'm gonna set south-southwest, about one-ninety, I guess. See how the tide is. Were you out yesterday?"

"Down the Figs. I got out late and didn't want to mess with that bar getting funky. I ran aground there the other night."

"The Figs? How was it?" Mike shook his head and made a thumbs-down gesture. "Yeah, I heard they were gone from there. When are you gonna set? Twenty of?"

"The later the better," Mike said. "No sense rushing it."

"We could probably set as late as six forty-five."

Mike nodded, signaling his agreement, just as another long-liner motored out of the fog to the north. He said to Ben, "We're at the south end of the grounds. This is a pretty busy spot."

The boat out of the north veered west and out of sight, but

another long-liner materialized on the western limit of our visibility and prepared to set. By then Mike had moved to the east of the *Seabag* and left enough room, he judged, to make a slightly more southerly run on his gear than Fred intended. Then it looked like derby fishing when, at precisely 6:45, all three boats rose from their skegs on parallel bearings and began loosing their anchors and lines over their transoms. Mike shouted once to Ben to unwrap a tangled buoy line from the pole of a high-flier. He eased the boat to six and a half knots, dropping his gear in thirty-two fathoms of water. He set two lines of three bundles each and took a call from Fred Bennett as he dropped his fourth high-flier.

"Mike, you gonna make another set here?"

"Thinking about it."

"I think you're going to be on top of me. It looks all right up there, but one end down here is swinging over in the tide."

"No problem — I'll set somewhere else." Mike called over his shoulder to Ben, who was still in the stern: "Ben, did we get anything on the west side the other day? Isn't that where we got plenty that time?"

Another voice came on the VHF, maybe from that third long-liner: "I think it's okay up at the north end. It's crowded all over down here, but I'm pretty sure it's okay up north."

Mike covered all his bases, making two sets on the west side of the grounds and one more on the north. At the south end, the first few hooks brought skates. They came twisting up out of the blackness like blown newspapers, like the past come back to haunt you, and broke the surface as clenched fists, their wings wrapped around the line, their tails furled under them. Mike threw them back as the next spate of hooks rose empty.

And then the cod came in. Out of a void, out of a medium at once as clear as glass and as black as pitch and without length or depth, the cod came as if summoned from nothingness: a white dwarf star elongating into a comet and then taking on color, becoming instead a golden ingot, and then assuming motion and capacity and animate spirit. It was Midas in reverse, watching that gold become flesh, or something in between, actually, a third state of being combining the graces of both. The cod rose goggle-eyed into the air, their gills intoxicated with oxygen, their orbs drowning in light. Eight out of ten, nine out of ten: nearly all were of legal size, twenty to twenty-six inches, five to fifteen pounds, fat and firm and mature, hardly any to throw back. They came not in droves or pestering hordes, but singly and in pairs for every several hooks. "Slow but steady," Mike said.

We took in 450 pounds on that first set and did better on the others, even on the set on the north side, where the line was cut, probably on a rock, so that Mike had to haul its pieces from opposite ends. Finally his bins were full of virtually nothing but cod: two haddock, one pollack, two or three wolf fish, and then two thousand pounds of the smaller scrod, the mid-sized markets, and several big twenty- to twenty-five-pound steakers, which would fetch the best price from Dave Carnes. In the thin, skeletal light of that February afternoon, the heaped fish seemed to glow like banked coals.

We were home by four P.M. with the fish all gutted. Mike eased the *Susan Lee* into a berth in front of the Chatham Fish Pier, where a fish packer named Eddie stood at a piling and gazed down into Mike's bins. "Hey, look at all them scrod," he said.

"Scrod, hell," Ben said. "Nothing but markets and steakers."

Mike beamed up at the packer and said, "Eddie, you got a shave. Did you get lucky too?"

≈14

Men's Lives

ON AN EARLY MORNING in June the road to Snow Shore is moist and foggy. Brian turns the Ford up Tonset Road, along shoulders snowy with Queen Anne's lace, house lots flecked with locust and hydrangea in bloom. He says that the recent poisoning of gulls on Monomoy by federal Fish and Wildlife agents is a perfect example of a too-much-money-to-spend sort of solution. "Half a million dollars to save fourteen plovers," he says with disgust.

One morning last month agents dropped thousands of toxic bread cubes into the nests of black-backed and herring gulls on South Monomoy Island, a part of the Monomoy National Wildlife Refuge. Their purpose was to create a 350-acre nesting area for two endangered shorebirds, the piping plover and the roseate tern, free of harassment by gulls. But several environmental groups contended that the feds were doing this only so the state could continue to relax restrictions on off-road vehicles on Nauset Beach, where the plovers and terns also nest. When a federal appeals court in Boston rejected the environmentalists' plea to halt the operation, demonstrators appeared on Monomoy the day of the poisoning with four hundred sandwiches of blackened Cape Cod tuna and activated charcoal pills, the bread and tuna to spoil the gulls' appetites for the bread cubes, the charcoal pills to absorb any poison.

"A fisherman is fined fifty dollars if he kills a seagull, so not to name any names, but if your boat is getting shat on all the time, all you have to do is kill a gull and leave it on the foredeck and the rest will stay away," Brian says. "All they had to do on Monomoy, I submit, was to leave one dead gull by each plover's nest." Brian swings left off Champlain Road to the narrow lane that leads down through a tangle of locust and sumac and beach plum to Snow Shore. "Well, God grant me the wisdom to accept the things I cannot change, and also enough ammo to hold them off till daylight."

At Snow Shore the boats lie still at their moorings, their reflections on the poured glass of the inlet as perfect as the artificial clarity of memory. The horizon beyond them, though, looks as though it were hung with linen, the fog at once amplifying and muffling the murmuring of bass fishermen and clam diggers on the other side of the harbor, the mewing of the gulls scouring the flats. "We ought to be hearing Mike on the radio today," Brian says. He drags his dinghy from above the tide line and slips his oars into their locks. "He's been going out for dogs a lot lately."

The sweep of Brian's oars through the water sounds like the rustling of silk. He seems to grow larger as he skims out into the inlet, as he moves in a *pas de deux* with his reflection to the bow of the skiff, as he unhooks the boat from its mooring, secures the dinghy, and then climbs aboard. The rattle of the outboard is not as thunderous as usual. Its sound is absorbed into the fog like oil into cloth. Brian assumes a ghostly aspect as he stands stiffly in the skiff's stern, and the boat seems to hover over the water, separated from its own sound, while it advances toward the beach. In the bed of the pickup Brian's totes of mackerel, salt porgies (a.k.a. scup), and thawing redfish smell like the mortal remains of souls who have waited too long for this crossing.

Brian says that a low-pressure area has gotten stuck over the Midwest and the Cape will be enduring unsettled weather for a while. We load the totes of bait into the skiff. Bait has been hard to find, and Brian scrounged the six-inch porgies, by now as brown as old apples, from somewhere down by the Cape Cod Canal. The mackerels and redfish are fresh from the oubliette, the latter as bright and hard as candy. The *Cap'n Toby* rests in a still life at its mooring as the skiff approaches, then shudders briefly as a breeze kicks up. Brian nods, ties the skiff off, clambers to the foredeck. The breeze skips with cat's paws and moonshadows across the harbor. "Yep, all you have to do is touch a boat, and the wind's up out of the east," Brian notes.

In the cockpit he pulls the cover from his new engine to open the seacocks. No longer does an oil-blackened plastic milk jug hang from his crankcase breather line. The V-8 Mercury Cruiser glistens as though wet in the soft light, its valve covers a pewtery silver, its lines and hoses an immaculate textured black. The engine's fifteen-knot cruising speed (compared to eleven knots for the Volvo diesel) allows Brian to haul around twenty-five pots an hour, five more than he could manage last year. Although he pays more for fuel than he used to, he calculates that he more than makes up the difference in gear efficiency — or would if his gear were close to being normally productive.

Saddled with a $10,000 debt on the new Mercury, Brian is struggling through another slow and mystifying spring. None of the charter-boat captains has been getting any bluefish yet, and Brian knows that his bait is sub-par. But if there aren't lobsters out there in the first place, then even bluefish will just rot in the bags. He has 440 traps in the water now, and another 60 waiting to go out. He's already decided that if the lines he's hauling today — three miles out, on the edge of the shipping lane out of Boston — come up empty, he's going to move them, though he doesn't know where. Nor does he know where he'll put those 60 new traps. This easterly wind that has just come up won't help with the fog; if anything, it will just pile it in thicker, and Brian is nervous about being in the way of big ships in foggy weather. "But maybe it'll bring the water temperature up a little," he says, "and that might help with the lobsters."

We fill the bait bags at the mooring, with a mix of mackerel and porgie in every bag, a piece of redfish in every third or fourth. At ten, with the tide running into the harbor through a new hole at the northeast end of the bar, Brian starts the engine and casts off. Two or three dozen lobster traps — more than last year, all being fished on recreational licenses — lie scattered between the boat channel and the clam flats off to starboard. More lie toward shore on the port side, their white floats like loaves of bread cast on the waters. On the beach to the southeast of the bar, a girl stands with her surf rod bent like a fishhook, its line taut to the water. Her companion looks on with the silvery lump of a striped bass lying at his feet.

Not all fish stories are unhappy ones, though it may be warranted that all fish stories now are difficult.

Like the Atlantic salmon, the striped bass is an anadromous fish, swimming from saltwater into fresh to spawn. It is believed to have spawned at one time in nearly every estuary on the Atlantic coast, and Captain John Smith reported seeing such multitudes that he imagined he could walk across a river "dry-shod" on their backs. Later this plentiful fish, with its fine, fat, delicate flesh, became no less important than the cod to the sustenance of the first Pilgrims of the Massachusetts Bay Colony. In 1639 the bass (along with the cod) was the object of the New World's first written conservation law, with an ordinance that forbade its use for fertilizer. In 1670 revenue from the sale of striped bass provided the means for the Plymouth Colony to open its first free public school.

Like the cod, and much more than the Atlantic salmon, the bass is tolerant and adaptable. Most partial to such small fish as mackerel, menhaden, flounder, sand lance, whiting, smelt, and silverside, it also feeds happily enough on lobsters, soft-shell clams, crabs, mussels, annelids, and squid. Anglers love its combination of beauty, power, savvy, and size. Its dorsal area ranges between tints of blue and olive green, with highlights of silver and brass, and its silver-white flanks are traversed by the parallel dusky bands from which the striper takes its name. Mature fish aged four to six years average between five and fifteen pounds, though the bass can live up to forty years and reach better than a hundred pounds in size. The Massachusetts record is seventy-three pounds, equaled on three occasions, the most recent off Nauset Beach in 1981.

Though tough and long-lived, the bass is not as prolific as the cod. Females produce fewer eggs, and the fact that stripers spawn in the shallow and volatile frontier between fresh and salt waters means that spawning success depends on a lot of mixed-up variables: rainfall, salinity, wind, current, water cleanliness, nutrient availability, and so on. Therefore the stripers' numbers have always been of the boom-and-bust variety, with the fish sometimes disappearing for as much as a decade, then reappearing again in great numbers, thanks to favorable circumstances and the exertions of only a relatively few female fish. The latter years of the nineteenth century, however, brought enough dams and dredging and industrial pollution to thwart even this hardy species, and thousands of estuaries, rivers, and creeks were removed from the stripers' spawning grounds. After 1897, no catches of striped bass were reported north of Boston for thirty years.

In 1928 a fifth dam was built on the Susquehanna River, which feeds into Chesapeake Bay. By then the Chesapeake was the major spawning ground for bass ranging the length of the East Coast, and this dam so reduced bass stocks there that it was feared the fish would become extinct. But the enlargement of the Chesapeake-Delaware Canal a few years earlier succeeded in flushing out much of the stagnation of the upper bay, and with a succession of good spawning years in the mid-1930s, striper numbers began to climb again.

In the 1940s, rod-and-reel sport fishermen decided that they could also do something to keep bass numbers high. "Schooling up in large lobbying aggregations," notes Peter Matthiessen in *Men's Lives,* his eulogy for Long Island's commercial fishermen, "they began to put pressure on the politicians for legislation to reserve the bass for recreational anglers, who could round up far more votes than the commercial men." So in 1945 the commercial netting of bass was prohibited in Massachusetts, and shortly afterward in Maine, New Hampshire, and Connecticut as well. Striped bass biologists said this was unnecessary, since commercial netting accounted for only a sliver of the annual take and there were plenty of bass anyway. "It is a curious anachronism," noted the magazine of the New York Conservation Society, "that the unusual abundance with which we have been blessed has, in a round-about way, resulted in frequent acrimonious disputes between commercial and sporting interests."

By then the Atlantic States Marine Fisheries Commission was conducting and publishing its own research on the striper question. The commission was not cheered by what it saw as the politicization of the bass fishery. "During the past thirty years there has been a growing trend toward social legislation in the marine fisheries of the several Atlantic states," the commission said in a 1953 statement. "Except in rare instances, such social legislation seeks to protect one fishery interest at the expense of another. . . . Such acquisitive attempts often claim conservation and sound management as their objectives. Rarely, however, is there sound scientific evidence to back these claims." Unless that trend was checked, the commission concluded, "and far greater consideration [was] given to scientific data and warranted conclusions, the longtime result may well be a gross mismanagement of our marine resources."

The ban on net fishing stayed, but bass numbers fell again in the late 1970s. Fishing pressure was certainly a factor, but what made that pressure suddenly excessive was declining water quality in Chesapeake Bay, where at the same time the shad fishery was dying and the oyster harvest was weakening. The Atlantic States Marine Fisheries Commission responded in 1981 with its own coastwide management plan for the bass, a regimen that recommended minimum size limits and restricted fishing on spawning grounds. When this became law in 1984 (the Atlantic Striped Bass Conservation Act), the plan was toughened with an increased size limit, closed seasons for commercial harvest, bag limits for recreational harvest, and a 50 percent reduction in overall harvest from 1981 levels. Some states were even tougher than that, declaring a complete moratorium on bass fishing.

Now, a decade later, the bass is back. Removal rates, which refer to the fraction of the total mature stock removed by fishing, averaged 65 percent during the 1970s; in the late 1980s, these rates were down to 10 percent. In 1990 the law was amended to allow increased levels of fishing in concert with monitoring programs that would hold removal rates below 20 percent. Meanwhile bass stocks have doubled in size, growing at a pace of 25 percent per year. Recreational anglers took 2700 metric tons of bass in 1993, while commercial rod-and-reel fishermen (who sell to the restaurant trade) took 800 tons, both harvests four times the respective takes of 1988. The striper is once again the mainstay of the Rock Harbor charter-boat fleet, and only Brian and other lobstermen who get bait from there are unhappy to find big racks of that sweet-flavored fish cooking in their barrels. Brian's lobsters like the oilier bluefish, but since stripers and blues feed on many of the same small fish, a rise in the size of bass schools feeding along the beaches is generally accompanied by a decline in the presence of blues.

With the stunning success of the fisheries commission's bass plan, however, and with the striper's return to abundance, the arguments, a curious anachronism, are back as well. While the allowed recreational harvest of bass has climbed in the 1990s at a pace in keeping with the increase in stock levels, the commercial harvest has been frozen at its 1990 level. Last spring the fisheries commission got the nod from its biologists to allow states to increase their levels of commercial harvest. This prompted the Massachusetts chapter of the New England Coast Conservation Association, a recreational fishermen's organization, to deliver a petition with five thousand signatures to the state's Division of Marine Fisheries. The petition demanded that all commercial sale of striped bass be banned in Massachusetts and that the fish be declared a game fish only, available solely to recreational fishermen.

The fisheries commission called for public hearings on the issue, a development that the Massachusetts DMF director, Phil Coates, bitterly resisted. "There are no resource issues here," Coates told the newspapers. "The issue is: Who gets the fish? Public hearings will not tell us anything that we do not already know, but will subject the public to a lot of acrimony." Coates personally opposed the ban and did not want the commission to be drawn into the resource allocation squabbles that have proven so enervating to the New England Fisheries Management Council. At its March meeting, however, the commission ordered Coates to proceed with the hearings, and last month the Cape Cod hearing was held at the high school in Sandwich.

Three hundred people were in attendance, and acrimony was in the air. "I see this move as pure greed," said Jerry Kissell, a Sandwich commercial fisherman. "Greed may be good in the stock market, but it makes no sense here." A NECCA representative, Whit Griswold, replied, "As long as there is a price on this fish's head, it will be harvested at its maximum limit." Other NECCA representatives suggested that out-of-state anglers would flock to the Cape if the state had its own exclusive game fish and that this would be of more economic benefit than the bass's modest commercial revenues. "Can anyone say *tourism?*" asked NECCA member Dean Clark.

Brian would enjoy toying with the logic of an argument that first pronounces it dangerous to put a price on the striper's head and next does precisely that. In either event, the scientific community agrees with Coates that the ban is unnecessary as a conservation measure; nor do all recreational fishermen support the proposal. Several Cape sportsmen's organizations have voted to back "professional wildlife management" and not endorse gamefish-only status. Greenpeace and the Massachusetts Audubon Society have also declined that endorsement, as has the Conservation Law Foundation.

Thoreau describes disputes over beached pilot whales and the rights to certain oyster beds. In his day, the Cape's oyster beds were stocked with seed from Chesapeake Bay, since the Cape's native oysters disappeared in 1770. According to superstition, this was the result of quarreling over the oysters by Wellfleet and other towns. But the Atlantic Striped Bass Conservation Act and the regional fisheries commission's work in support of that law have been a model for what can be accomplished without quarreling — instead with patience, real enforcement, and state governments and a regional organization working in concert. That success has kindled hope in the hearts of all who would like to accomplish the same for the much-quarreled-over cod and haddock. It has also, with the failure of the New England Fisheries Management Council to arrive at a lobster management plan, swept the Atlantic State Marine Fisheries Commission into that regulatory breach. Of course, the bass plan deals largely with the pastimes of recreational fishermen, who have other game fish in other places to turn to, if need be. The fisheries commission's lobster plan will deal with the livelihoods of commercial fishermen, who have nowhere else to turn. Brian recommends foul-weather gear.

Meanwhile, the bait barrels at Rock Harbor fill through the summer with fragrant racks of bass. Another generation of youngsters is working long hours for the vociferous Stu Finlay and other charter-boat captains. And abundance, no less than scarcity, proffers enticements to human greed.

Call it an atmospheric distortion. The imminence of Amendment Seven, now less than two weeks away, has magnified events in the lives not only of groundfishermen like Carl and Mike, but also of lobstermen and shellfishermen, lending the developments of the winter and spring a fatefulness not necessarily inherent in the events themselves. They cohere like the meteorology of storms. They loom large and throw off sparks.

In January an Orleans district court judge considered the case of three Provincetown lobstermen caught with eighty-three banded, egg-bearing female lobsters aboard their boat. The lobstermen told the judge that they were only going to use the eggers to lure male lobsters to their traps, and the judge said okay and duly acquitted them. Other lobstermen, acquainted with the murderous aggressiveness of eggers and the fact that females emit pheromones only after molting, were outraged. So were industry observers and other fishermen. "It's just a giant 'Screw you' to fishermen — all fishermen," said Carl Johnston. "It's set us back eight to ten years."

Brian was no less angry. "There is no longer any element of commonwealth in the Massachusetts lobster fishery," he told me. "This was the first fishery in the state to go limited-entry, and it's now the private domain of thirteen hundred license-holders. So those of us who have that privilege, it would seem to me, have even higher standards to meet in taking care of the fishery. But there are some guys who take advantage of whatever system you have, whether it's natural or manmade." He described the affair to the *Cape Codder* as "a knife in the heart of fisheries management. . . . While fisheries violations may not seem as bad as murder, wife-beating, or dealing crack, the judicial system needs to put on a different hat than the one it wears now when dealing with resource issues."

Other systems besides the lobster industry's are up for grabs, including the family-sized aquaculture grant system in Orleans that Brian helped design. Brian has recently rejoined the town's shellfish advisory committee and is among those debating the future of the quahog grants in Pleasant Bay: whether they will indeed serve as a public safety net for a few laid-off groundfishermen, especially now that Amendment Seven dictates the certainty of layoffs there, or whether they will be a launching pad for some aggressive entrepreneur's success story or some corporate Rockefeller's cornering of the market. There are now twenty-six grantholders in Pleasant Bay, eighteen of them — Carl Johnston, Dan Howes, and Mike Russo included — packed into the single bar between Sampson and Pochet Islands. Because of demand for space, only five grants have expanded to their two-acre maximum. Jay Harrington has voluntarily returned one of his two acres to the town.

At the same time, the Massachusetts Aquaculture Association

is pushing a plan to allow grant-holders in Orleans and other towns to sell their leases if at any point they want out of the business, and others are telling Brian and other commonists just to get out of the way. Last March, Victoria Ogden, the publisher of the *Cape Codder* and a newcomer to the area, wrote an account of her visit to the International Boston Seafood Show, where the stars of the exhibition were dishes prepared with salmon and catfish raised on farms around the world. A few samples of these convinced her that Cape fishermen were shortsighted: "Instead of fighting the few entrepreneurs who have set up beds, why not make aquaculture the next growth industry on Cape Cod?"

In her next column, Ogden reported that her phone had been jumping off the hook. One call was from Jim Harrington, a lobsterman and Jay Harrington's brother. She wrote,

> Jim Harrington told me, up front, he was not against aquaculture, and, in fact, he has his own one-acre bed. His fear is that families and then corporations will control or be able to buy up all the bottom, leaving no chance for an individual to make his living either in the wild shellfishery or as a small aquaculturist. "I have one acre and some of these families have four. What's to stop corporations from buying up these parcels, versus the young man on the waiting list? . . . The traditional fishery has helped me raise my family and build my house," Jim said, "and others deserve that opportunity."

What Harrington fears has happened before. The lease sales of shellfish grants around Long Island to private corporations all but ended oystering in open waters there several decades ago, and was the first punishing blow struck against the vitality of that small-boat fishing community.

Brian called as well, and also wrote another letter to the editor, pointing out that there is such a thing as municipal aquaculture. Earlier this month he made a practice of it, joining other Nauset Fishermen's Association members in planting about a million seed quahogs in Little Pleasant Bay and the Nauset Harbor estuary. The seed was purchased with the association's $84,000 FIG grant, and was larger and more expensive than the seed usually used by aquaculturists. But the NFA has its eyes on the survival rate; estimating that conservatively, at 75 percent, the association believes the seed will be worth $200,000 to local shellfishermen.

Carl Johnston says his own grant is looking good and seems to have come through the winter with little ice damage or mortality. He himself got through the winter all right with Mark, though it wasn't easy. Once the *Honi-Do* was repaired, Mark took it out dragging for shrimp again, only to discover that Dave Carnes at Chatham Fish & Lobster wouldn't buy any of his shrimp. Carl thinks that Carnes was mad about all the phone calls Carl made to other buyers on the night he couldn't get in early enough to sell to Carnes, and about his eventual sale to Chris King, a competitor. That was enough to get Mark mad as well. He told Carnes to fuck off, Carl says, and then geared the boat for groundfish. They went on "Toby tows" along the beach, catching blackback and yellowtail flounder and even an occasional halibut, all of which they sold to Chris King, who took the fish down for auction in New Bedford and paid Mark according to the price they fetched there. By and large, Mark liked what his fish were fetching.

One afternoon, however, King's fish packer said it would be three hours before he could unload the *Honi-Do*'s cargo of gray sole caught that day at Joe Bragg's Ridge. After Carnes's packers said that they would have to ask Dave before they could buy and unload Mark's fish, Mark went down to the other end of the pier and sold to a third buyer, Chatham's Finest. On another occasion the *Honi-Do* was approaching Chatham with four hundred pounds of dogfish along with a good load of gray sole. Carl knew that King liked to deal in large volumes and, unlike Carnes, normally wouldn't take anything under two thousand pounds. He advised Mark to call ahead to King's packer to see if they would take the dogfish, but Mark said, no, it's all right, they'll take them. He and Mark ended up having to dump all the dogs over the side.

On May 1 each year draggers are pushed away from the beach, banished to federal waters so as not to interfere with the strings of lobster pots being set inshore. Mark took the *Honi-Do* out to deep water, 100 to 115 fathoms, as much as 18 miles out, going for cod and gray sole and plaice. They got good loads of cod, but prices were generally low, sixty to eighty-five cents per pound. By then Mark was persistently annoyed by how often King's truck was unavailable when he wanted to unload, and he was also noticing that the prices Carnes was offering were better now than the prices he was getting at auction.

So at last Mark and Carnes had a talk and renewed their professional association, though relations have remained prickly between Carl and Carnes's packers. "I was in the freezer there the next day to buy some ice, and Ray Durkey was looking over my shoulder," Carl told me. "I had both doors open so I could find the softest ice to shovel, and Ray was busting my chops about that, saying I was taking all the best ice and they were going to charge me more. So I just built a pyramid in that barrow, stacked it as high as it would go, and let as much fall on their floor as wanted to fall on the way out. Point made."

With money getting tighter and boats such as the *Honi-Do* being sharply curtailed in the number of days they can work, relations between fishermen and buyers are only going to get pricklier. Carl guesses that Amendment Seven isn't going to hurt the *Honi-Do* too much this first year. "And I'll be surprised if it actually comes down on July first. There's always a court order or something that pushes it back. That second year, though, is going to be a major hurt." That will be when the 136 days at sea fishing for cod and haddock and yellowtail which the *Honi-Do* will be allowed during the first regulatory year will shrink to 88. Carl estimates that of the 180 days or so that the *Honi-Do* works now in an average year, anywhere from 120 to 130 are devoted to those species. So this summer it will be business as usual, for the most part. But afterward? "And then there's that reduced quota, that total allowable catch," he says. "They haven't really defined what they're going to do once that's reached." With a boat that's largely paid for, Mark is in better shape than a lot of dragger skippers. Maybe Mark can afford more time ashore than he already takes, and maybe not. In either event, Carl can't.

Nor can Mike, who as a boat owner is carrying a lot more debt than Carl. At the beginning of May, he hauled his boat out again for repairs: cleaning, painting, a new rudder, a new propeller, and a reconditioning for the worn-out sheaves on his hauler. The *Susan Lee* was back in the water within five days and handling like a new boat. But with cod prices as low as they were and dogs nowhere to be found, Mike was lucky to break even if he went out. "All ahead stop," he said as he described the low yields and the rock-bottom prices of what was shaping up to be his leanest spring in the last five years.

All ahead stop on other fronts as well. The deal on the bait shanty fell through when the building's current tenant exercised his right of first refusal and bought the shanty in April. So Mike will just have to go on paying rent and putting up with the balky freezer at the shanty he's been using. The good news in April was that he found an experienced crewman who was ready to work with him through the summer. Twenty-four-year-old Sean St. Pierre went out with Mike once that month, but died only a few days later when his truck rolled over at an intersection in Harwich. Mike continues to fish alone, his ass dragging.

I last talked to Mike at the beginning of this month. Commerce Secretary Mickey Kantor, appointed in April after the death of Ron Brown in an airplane crash in Bosnia, had just announced his approval of Amendment Seven, and the law was still on track for a July 1 implementation. Mike awaited a box of paperwork from the federal government, including the toll-free number he'll have to punch into his cell phone every time he leaves the harbor from July 1 on. In the meantime, with cod scarce and the dogs still missing, he and another Chatham long-liner were thinking of outflanking the competition with a run on that skipper's boat all the way out to the Northern Edge, on Georges Bank. "Nobody from Chatham's been out there for a couple of years," Mike said, "but I hear that's where the cod are now. The big draggers have really been whacking 'em out there. It's a ten-hour steam if we take it easy on the fuel. So we'd leave at six or seven P.M., then fish the next morning. And if you're not working inshore, then you don't have any of this territory bullshit about who's got a claim to work where. Out in the channel, out on Georges, no problem, everybody's pretty good. Guys get better the farther east you go. That's where it's at. So I don't know. I'd like to try it."

Dan Howes is content to be an observer and not a player in the travails of the groundfishery. Unfortunately, his sojourn in Maryland this winter also involved a lot of observing. He ended up with a site on the *I'm Alone,* a forty-eight-foot Chatham boat working for the winter out of Ocean City. But the big gill-netter hardly ever got out of the cove. "In a normal year you can fish five, six, seven days a week down there," said Dan. "But this year we just sat around and watched the wind blow. It was hard to put any days together. We averaged maybe eight or nine days per month. There were fish out there, but we couldn't get at 'em."

He came home in the middle of April and put the *Last Resort* back in the water at Sesuit Harbor. Since then he has been "living on the island," as he says, running his tows off Billingsgate. "It's been good. I've been getting about eighteen or twenty bags a day, about a third of that small stuff. There's no grass out there at all now, so you have to be right on top of it, knowing where that edge is."

Dan remains unfazed by the prospect of large numbers of refugees from Amendment Seven pouring into the wild quahog fishery. "I'm not going to show anybody anything out there," he says. "I hate to be like that, but nobody showed me anything. It's not cheap to rig up to do it right, and then you've got to work at it to learn where those edges are." He is convinced that the whole affair is just a charade, just workfare for bureaucrats and scientists. "It's all political bullshit. Amendment Seven is totally unnecessary. I guess scientists need the work. But all they had to do was close the spawning grounds, which they've done, and then just wait for the fish to come back, which they are. There are a lot more cod out there than they say there are. But cod's a fresh market item, and with Canada opening up their cod fisheries again, and all the pollack coming in from Alaska, the prices are going to be volatile. I'm glad I'm watching."

Brian told me that he wasn't catching anything this spring, and when the first two wire traps come up empty, he says he's glad, it proves he was right. "On the other hand," I tell him, "if there were lobsters in there, it'd prove you were even smarter than you thought you were."

"Well, that's almost impossible."

He drops the second trap back into the water and watches its tarred warp snake out over the transom of the *Cap'n Toby* as the boat steams north, against the tide, up his line of traps. "So much for salted porgies for bait," he says.

The seas are running one to two feet, building to three. "So far, so good," he said on the way out. "I thought it was going to be rougher out here. It's pretty lumpy, but there's no wind. The tide turns at one o'clock today. Might smooth out a little more then." The fog has unexpectedly dissipated as well, and the white cliffs of Eastham and Truro run like a chalk line over the western horizon. We're hauling in a hundred feet of water, which slaps against the sides and humps under the keel in glassy, sun-spangled escarpments. The boat rocks and shivers beneath our heels.

We haven't heard Mike Russo on the radio this morning, but the airwaves have been busy nonetheless. Brian's friend Bob Maraghy is out hauling traps in the *Severance* today, and we overhear a call Maraghy receives from the Coast Guard's Chatham station. "We had a request for an inquiry," the radio operator says. "Everything all right?"

"Everything's fine," answers a puzzled Maraghy. "This is the *Severance,* standing by on channel six."

Brian picks up his microphone as we close in on this string's third trap. "Hey, Bob, it's nice of your loan officer to be checking on you."

The third trap contains a starfish, a rock crab, and a dying mud hake, its belly heaved up. Brian keeps the hake for bait. The sun shines like a projector lamp through the head of a porgie in the trap's bait bag. The skin is translucent. Inside, sand fleas swarm through the canyons and tunnels of its skull.

The fourth trap is empty. The fifth has three lobsters: two shorts and one keeper, a one-pound chick. The sixth has three lobsters as well, but they're all eggers and are thrown back. The next four yield a smattering of shorts, chicks, and three small deuces, along with another mud hake and some cunners, a perchlike fish, with their eyes blown out of their sockets. These few lobsters are at least encouraging to Brian. "After those first four traps, I was thinking there were no lobsters here whatsoever," he says. But only slightly; when the next several traps offer just two keepers and a cull, Brian says, "This really is piss-poor."

Some of the traps come up with spiky gardens of green sea urchins sprouting oddly from their wire floors. Brian likes to use the factory-made wire traps out here, because they're a lot lighter than his wooden ones when they get waterlogged and they seem to fish better in deeper water. But he doesn't like their cull rates, which is to say the frequency with which lobsters seize the wire mesh while being removed and lose a claw when he tries to persuade them to let go. He struggles with one obstinate animal, which may or may not be a keeper, and finally the lobster comes loose. Its carapace is measured; it's a short, and he tosses it back over the side. By the time he finishes his second string, having cleaned and rebaited twenty traps, there are maybe a dozen legal-sized banded lobsters scratching in the tote at my feet.

"Can't really tell what the best thing to do is," Brian says as he takes a peanut butter sandwich and a Diet Pepsi from his cooler for lunch. "The bait might be bad, or it might be too early for lobsters this year in these shoal waters, or maybe there just aren't any more of 'em anywhere." He shuts off the engine, lets the boat drift. A sooty raft of storm petrels has collected in the boat's wake, bobbing in the chop. "I ought to move these traps next week. The problem is, I just don't know where."

Over lunch Brian says that the biggest lobster he ever caught, a fifteen-pounder, was taken in this area. "Harry Hunt never caught big lobsters like that, fishing down around Nantucket. He liked to keep his parlor heads tight, and the really big ones didn't seem so comfortable with that. But he made up the difference in quantity. Back in the seventies we commonly hauled up pots with fifteen to eighteen lobsters in them. We'd haul three hundred and fifty traps per day, get in late, go out again early, live on peanut butter and Ritz crackers."

An anniversary of sorts is on the horizon. Brian says that next summer will mark twenty-five years since he cast his lot with Harry, since he decided to make his living on the water, since the town and bays and beaches that had always been a place of retreat for his sports-minded grandfather and bookish mother — physically, spiritually, economically — became instead Brian's front line of engagement with all that Hunt claimed opposed him: the wind, the tide, the weather, and the envy and malice of every other man out there.

"The mistake I made as a fisherman," Brian says, "was taking on faith everything that Harry taught me. He fished until the age of seventy-nine, but his heyday was in the late thirties through to the sixties. He developed his strategies during those years as to how to bait, what kind of bait to use, where to fish, and so forth. He was successful, but it was a time when you had very little competition as compared to now. And he had such a strong personality that he threw a lot of guys off. He'd be down there at the landing every day telling 'em what to do and where to do it. One year Smitty — Steve Smith — bought a lot of new pots from Anderson's, and Harry told Smitty that the parlor heads were all fucked up but he'd tie 'em up right for him. So he did, and Smitty didn't catch shit with those pots. The next year Smitty bought a new batch of pots from Anderson's, and he only let Harry tie up a few of the heads on those. The rest he left as they were, and they consistently outfished Harry's pots. Harry also didn't like pots where the heads were nailed to the bottom runner. He insisted you needed a brick step-up into the head. If a different design was outfishing his own, he'd swear his traps were being robbed. Everybody said I'd learn a lot from Harry, and I did, sure, but that also meant that some of my decisions when I went out on my own were based on faulty premises. It's taken me a long time to figure that out. That, and I never had much money to really make a go of this business."

In accounting for the day's disappointing run so far, Brian is not currently considering the possibility that his traps have been pilfered. I ask about his suspicions last spring, and he allows that the robbing then probably wasn't chronic, as he first believed it to be, but just something that happened one day early that season, for whatever reason. Brian sits on an overturned bucket in front of the console of the *Toby,* sipping his Pepsi, haunted by ghosts, filled with the past. I see him now inside the pages of a Philip K. Dick novel, heading out on a highway at night in a stolen pickup in order to see what lies beyond the town limits, resolute on breaking through to whatever otherworldly reality lies behind the town's spurious facade. The suspicion settles on him that the truck is not moving at all, that the wheels are spinning uselessly in an illusion of movement. He contemplates stopping the truck to search on foot: "The hell with that, he thought. At least he was safe here in the truck. Something around him. Shell of metal. Dashboard before him, seat under him. Dials, wheel, foot-pedals, knobs. Better than the emptiness outside." Beyond the hood of his pickup, the darkness is that of the blind.

I think Brian regrets that he doesn't have now what Harry Hunt had: a hungry piss-and-vinegar sort of sternman such as he once was, to whom he could pass on his premises, faulty or otherwise. I suspect a part of him has always dreamed that his son, Mike, might be that sort of sternman, that his permit might one day pass from father to son, as it has with the Bergquists in

Chatham, and that the modern privilege and the ancient lore and the timeless beauty of lobstering would stay in the family. But Mike hasn't inclined that way. Brian feels helpless in communicating to him what he sees and feels out on the water, when everything's lying just like that and the lobsters are coming in two to three pounds per pot. These days he thinks it's probably just as well.

Earlier today he laughed harshly, saying, See, I'm not the only cynical bastard out here, on overhearing a radio exchange between two other lobstermen enduring terminator runs. "Well, this is a real profitable day," said the first. "I should have known better." "I still don't know any better," said the second.

Brian keeps on trying, out here and in the committee room, dreading the day when he might finally know better. Before Christmas he represented the Nauset Fishermen's Association as one of three speakers in a luncheon event organized by the Lower Cape Community Development Corporation. Other speakers were Bob Ahearn, the owner of a Dennisport fish market and restaurant, representing the Cape Cod Commercial Hook Fishermen's Association, and Andy Rosenberg, the Northeast regional director of NMFS. Brian looked forward to appearing on the same docket with Rosenberg, whom he respects, but Rosenberg failed to show. His place was taken by Terry Smith, the chairman of NMFS's research science center at Woods Hole.

The luncheon was at Orleans's Jailhouse Tavern with a group comprising, said Brian, "politicians, bureaucrats, and social-service types with degrees from Wellesley." Brian described what the small boats provide the Cape economically — two to five jobs for each of the Cape's four hundred groundfish permit-holders, an additional four to five jobs each in fish processing and wholesaling — and stressed how important it is that local youngsters can enter the industry. Restricted access or limited entry should be tools only for the rebuilding process, he said. Small boats run by local owner-operators, he added, have less of a negative impact on fish stocks, whatever the stock, and more of a positive impact on local economies than large boats. Bob Ahearn concurred, and expressed the CCHFA's enduring anger that long-liners and other hook fishermen, small-boat fishermen all, were not exempted from Amendment Seven. But Terry Smith countered that the port of Chatham, with its many small long-liners and gill-netters, accounted for a tenth of New England's 1993 cod harvest. With Amendment Seven allowing a total catch of only 4.1 million pounds of cod (versus the 32 million pounds taken in 1993), Smith felt that it was only right that the Cape's long-liners and small owner-operators of whatever persuasion take the hit along with everybody else.

There is no refuge ashore. The hits come from both directions as the gentrification that withered Long Island's fishing fleets continues apace on Cape Cod. In the Orleans library I came across a 1983 editorial column in the *Cape Codder* commenting on two recent events. First, a Harwich lobsterman was prevented from storing pots in his yard by his neighbors. Second, an Orleans resolution to build a pier for fishing boats at Snow Shore was bled to death by a lawsuit filed by abutting landowners. The editorial noted that "there have always been those who have moved here expecting Cape Cod to conform neatly with the values they brought from more suburban or urban areas. Usually they have been disappointed. But rapid development has attracted more people whose concern for the Cape and its character extends no farther than their condominium or subdivision lawn, whose interest in the community stops at their tax bill." The succeeding years have brought even less disappointment for those who prefer wine to brine, and have not only enriched the coffers but exacerbated the myopia of those assessors and developers who, as Robert Finch puts it, "can no longer see the forest for the fees."

What Brian remembers of forest is gone: the woodlots, the silent sandy roads, the long beach plum entanglements overlooking Rock Harbor. What was a place of retreat for two previous generations of Wiggins and Gibbonses has indeed become the

a complaint
're buyin', it's
wich's plastic
oth hands, the
t sheen of the
rd window as
ually from the
cross another
g against the
ial sweetener,
ike forearms,
ven now as a
d metabolism.
arteries of the
on to Brian's
ure, all he sees

f he had to do
an concluded
opinion that
Barbara Bush
gh the disclo-
late the terms
f you're really
d after Hagen
in the murky

es the bucket
de has turned,
sky over the
g on high to a
deep as time.
bears south,
ugh the water
n either side,

quering civilization —
d importunities swal-
an West. Thoreau said
put all America behind
nce, seeking in particu-
rely of the soil on the
arbor, where Dr. Ralph
s surrounding Pleasant
lusive access to water
narguable assertion of
imited entry. The food
eir wine, is gotten else-
ge. At the same time,
s of the bridge, food is
ial commodity distinct
t also from the oceanic
hunters, farmers, and
Fishermen watch with
as People for the Ethi-
erating" lobsters from
l, New York, to change

ies *News* titled "Family
portunities" no doubt

ndividuals. They are hard
lo," said Paula Roderick,
lford Fishermen's Family
that when her fisherman
er choices, most succeed
," she said. Some prime
en Odd Johanssen and Al
m the Bay State School of
ork in the air condition-

A character in a Walter Scott novel, receiv
about the cost of some fish, replies, "It's no fish
men's lives." Brian sits on his bucket with his s
wrap wadded at his feet, his Diet Pepsi cradled in
radio murmuring behind him, the cellophane-br
sea winking in and out of the wheelhouse's start
the boat softly rocks. In the silence we hear occa
tote near the pegging board the scraping of a cla
lobster's carapace, the scratching of a pincere
tote's plastic. Brian likes the Diet Pepsi for its art
its lack of sugar. In his knotty hands, his keels
the blood vessels are hardening and narrowing
long-term degenerative effect of his body's skev
Particularly susceptible to diabetes are the hairli
eyes and kidneys. This provides a literal dime
admission to me once that when he looks at the f
is a black hole.

And yet, almost in the same breath, he said tha
it all over again, he would do much the same.
the short autobiography he wrote for me with
really his life was pretty humdrum, that the lives
and Shirley MacLaine make a better read, altho
sure of certain elements of his own story would
of the Federal Witness Protection Program. "Bu
interested," he wrote, "I'll tell you what happe
sequestered Siegfried's treasure-trove by sinking
Rhine at Worms."

He stands stiffly, stretches out the years, sl
against the bulkhead, and goes to the wheel. The
and the seas, finally, are lying just like that. T
eastern horizon is the color of skim milk, deepen
pale and glaucous blue, webbed with clouds and
The engine fires and hums, and the *Cap'n To*
climbing to its skeg and clipping so cleanly thr
that it seems as though it were standing still.

unsequestered, glinting in commonwealth from the crest of each wavelet, is spread the treasure-trove of Siegfried's gold.

The fatal flaw of the 1976 Magnuson Act was that it was divided in its soul. Returning to the site of the nation's first treasure-trove and surveying its depletion, the law sought to restore that treasure by banishing the big foreign ships from Georges Bank and extending the length of America's commons to two hundred miles out to sea. But in envying the technical efficiency of those great ships and the wealth that they took from Georges, the law left itself incapable of defending from U.S. citizens what it had taken from other nations.

Throughout the 1980s the Magnuson Act was routinely reauthorized by Congress, while those scientists, conservationists, and fishermen who saw disaster in the offing found themselves helpless to arrest that momentum. The first real stab of the brakes on this process, the Conservation Law Foundation's 1991 suit against the federal government, was followed by a gloomy 1992 NMFS report on the status of marine resources. "The good news is that the act has fulfilled its economic goals through the Americanization of our fisheries," commented Gerry Studds. "The bad news is that the act has not been nearly as successful in ensuring the sound management and conservation of our fisheries." When Magnuson came up for reauthorization again in 1993, Congress had to go back to work on it, and the direction of American fisheries management into the next century was once again — and still is, as of this day in June 1996 — up for grabs.

Brian says that as early as the 1960s, oil companies were prospecting around Georges Bank and in the Great South Channel. He heard stories of wonder-struck Chatham fishermen steaming through long stretches of water carpeted with belly-up cod and haddock, all killed by geologists' depth charges, all finally gaffed into those fishing boats. The sale of oil leases on Georges Bank would have been squarely in the federal tradition of the private disposal of public resources, but an unprecedented alliance of fishermen and conservationists in the late 1970s, along with protests from the Commonwealth of Massachusetts and a lawsuit by the Conservation Law Foundation, succeeded in heading off such sales. Those strange bedfellows, fishermen and conservationists, separated the next morning, no doubt scrubbing themselves with lye as they did so.

But in the latter part of these Magnuson reauthorization hearings, these old playmates — or at least some fishermen, some conservationists — have found each other again. One-time lobsterman Jim O'Malley, the director of the East Coast Fisheries Federation and now also a New England Fisheries Management Council member, is glad he had a candle in the window. "Nobody gave a damn about what we were saying until Greenpeace showed up in 1994," he said. "That was a watershed moment for us." Greenpeace put fishermen — small-boat owner-operators, most of them from the Pacific Northwest — in front of congressmen who were hopeful this time around to ensure sound management. For their part, the fishermen found that their words carried further when amplified by a congressional microphone and Greenpeace's media-savvy lobbying tactics.

What has been strange for O'Malley and other New Englanders, however, is to find themselves deployed out on the flanks, not in the center, of this particular battle. Most of the amendments to the law contained in the bill approved (overwhelmingly, 387–37) by the House of Representatives last October represented ordained outcomes to battles that the fishing industry could not have reversed even if they had wished to fight: requirements for each regional council to define overfishing and remedy it; a requirement for the secretary of commerce to do so in the breach; requirements for measures minimizing bycatch, protecting habitat, and managing species on more of an ecosystem basis; and a replacement of the phrase "optimum yield," including its consideration of the social and economic consequences of harvest limits, with the phrase "maximum sustainable yield," which considers only the welfare of the fish stocks. O'Malley knew that the amendments on bycatch and habitat could turn into clubs to use against the dragger fleet, that "maximum sustainable yield" meant that now NMFS, not the councils, would be making the bottom-line decisions on what fishermen could catch. That was too bad, but these were measures that Greenpeace and other environmentalists supported. Since the Magnuson reauthorization was the only important piece of environmental legislation addressed by the 104th Congress, and since the fishing industry by itself had little political clout anymore, neither Democrats nor Republicans were willing to bend on these issues.

O'Malley is more interested in the work of the Devil, who has lately been walking up and down the Gulf of Alaska rather than Georges Bank. Georges is walled off and nearly bankrupt; meanwhile, the New England council, fortified by O'Malley's Manichaean vision on this issue, continues successfully to resist NMFS's urging to implement ITQs in its groundfisheries. But there is still plenty of pollack off Alaska, along with a fleet of sixty Seattle-based factory trawlers, ships that harvested their entire offshore quota of pollack in just eighty-five days in 1993. The owners of those ships have to keep them working in order to service their debts and want to see the ITQ systems used in other North Pacific fisheries extended to the pollack fishery in particular, to all American fisheries in general. The purchase of ITQ shares, then, in a restored Georges Bank, say, would relieve trawler congestion off Alaska and bring the benefits of those ships' efficiencies to other sectors of the seafood market.

A war was in the offing, but first the generals had to be chosen. The old order started to crumble with the Republicans' 1994 election-day victories in the House and Senate. Then it became quite apparent where the money and the power now lay in American fisheries. The Republican House majority immediately dismantled Gerry Studds's Merchant Marine and Fisheries Committee and handed over its business to a refurbished Public Lands and Resources Committee, chaired by Don Young, of Alaska.

Later Studds, who had served twelve consecutive terms in the House, learned Portuguese in order to communicate better with his New Bedford constituents, and pioneered the first alliance between environmentalists and fishermen, announced he would not seek reelection. Similarly, Massachusetts Senator John Kerry saw his chairmanship of the National Ocean Policy Study Group disappear. That group became instead the Subcommittee on Oceans and Fisheries, under the leadership of Alaska's Ted Stevens.

Studds's chief contribution to the Magnuson reauthorization, the $25 million vessel buyout program for the Northeast, survived, but with the loss of those chairmanships, the New England representatives have largely become spectators to a high-seas showdown between the Washington delegation, which has the ships, and the Alaska delegation, which has the fish, and its outcome will have much to say about the fate of New England waters. The House version of the bill is actually something that a commonist such as Brian could love, thanks to a pair of amendments attached last year by Democrat George Miller, of California. The first prohibits the transfer of quota shares in any subsequent quota-based management programs, and requires that those holding shares not be absentee corporate owners but actually work on their boats. The second redefines the law's use of the word *efficiency* to mean not massive fishing and processing power but rather "fishing which yields the greatest economic value of the fishery with the minimum practicable amount of bycatch and provides the maximum economic opportunity for, and participation of, local community-based fleets and the coastal communities which those fleets support." Don Young, whose Alaskan constituents feel like Yankee fishermen once did when staring up at those trawlers, voiced strong support of the Miller amendments, as did Gerry Studds.

But not all fishermen were happy with that: certainly not the Seattle trawler fleet, nor other West Coast fishermen already vested in ITQ programs and hoping to expand into more. Others resented and feared the role of the environmental lobby in shaping the bill, though conservationists were not unanimously behind it. Both the Environmental Defense Fund and the Center for Marine Conservation support ITQs as tools to reduce bycatch and overfishing. But Gerry Leape, the director of Greenpeace's fisheries campaign, was jubilant with the House vote on the bill: "There's been a turn in the road here. Congress is shifting the Magnuson law away from developing the fishing fleet, as it did in the 1970s, to creating a sustainable fishery, and therefore a sustainable fishing industry."

But this year, as the Senate prepares its own version of the bill, a counterattack has been mounted by Republican Senator Slade Gorton, of Washington. "There is, I believe, a perception that an attack on efficiency is a triumph for small vessels and a blow to what are perceived to be the larger, more cost-effective vessels such as those in Washington's factory trawler fleet," Gorton said. "This perception reveals a disturbing trend toward unfairly demonizing more productive, more efficient fleets."

Ted Stevens has crafted the bill's Senate version, now known as the Sustainable Fisheries Act, to be even more galling to Gorton than the House version. In addition to the Miller amendments, Stevens is pushing for a five-year moratorium on new ITQ programs while the federal government studies the impact of existing ones. Alaskan small-boat operators were not reassured, for example, by the results of an ITQ program instituted early this year in the North Pacific halibut and black cod fishery. There, $100 million worth of quota shares were conferred on only forty vessel-owners. The Seattle trawlers would command a similar lion's share of the pollack fishery, and the Alaskans are now attacking the big ships at their most vulnerable point, their volume of bycatch, pointing out that in 1994 trawlers off Alaska discarded 580 million pounds of fish too small or of the wrong type for their processing machines: groundfish, salmon, halibut, herring, and crab. This was three times the combined bycatch of two thousand other boats that fished groundfish that year. "We hope," said Stevens, "that this bill will bring an end to the inexcusable amount of waste that is occurring in the fisheries off Alaska and in other parts of the country."

Slade Gorton has fought this moratorium on every front, though in a March Senate Commerce Committee markup of the bill, he could do no better than whittle the moratorium down to four and a half years. But Stevens may have overplayed his hand. Gorton is threatening a filibuster against any attempt to bring the bill to the Senate floor for debate this summer. Stevens believes he has more than the necessary sixty signatures to override the filibuster, but with the Senate recessing for August and then adjourning at the end of October, Gorton's threat alone may be enough to stall the bill during the Senate's hectic last weeks. If the bill fails to pass this summer, it could take up to two years to get it to the Senate floor again. And if that happens, the Magnuson Act would probably be reauthorized again in its present form. Gerry Studds would lose his buyout program, regional councils would continue to grapple painfully with the social and economic consequences of harvest restrictions, and ITQs would remain a management-style presence in any fishery.

Also lost, at least for that time, would be any opportunity to fulfill the original promise of the Magnuson Act: finding the means, finally, after centuries of boom and bust on the seas and in the bays, after a national lifetime of accelerating growth and consumption, to align those two disparate graphs into the sort of smoothed-out stasis that promises sustainability; the means, in other words, to live on this earth, to draw both fruit and solace from its wilderness.

Jim O'Malley, from his front-row seat, compares the events of the spring to a debate on the giveaway of the Grand Canyon. Last month, meanwhile, Greenpeace activists wearing masks in the likenesses of Slade Gorton, Washington Senator Patty Murray, and Tyson Foods president Don Tyson appeared on the Senate lawn. With Gorton and Murray brandishing fistfuls of dollars, the Washington delegation joined Tyson in a bed that had been set up on the lawn in view of the Capitol Building. In place of satin sheets, those enticements to sin, the bed was strewn with fishnets.

Voices are in the air. From somewhere over the horizon, a Coast Guard cutter is hailing the *Weymouth,* a boat that Brian knows. He says it's an old eighty-five-foot eastern-rig scalloper. "Where are you from?" the Coast Guard asks.

"New Bedford."

"How much do you have aboard right now?"

"About four thousand pounds."

"How many in your crew?"

"Seven."

"When will you be in port again?"

"Probably Saturday."

"May we come aboard?"

"Help yourself."

Brian says that this is probably a routine boarding. The *Cap'n Toby* is only three fourths of a mile off the beach now, racing to a line of traps sunk in forty feet of water off Marconi's radio station. The terns are back, nesting in droves on Monomoy and Tern Island and Nauset Beach, and a flock of them are pinwheeling over a stretch of water between the boat and the beach. The air streaming by the wheelhouse sparkles like champagne, like the first bright day of spring, though this is a quality that the air on Cape Cod retains until deep in the summer. To the east the flags of other lobstermen's trap lines jab the horizon like pennants around a stadium's rim.

The day after tomorrow is the Eve of Saint John, Midsummer Night's Eve, on old calendars. According to Cape fable, that is the night on which all the lost ships rise to the surface to resume their interrupted journeys: the *Sparrowhawk,* bound for Virginia but aground off Eastham in 1623, repaired by William Bradford, who learned of the wreck from Nauset Indian couriers, but wrecked for good by a second storm before she could sail; the pirate ship *Whydah*, aground with 145 lost off Wellfleet in 1717, whose drowned captain, Black Sam Bellamy, pronounced that he preferred to rob the rich under the cover of his own courage than the poor under the cover of the law; the *Confidence*, skippered by Isaiah Knowles, whose vessel floated into Cape Cod Bay on her beam ends in 1806 without a living soul aboard; the *Pomona*, lost with six other Truro vessels on Georges Bank in the October gale of 1841, which finally drifted bottom-up into Nauset Harbor with three drowned boys in her cabin; the *A. Roger Hickey*, a two-masted motor-auxiliary fishing schooner aground in 1927 near this spot Brian is working now, her distress described by Henry Beston during his winter on Nauset Beach; the S.S. *James Longstreet*, bound for Europe with an arsenal in the hold to throw against Hitler; the *Patricia Marie*, with Cap'n Crunch, Billy King, still at the wheel, Dicky Oldenquist at his side, and her decks still loaded to the scuppers with scallops; the *Banshee*, cut in half by the *Hawkeye* in the fog off Monomoy last summer; and thousands more, from a sixty-four-gun British man-of-war to dories pitchpoled in the surf. Their timbers stir even now as they begin, with movement as slow as the hands of a clock, to wrest themselves free of the mud and sand. They throw off bubbles like pearls with each increment gained toward the surface.

Brian is fishing his wooden half-round traps here in five fathoms of water, and so far these are fishing better than his wire traps offshore. One trap comes up bursting with five good-sized keepers, a deuce and four chicks: "Now if every trap was like that . . ." He says he still wishes he had some bluefish for bait, and with that, time slips out of joint, like an elevator moving between floors, and the cliffs of Wellfleet entirely dissolve. "How come there aren't any blues?" Brian asks.

He stands at the bed of his pickup at the top of the dock ramps at Rock Harbor. Two totes of fresh striper racks are in the bed. A jowled and graying man in a Goose Hummock baseball cap, a charter-boat captain, has his head out the window of a pickup pulled off to the side of the street across from Brian. A sluggish southwest breeze is moving like hot engine oil between the black grit of the parking lot and the aluminum hardpan of the cloud cover overhead. Thunder rumbles from over the horizon, down toward Chatham.

The charter captain says the blues are too far up, near Truro now; he caught all his bass near Brewster.

"Well, nobody's happy unless he's got something to bitch about," says Brian. "So I guess I'm happy."

The charter captain is happy too. "Look, I don't want to make trouble, but one of your competitors brought me a load of dead lobsters the other day. He's just biting the hand that feeds him. I don't want that shit."

"Maybe he didn't know they were dead. His sternman might've just left 'em in the tank too long."

"Well, I didn't like it. Maybe you can let him know that."

"Sorry I haven't brought you more lobsters myself. It's been a bad season so far."

"That's all right. Hey, let's you and me go fishing down around Monomoy the next time it's foggy."

"Yeah. We can wait there for the fast boats to go by."

"You hear that one guy had his leg cut off? The propeller got him. Boy, that fella on the *Hawkeye* better have some deep pockets. They're gonna rake him over the coals. You think he ever checked his radar?"

"I don't know. Sometimes down there around the rip you can lose boats in the sea clutter."

"Yeah. And you got that high prow on those Hatteras boats. You get a little speed and you really get up in the water. Then you can't see a damned thing ahead of you."

The charter captain waves, pulls into traffic, getting a little speed, and then Dan Howes's pickup pulls into the space he vacated. Brian asks Dan if he's going out. "This afternoon," Dan says. "Just the high tide. I can do as well working just the high tide as I can the whole day."

"Well, don't get struck by lightning."

"No, I'll stay out near Andy's boat. He's got that nice tall mast."

By the time we get down to Snow Shore, the clouds have turned to slicked carbon. Nick Leroy, the clam digger, is standing on the beach and staring across the harbor at a pair of other diggers still working the exposed bar. "Those guys have either got balls or no brains," he says. At that instant a bolt of lightning snakes down from the clouds behind the bar, and a clap of thunder comes rolling in a wave off the ocean. Both diggers sprint with their rakes to their skiffs as the first drops of rain pock the flats.

We sit in the cab of the pickup, waiting to go out, while a fine, fulminating thunderstorm crackles around us. Brian worries about getting his traps hauled and his ass back into the harbor before the tide has run out too far. He mentions that his son, who has been working on the *Blue Heron,* is scheduled to be the guide on a trip around the area this afternoon. "I wonder if that'll get canceled," he muses.

Twenty minutes later the sky in the west is breaking up into laundered sheets of blue and gray and white. Nick says the fresh water is sure to keep the clams deep in their beds, but he's going out to scratch some anyway, and he gives Brian a ride out to the skiff. We load the *Cap'n Toby,* fill the bait bags with a mixture of bass from Rock Harbor, cod from Esther at the Nauset Fish & Lobster Pool, and sand dabs from Chatham Fish & Lobster. Then we steer out of the harbor through sandbars already starting to dry and harden beneath the midsummer sun.

Outside, the tide is running north with the wind. The waves are smooth and low and dappled with light, the clouds soaring in pale flying buttresses beneath a blue cathedral sky. We head north to Wellfleet, out to the Can, with the cliff-abutted beaches off our port side already starting to fill with sunbathers.

The lobsters come up in bunches or not at all; come up, when they do, in skittering, angry pods: shorts, chicks, one-clawed culls, clawless pistols, stone-clawed deuces, and eggers — green eggers and brown eggers and blond eggers, the whole color wheel of eggs as they mature through the year in their grapelike clusters on the females' pleopods, and spent eggers, with just a smattering of leftover caviar clinging to the pleopods' fine hairs. "I like seeing all these eggers," Brian ventures, "but I'd rather handle money." The fifth trap of the first string comes up with a green egger, a cull, and a half-eaten chick, cannibalized by the egger. "I hope that's not a sign of things to come." The wooden traps, both square and half-round, rise slimy with red beard sponge, their bait bags maggoty with sand fleas. A prodigious six-pounder has claws that are too big for the elastic bands, that have to be pegged instead. It straddles the pegging board like a crucifix. It raises its pegged claws as if displaying stigmata.

There are two other keepers in the trap with the six-pounder. Brian says, "I caught these with blues, and now I'm rebaiting with cod. Entropy." A lath is loose on the top of one trap, and it has to be renailed. A parlor head has come unstrung in another and has to be retied. Repairs are made hastily, with the clock ticking, the tide running.

The lobsters are beautiful and monstrous, sunstruck and sea-slicked. Their claws and carapaces are mottled and marbled, green and orange and blue and calico, camouflage and carotenoid, their colors layered in rock and weed and light, their margins peppered with dusk and darkness. Their claws, crusher and cutter, lift and open as if inviting an embrace. One keeper, a chick, has had one eye stalk sheared off, its socket cleaved. It's an old wound, and now a half-grown mussel is sprouting like a fly's eye from the canyon in the skull. The one-eyed lobster has sunk the teeth of its cutting claw into the mesh of a bait bag and won't let go.

Brian finds four lobsters in the next trap, all of them shorts: "Short city." The tide has turned, and the traps feel like anvils at the end of their warps. "There are a lot of small lobsters here, a whole herd of 'em." Rock crabs go skittering down the starboard gunnel. Herring gulls hang suspended in the freshening breeze, like wind chimes above the boat. One gull has landed on the transom and is regarding the shack box, with its stripped fish skulls and empty racks, with an investor's speculative eye. Brian holds a squirming keeper up against the sky: "That's a prime Nauset lobster — two or three pounds, a good hard shell full of meat. That's the best eating lobster in the world." A snapping five-pounder comes up in the next trap, the last in the string, which has a rack of new laths across its top and is the trap that Brian believes was maliciously stove in and robbed this spring. "Eleven lobsters for ten traps. That's not bad, these days."

The wind is up to fifteen knots and working against the tide, kicking up the seas like a wood plane scrubbing against the grain. An empty fish tote blows overboard, tumbling over the gunnel, then floats belly-down and is retrieved with a boat hook. The *Cap'n Toby* is working strings tighter to the shore now and is in the lee of the Wellfleet cliffs. "We wouldn't be out here if the wind was from the northeast," Brian says. "Two shorts. That's more like it. Typical day." The next keeper is a pistol, missing both claws. The regenerating limbs are candy-red and thumbed, like tiny mittens. "I used to be able to jig in here for cod in the winter, catch five hundred pounds a day. You could even surf-cast for them, using clams and quahogs for bait. Lotta 'used to' in this business." The next trap has four lobsters in it. "Three keepers. Wahoo!" The next one, the ninth in the twelfth string Brian has hauled, drops to the gunnel with four keepers inside. "If the fishing was like this all day, we'd be rich. But it wasn't, and we're not." The rebaited trap slides over the side. The warp uncoils from a pile at Brian's boots and disappears over the transom like the last whisper of regret, like the final few notes of a horn solo.

Twelve strings are all that Brian has time for. The rest will have to wait for another day if he hopes to get into the harbor before the tide is out. The sky is scrubbed clean, except for some clouds spread like bunting in the south, toward Chatham and Nantucket. A glistening shard of rainbow hangs suspended in the bow spray of the *Cap'n Toby.* The water lies easy in Cape Cod

Bay, where Dan Howes no doubt is living on the island again. At the four-lane bottleneck of the Bourne Bridge, the weekend traffic has slowed to a standstill, but herring gulls soar easily overhead. They float up on thermals thrown off by the pavement, viewing from above the Mount Pisgah of the bridge's stanchions the western prospect of that more goodly land dreamed by Bradford and his party from their landfall on Cape Cod. Then the gulls plummet down into the canal and scatter, scavenging, into the bay.

In the harbor Bob Maraghy's *Severance* and Steve Smith's *Murrelet* are resting at their moorings beneath Nauset Heights. Brian ties up and checks my sorting of the day's catch into the stern holding tank, deuces into the starboard side of the tank's partition, chicks into the port. The banded lobsters rest in layers in the tank. They move with a drowned nickering of pincer and carapace as Brian's gloved hands search among them.

The *Blue Heron,* with young Mike Gibbons in the bow and on the microphone, is moving into the harbor at just this moment, approaching the *Cap'n Toby.* The bargelike party boat is three-quarters full, its passengers in sandals and cotton shorts and sunglasses. They rise from their seats and stand beneath its sunroof with cameras lifted to their faces. Brian has plucked the six-pound deuce out of the holding tank and is proffering it like a talisman to his son. He stands in the stern amid the scissoring of camera shutters, the pleased murmuring of strangers. The light from over the heights, cascading in bolts out of the west, falls around him like glory.

Epilogue

My work with Carl Johnston, Dan Howes, and Mike Russo ended in May 1996. I continued with Brian for another three weeks, and it was on June 21, the same day that I last hauled traps with Brian, that Mike Russo lost the *Susan Lee.*

Mike had been fishing off Peaked Hill, on the north shore of Provincetown. He and some other Chatham long-liners had found a pocket of warmer water along that stretch of the Cape — fifty-eight degrees, as opposed to temperatures elsewhere in the low fifties — and he had been rewarded that day with a bonanza load of dogfish, more than five thousand pounds. He was steaming home at eleven in the morning at the end of his sixth straight day on the water, running a swift nine knots past the tall dunes of Truro, steering the *Susan Lee* inside the lobster traps off the beach. The tide was running north, the seas light at two to three feet, and the wind out of the southeast at fifteen knots.

"There were five of us, me and four other boats, in a wagon train going back to Chatham," he told me later. "I was second in line. I had my times down good for when I'd make the Chatham bar, and I was thinking of just one thing — staying in line and not losing my berth at the dock. Otherwise I would've had to wait an extra hour to unload."

Once inside that lobster gear, however, he noticed that his wash-down hose had dropped into the boat and was draining water into his cockpit and then out the scuppers instead of over the rail. It was probably no big thing, but he didn't want any extra weight in the cockpit with all the fish he already had there, and as a matter of seamanship, he didn't like the idea of an unsecured hose back there anyway. He put the *Susan Lee* on automatic pilot, bearing 160 degrees south-southeast, and fought his way aft over dogfish heaped waist-high in the bins behind the wheelhouse. He secured the hose, contemplated traversing that mountain of fish again, and decided instead to go around, climbing forward along the starboard rail.

"It was poor judgment, strictly the result of fatigue," Mike said. "I'd hauled thirty-six thousand pounds of dogs in just that past week, and my whole priority that day was getting unloaded on time. I didn't want to stop, and I never gave a thought to the risk factor. I was just too tired."

The risk factor clicked like a cartridge into the barrel chamber of the fate factor. Mike mounted the rail and began clambering forward. The boat rolled unexpectedly to port, and he lost his grip. He tried to grab the rail again on his way down, is thankful now that he didn't hit his head on it. In his amazement he hardly felt the icy shock of the water as it closed around him, as it smothered his forward momentum. He broke back to the surface to behold the dwindling transom of the *Susan Lee* as the boat sped away from him. His first thought? "You gotta be kidding me."

The beach was about two hundred yards away, and at first, once the boat was gone and it was just him and the waves, the sudden silence and his blank incredulity, Mike was able to keep the panic at bay. He always wore his boots loose; that day he had on his winter boots without the thick socks he usually wore with that pair. He kicked them off easily, moving as if in a dream, and they dropped like stones to the bottom, five fathoms down. He felt himself foundering in his oilskin pants, but discovered that as long as he didn't kick, if he just held his legs rigid as he did the breaststroke, then he could get some measure of buoyancy out of his legs. "There's the beach," he told himself. "It's time to get going." He stretched out on the surface, his arms reaching for the beach. The rolling water that intervened between him and dry land stretched away like an unabbreviated lifetime.

Mike harbors no illusions about his meager abilities as a swimmer. He was surprised nonetheless at how quickly he tired as he stroked for the beach. He didn't know if it was just the cold or the effort of hauling his legs as dead weight behind him. He had been swimming for about fifteen minutes when the next boat in the wagon train home, the *Never Enough,* skippered by Bruce Kaminski, came running up the beach. Mike let his legs drop beneath him, his body going vertical in the water, as he waved and hollered. But Kaminski was busy with his crewman, scrubbing his boat down. The vessel flew by, its occupants oblivious of the man in the water.

Mike watched the *Never Enough* fade into the distance and then felt his legs, still wrapped in their oilskin pants, heavy beneath him. He strained to raise them back to the surface so he could float again, felt them instead pulling him down to the bottom like ballast. That was when the panic started to build, when the chill of his own mortality began to grow like ice crystals in his bloodstream. He fumbled at the clips on his shoulder straps, contrived at last to undo them, and finally, with an exhausting sort of swimming, wriggling, shimmying motion, managed to fight free of his oilskins.

He was surprised at how much lighter he felt, how much faster he could move with his legs kicking behind him, and then no less surprised at how quickly he tired again as he plowed his way through the chop. When Terry Picard in the *Wendy-Jean* (the same boat that the disabled *Honi-Do* dinged last February when it was towed into Aunt Lydia's Cove) came running up the beach, Mike dropped his legs beneath him again, waved and hollered once more. "But he had all his doors closed. No way he could see or hear me," he said. The *Wendy-Jean* disappeared over the same narrow horizon that had swallowed the *Susan Lee* and then the *Never Enough.*

Mike's legs had gone leaden again, only this time there was nothing he could shed to lighten them. Fatigue, cold, disappointment, despair, and the rising shrillness of his panic: they clung to him like a second set of waterlogged oilskins, like concrete overshoes, and he felt himself swallowing seawater, felt the waves closing over his head like a trap door as he went under. Mike remembers opening his eyes as he sank, looking around at the green darkness when he was four or five feet beneath the surface. He remembers his next thought, remembers seeing in his mind the faces of Susan and Abigail, and recalls a statement of policy taking shape there in the soft void illuminated by those faces: "Fuck this, I ain't dying."

Anger and love drove him from the darkness, lifted him back to the surface, and then put the panic back in its box. Or maybe it was simply his wedding ring, something he became conscious of only after he'd broken back to the top: "I had this feeling somehow like my wedding ring was about to slip off my finger, and that I had to keep it on or else I'd drown. As long as I had it on, though, I'd be all right."

He was all right at that moment; he felt an unexpected sense of presence and well-being as he broke through the surface. "I just kept thinking about my family and focusing on my breathing and started hauling for the beach again. I couldn't kick anymore, but if I kept my ankles crossed, I could keep my legs floating on the top okay." He got a clear — and exultant — view of the beach from the crest of a wave: "Holy shit, I'm making it. I'm getting somewhere." He breathed and reached, breathed and reached, not panicking, not hurrying; he had, after all, the rest of his life to get there.

But finally the stretch to the beach became a race with hypothermia, with the headache that went off in his head like a concussion grenade as the heat drained out of his core. Then he had to fight the waves' backwash as he neared the shore. "When my toes touched that gravel," he said, "they just wrapped themselves around the sand down there. They never felt anything so good." He fought clear of the surf and collapsed under a line of the highest dunes in Truro. Two miles to the south, at Wellfleet's Newcomb Hollow Beach, sunbathers looked like ants. He had been swimming for an hour and his body temperature was down to ninety-two degrees. He lay shuddering on the gravel, sand fleas leaping about him as though he were carrion, his teeth rattling in his skull with the seismic bursts of his headache and that shuddering.

He got up in order to tell people about his boat, to start working on getting her back. He saw the last long-liner in the train, Mike Anderson's *Bad Dog,* approach from the north, also running inside the lobster traps. Mike tried to raise his arms but found he could barely lift them. He saw Anderson turn and look at him and then continue south. (Later Anderson told Mike, "I was wondering who the hell that nut was on the beach.") He trudged on, was brought up short by a great bird he'd never seen before around the Cape — a bald eagle, standing on the sand as it tore up the carcass of a seagull. He had a vision of the eagle tearing at his own carcass, preparing a way for the sand fleas, and he gave it a wide berth as he staggered toward the sunbathers in Wellfleet.

After the accident, Mike had to give up his bait shanty and let his crew of baiters go. Later he needed to haul some seed quahogs out to his grant in Pleasant Bay. A cold sweat broke out on his face as he approached his skiff that day. He forced himself into the boat, moving as though his feet were on thin ice, as though he might at any instant plunge through the skiff's bottom and into the cold water again. Once he was inside and still dry, he was all right and could make the trip.

For a few weeks in July he ran somebody else's boat, a gill-netter that had been geared for long-lining instead. Every morning before leaving dock he called NMFS's 800 number, gave a bored-sounding operator the owner's name and identification number, the boat's port of call and destination, and copied down the eleven-digit number encoding the month, day, and time of his departure, which he was required to have at hand in the event of a Coast Guard boarding. Mike hated doing that, but not as much as he hated running somebody else's boat for a crewman's wages. Prices were low, and it was probably too soon for him to be behind the wheel of any sort of boat out of sight of land. "I'd just lost my nerve," he told me. "I had the heebie-jeebies out there all the time."

He believes now that while he wasn't fated to die that day — or at least he and the fate factor reached an accommodation in that respect — he was simply fated not to recover his boat. "Clusterfuck," Mike said, considering all the things that went wrong. Because of her automatic pilot, the *Susan Lee* sped by other boats as if her skipper were at the helm, and Mike was thwarted on every front in his efforts to alert other fishermen otherwise or get salvage help. The first words out of his mouth that day to the team of Wellfleet paramedics who packed him into an ambulance were the *Susan Lee*'s speed and bearing. He told them to relay the news to the Coast Guard, along with instructions to broadcast a bulletin on VHF channel six, which is used by Chatham fishermen to communicate between boats and to make bar reports to the Coast Guard.

But the Chatham station had just been downsized, and the Coast Guard station at Woods Hole was then in charge of communications for that part of the Cape. According to the Coast Guard's procedures manual, channel sixteen is the frequency to use for vessel emergencies, though few Chatham boats monitor that channel. Woods Hole went by the book, and the bulletin about an unmanned boat steaming south-southeast went out into radio silence. Brian and I heard nothing that day aboard the *Cap'n Toby.* Nor did Woods Hole call the Chatham harbormaster, who would also have been able to mobilize fishermen in a search for the vessel. Finally, the Chatham station's inflatable rescue boat — the same craft that had been sent out for the *Honi-Do* in February, which was dispatched that day to intercept the *Susan Lee* — broke down at sea and required rescuing itself.

Meanwhile the *Susan Lee* nearly swamped a small boat coming into Chatham from jigging cod; the skipper of that craft said he thought Mike was down in the cabin making coffee or something. She was never seen again, disappearing into fog around the Pollock Rip, where the *Patricia Marie* had harvested her last load of scallops. The Coast Guard suspended the search that night. The next day, after Mike got out of the Cape Cod Hospital, he went with Roger Horne on the *William Gregory,* along with another boat and a tuna spotter in a small airplane, to crisscross the waters of the shoals east of Nantucket. This was to no avail.

Mike imagines the final moments of the *Susan Lee,* which was insured for only half her value, with almost a parent's peculiar agony. "Who knows how long she floated down there?" he wonders now. "There was a sloppy northeast sea going that day, and once she ran out of fuel, she would have started shipping water through the scuppers. The pumps would have run. Eventually the batteries would have died. Once the pumps stopped running, with all that fish aboard — that would have been the end of the story. Or maybe she took a sea over the rail and went down fast. Who knows?"

Several days later a New Bedford dragger found her lifeboat, released automatically when a vessel sinks, thirty miles east of Nantucket. "They must've gaffed it," Mike says. "I've got it in my garage. It doesn't hold air anymore."

In July, when Amendment Seven went into effect and Mike was running the converted gill-netter, the Massachusetts Marine Fisheries Commission rejected the New England Coast Conservation Association's petition to declare the striped bass a game fish only. In Washington, Ted Stevens failed to get the Senate version of the Magnuson reauthorization to the floor before Congress's summer recess, but he and John Kerry did secure the signatures of seventy-three senators supporting cloture in the event of a filibuster in September from Slade Gorton. This finally led to a compromise with Gorton: the moratorium on new ITQ programs was chopped from four and a half years to four, and Stevens's proposal that regional councils approve ITQs by a two-thirds vote was waived. This led to the Senate's passage of the Sustainable Fisheries Act of 1996 by a 100–0 vote. President Clinton signed the bill into law on October 11.

"We can and should thank the environmental community for helping to bring conservation issues to the foreground," wrote the editor-in-chief of the *National Fisherman,* Sam Smith. "Focusing on ecosystem management, for instance, could be a great benefit. Although our regional councils look at how species interact, especially when dealing with bycatch issues, the idea of managing on an ecosystem basis — a Greenpeace *cause célèbre* — rather than species has never been a formalized management doctrine. And as fishermen continue in greater numbers to diversify — fishing for crab, halibut, salmon, and groundfish all in the same year and all from the same boat — how one fishery impacts another, for instance, is a big financial issue."

Meanwhile, Mike is diversifying: doing some shellfishing, running boats here and there for other skippers, and otherwise wrestling with his grief. Friends, neighbors, his family, and fellow fishermen have been a comfort to him. "My phone rang off the hook for nearly two weeks after that happened," he told me. "The Nauset Fishermen's Association offered to help pay my bills. If I was physically hurt, I might've taken the money. But I had too much pride. It helped to know they were there, though. I've got them and a lot of other people to thank."

He appreciates how lucky he is to be alive, how rarely a longliner ever runs as close to the beach as he was, and for that reason he is perplexed even now by the incident's unholy marriage of catastrophe and great good fortune. Reports of fatalities elsewhere in the fishing industry trouble him in his soul, fill him with survivor's guilt. Other survivors have come forward, men who similarly cheated death, falling out of their boats or otherwise. They share their experiences like stories from the war. They take solace in speaking with someone who, like them, knows more than he wanted to know, is too well acquainted with the remorseless working of things. They warn him that the blues will get worse before they get better. They promise him that eventually they'll get better.

To hasten that day, and to help himself forget about how much this has cost him and how far it has set him back, he makes his plans for getting back in the game. He says his next boat will be a little bit bigger, thirty-five feet. He allows that it might be slower than the *Susan Lee,* but it'll be more comfortable, and that will help him to find and hold a good crewman. "I might have a boat custom-built," he says. "I'm talking to Andy McGeoch, a good builder at a yard in Dennis. I've got a track record now, and I've got backing. If I put up fifteen thousand dollars out of my savings, I got another guy who's going to put up the rest. I'd rather pay off people who believe in me than pay off the bank." He pauses, mulling things over. "Everything happens for a reason. I've got to believe that. I wanted a new boat anyway, but not exactly under these circumstances. Susie hasn't said anything. She knows I'm going to fish."

I read about Mike's accident in the August issue of *Commercial Fisheries News:* "A little mishap — one that could happen to anyone — nearly bought the farm for Mike Russo on June 21." That same column, "Along the Coast," reported that crewmen on the New Bedford–based dragger *Atlantic Star* were astonished to find a golden haddock in their nets. The product of two parents with unusual recessive genes, the fish lacked the shoulder patch and other silvery markings characteristic of haddock, and had a skin tone more like a cod's.

Older fishermen interpret a golden haddock as a sign of good times to come, of prosperity and plentiful catches. It takes no more than that. From such small synchronies of beauty and fortuity, hope breathes like a freshening breeze across the slips and moorings of another generation of voyagers.

Glossary
Notes
Bibliography
Acknowledgments

Glossary

Backstraps. The chains fastened to a dragger's doors, or otterboards, used to connect the doors to the boat's trawl net.

Bottom boxes. Square or rectangular wooden frames, usually screened on both sides, that are filled with sand and seed shellfish and buried in shoal bottom; the screens protect the seed shellfish from predation.

Bullrake. A long, flexible rake used to dig quahogs when working from a small boat.

Catboat. A broad-beamed, sloop-rigged sailboat with its mast stepped into the bow; used to transport cargo and to dredge shellfish around Cape Cod before the advent of outboard motors.

Cod end. The tail end of a trawl net, where cod or any other targeted species are collected as the net is hauled, and from which the catch is emptied onto the deck.

Cowbell. A device at the tip of a trawl net's cod end that opens and closes the net.

Doghouse. A small wheelhouse set amidships in a small boat.

Doors. The heavy slabs of iron and wood or steel towed on either side of a trawl net to keep it open; also known as otterboards.

Dragger. A boat that drags a trawl net behind it.

Eastern rig. A dragger that has its wheelhouse aft and hauls its trawl net in over the side rather than up a ramp in the stern; also known as a side-trawler.

Gangion. A short line connecting one of a series of hooks to a longer fishing line.

Gill-netter. A boat that uses gill nets, which capture fish by snaring their gills.

High-flier. A buoy that floats a six-foot flagpole and flag, used by long-liners and gill-netters to mark the ends of their lines or nets.

Idler chains. The chains that connect a trawl net to its net reel; these are disconnected when the net is attached to the boat's door and set and are reconnected when the net is hauled back.

Jigging. Fishing with a hooked artificial lure that is jerked up and down, whether with a hand line or a rod and reel.

Landing report. A document specifying the species and volume of fish delivered to a dock by a fisherman on a given day.

Lapstreak. In a wooden boat, hull planks that overlap like clapboards.

Lazarette. A compartment in the stern housing a boat's steering mechanism and sometimes used for storage.

Long-lining. Fishing for groundfish by setting long lines of baited hooks anchored to the bottom; also known as tub-trawling.

Net reel. The spool around which a trawl net is wound on the gallows frame of a dragger.

Nordmore grate. A grate positioned in front of the cod end of a trawl net and below an escape vent, designed to guide the target species into the cod end.

Otterboards. See *Doors.*

Otter trawler. A dragger that uses doors, or otterboards, to hold its net open under water.

Paravane. A stabilizing float on the end of a boat's outriggers.

Parlor head. A funnel of netting inside a lobster trap that allows lobsters to enter easily but prevents them from leaving.

Pegging board. The table on a lobster boat where lobsters' claws are banded with thick rubber bands; named for the wooden pegs that used to be used.

Pen board. A board fixed anywhere across a boat's deck to hold or confine catch.

Pleopod. A lobster's abdominal swimming limb; also known as a swimmeret.

Purse seiner. A boat whose net can be closed to encircle fish and draw them out of the water in a purse; also known as a Scottish seiner.

Side-trawler. See *Eastern rig.*

Scottish seiner. See *Purse seiner.*

Seed. Shellfish under the legal minimum size for harvest and sale.

Skeg. The part of a single-screw vessel connecting the keel to the rudder post.

Tickler. A concrete block or similar device towed ahead of a small dredge to clear vegetation and create turbulence.

V-notching. The practice of notching the tails of egg-bearing lobsters before returning them to the sea, thereby marking them as egg producers not to be taken; required in Maine, but not in New Hampshire or Massachusetts.

Warp. A line used for hauling an object, such as a lobster trap.

Notes

Prologue

8 "You can't harvest": Bill McKibben, *The Age of Missing Information* (New York: Plume, 1993), p. 26.

1. Siegfried's Fabulous Horde

17 "the most self-serving": Brian Gibbons to RAC, 1994.
21 "shoales of codde-fishe": Quoted in Robert Finch, ed., *A Place Apart: A Cape Cod Reader* (New York: Norton, 1993), p. 365.
25 "In my early": Gibbons to RAC.
26 "By 1971": Ibid.
32 "The 15th day": Quoted in Henry David Thoreau, *Cape Cod* (Hyannis, Mass.: Parnassus, 1961), p. 284.
32 "We had pestered": Ibid., p. 286.
32 "Puritan Massachusetts": Samuel Eliot Morison, *The Maritime History of Massachusetts 1783–1860* (Boston: Northeastern University Press, 1979), p. 14.
35 "Tub-trawling": Gibbons to RAC.
37 "Every Cape man": Thoreau, *Cape Cod,* p. 142.
38 "Boldly he sailed": Henry C. Kittredge, *Cape Cod: Its People and Their History* (Orleans, Mass.: Parnassus, 1930), p. 188.
39 "There were some": Gibbons to RAC.
39 "Animals": Quoted in Carl Safina, "The World's Imperiled Fish," *Scientific American* 273, no. 5 (Nov. 1995): 46.

2. Ivy Day in the Committee Room

47 "We used the model": New England Fisheries Management Council Lobster Technical Core Group, "Estimating the Effectiveness of Trap Reductions," report, May 11, 1995, p. 1.

51 "Picture a pasture": Garrett Hardin, "The Tragedy of the Commons," *Bioscience* 162 (1968): 36.
51 "As a rational being": Ibid.
52 "The essence": Quoted in ibid., p. 36.
54 "The establishment": Gibbons to RAC.

3. Not as Bad as They're Crying It to Be

83 "is destined": "Popularity of Frozen Foods Will Increase Fish Demand," *Atlantic Fisherman* 26, no. 11 (Dec. 1945): p. 23.
85 "The fresh groundfish": Margaret E. Dewar, *An Industry in Trouble: The Federal Government and the New England Fisheries* (Philadelphia: Temple University Press, 1983), p. 51.
86 "it would look": Ibid., p. 108.
86 "We were formerly"; Ibid., p. 69.
87 "Since the 1940s": Ibid.

4. The Ghost of Harry Hunt

102 "All the alternatives": *Commercial Fisheries News*, June 1995, p. 8A.
107 "What good": Gibbons to RAC.
107 "a fatality": André Malraux, *Man's Fate* (New York: Modern Library, 1934), p. 322.
107 "'Things are bad!": Jean-Paul Sartre, *Nausea* (New York: New Directions, 1964), p. 29.
110 "great lopsters": Henry O. Thayer, *The Sagadahoc Colony, Comprising the Relation of a Voyage into New England* (Portland, Me.: Gorges Society, 1892), p. 45.
110 "sweet, restorative": R. Brookes, *The Art of Angling*, 2d ed. (London: 1740), cited by Francis Hobart Herrick, *Natural History of the American Lobster* (New York: Arno, 1977), p. 173.
110 "the least boy": Alice Morse Earle, *Home Life in Colonial Days* (New York: Tuttle, 1898), p. 117.
110 "The animals": Herrick, *Natural History*, p. 368.
110 "Period of plenty": Ibid., p. 171.
111 "By 1880": Ibid.
111 "rapid extension": Ibid.
113 "In the fall": Quoted in George Brown Goode et al., *The Fisheries and Fishery Industries of the United States. Section V: History and Methods of the Fisheries*, vol. 2 (Washington, D.C.: Government Printing Office, 1887), p. 731.

113 "Period of real decline": Herrick, *Natural History,* p. 171.
113 "It is an easy": Ibid., pp. 370–71.
114 "General decrease": Ibid., p. 171.

5. *Wampum*

126 "a Yankee shipmaster": Morison, *Maritime History,* pp. 260, 285.
127 "The clippers": Kittredge, *Cape Cod,* p. 248.
127 "came into competition": Morison, *Maritime History,* p. 359.
128 "An increase": Ibid., p. 353.
128 "the long-suppressed": Ibid., pp. 370, 369.
129 "We ceased": Kittredge, *Cape Cod,* p. 256.
130 "One such man": Ibid.

6. *Piss and Vinegar*

140 "No natural boundary": Alexis de Tocqueville, *Democracy in America,* vol. 1 (1835), ch. 18.

7. *The Solace of Outward Objects*

149 "fell in with": Quoted by John Hay, *The Great Beach* (New York: Norton, 1963), p. 2.
150 "God [hath]": Quoted by Russell Bourne, *The Red King's Rebellion: Racial Politics in New England 1675–1678* (New York: Atheneum, 1990), p. 19.
150 "only dish": Quoted by Mark Kurlansky, *Cod: A Biography of the Fish That Changed the World* (New York: Walker, 1997) p. 69.
150 "But freedom": Kittredge, *Cape Cod,* pp. 59, 12.
151 "It is a wild": Thoreau, *Cape Cod,* p. 217.
151 "The time must come": Ibid., p. 318.
152 "Abstracted": Hay, *Great Beach,* p. 6.
152 "sick to its": Henry Beston, *The Outermost House: A Year of Life on the Great Beach of Cape Cod* (New York: Henry Holt, 1928), p. 10.
153 "for Orleans is": Thoreau, *Cape Cod,* p. 40.

8. *Terminator Run*

166 "When are you": Kenneth R. Martin and Nathan R. Lipfert, *Lobstering and the Maine Coast* (Bath: Maine Maritime Museum, 1985), p. 100.

168 "The conservation": Ibid., p. 55.
172 "The compliance rate": "Lobster Fishery," *Commercial Fisheries News,* May 1996, p. 2B.
174 "About Long Point": Thoreau, *Cape Cod,* p. 304.

9. Always Hopefully

182 "Scarring of the bottom": "Geology and the Fishery of Georges Bank," *Public Issues in Energy and Marine Geology* (July 1992), n.p.
183 "This was nothing": *Cape Codder,* Mar. 24, 1995.
188 "Every step": "Deadline Slips," *Commercial Fisheries News,* Aug. 1995, p. 14A.
190 "There are crowded days": Beston, *Outermost House,* p. 198.
190 "boisterous shore": Thoreau, *Cape Cod,* p. 81.
192 "evidence that the New England": "Commerce Adds $25 Million," *Commercial Fisheries News,* Sept. 1995, p. 8A.
192 "I have no desire": "Greenpeace Protests IFQs," *National Fisherman,* Oct. 1995, p. 11.
193 "thieves of summer": Robert Finch, *Common Ground: A Naturalist's Cape Cod* (New York: Norton, 1981), p. 83.
193 "monstrous": Beston, *Outermost House,* p. 44.

10. A Sea of Troubles

207 "Blaming NMFS": Gibbons to RAC.
208 "I expected that": "Amaru Agonistes," *Cape Cod Times,* Nov. 7, 1995, p. C1.
213 "Scallops scarce": *Cape Cod Times,* Nov. 2, 1995.

11. Another Fire Drill

219 "a new Nordic": Morison, *Maritime History,* p. 22.
219 "the lighted necklace": Finch, *Common Ground,* p. 114.
224 "bits of fog": Marge Piercy, "How grey, how wet, how cold," quoted in Finch, *Common Ground,* p. 225.
225 "At Nauset": Kittredge, *Cape Cod,* p. 60.
225 "In the late summer": Warren S. Darling, *Quahoging Out of Rock Harbor: 1890–1930* (Orleans, Mass.: self-published, n.d.), p. 18.
228 "Cape fishermen": "Fisheries Chief Says Closure Needed," *Cape Codder,* Dec. 5, 1995, p. 22.
231 "Dick Allen": "ASMFC Takes on Coastwide Lobster Management," *Commercial Fisheries News,* Jan. 1996, p. 3B.

233 "eat . . .": Quoted by Benjamin Schwarz, "What Jefferson Helps to Explain," *Atlantic Monthly*, Mar. 1997, p. 63.
234 "In the quarter century": Quoted in "Arvin and Dinuba: A Tale of Two Towns," *Harrowsmith*, May-June 1991, p. 35.
235 "Parenting becomes": Clifford Cobb, Ted Halstead, and Jonathan Rowe, "If the GDP Is Up, Why Is America Down?" *Atlantic Monthly*, Oct. 1995, p. 67.
239 "One of the most": Darling, *Quahoging*, p. 2.
240 "Beginning well aft": Kittredge, *Cape Cod*, p. 202.

12. *The True Atlantic House*

248 "that seashore": Thoreau, *Cape Cod*, p. 74.
264 "The act": Mark Leach, "The Real Solution: Limit Mobile Gear Fishing," *Commercial Fisheries News*, Feb. 1996, p. 8A.
265 "Some fishermen": Quoted by Michael Crowley, "When Trawling Meant the Guillotine," *National Fisherman*, Apr. 1996, p. 40.
265 "a net": Ibid.
268 "Individual lockers": Quoted by William W. Warner, *Distant Water: The Fate of the North Atlantic Fisherman* (Boston: Atlantic–Little, Brown, 1977), p. 37.
269 "I couldn't believe": Ibid., pp. 47–48.
269 "Thus occurred": Ibid., p. 52.
271 "contributed so much": Quoted by Dewar, *Industry in Trouble*, p. 76.
272 "We hear": Ibid., p. 70.
272 "I tried": Ibid., p. 71.
273 "It's going to fail": Ibid., p. 72.
274 "I've been": Ibid.
274 "Huge as the total": Warner, *Distant Water*, p. 58.
276 "a harmonious": Ibid., p. 70.
276 "but there should be": Quoted in Michael Crowley, "Does Dragging Harm the Habitat?" *National Fisherman*, Apr. 1996, p. 39.
283 "the great northeast storm": Beston, *Outermost House*, pp. 80, 88.

13. *In Cod We Trust*

285 "one of the rarest": Beston, *Outermost House*, p. 150.
285 "hideous wilderness": Quoted in John Hay, *The Great Beach* (New York: Norton, 1963), p. 1.
285 "Though there were": Thoreau, *Cape Cod*, p. 141.
288 "This plan will have": "Amendment 7 a Go," *Commercial Fisheries News*, Mar. 1996, p. 15A.

290 "he who waits": Ibid., p. 247.
292 "the curse": Warner, *Distant Waters,* p. 325.
292 "The surf was alive": Beston, *Outermost House,* p. 176.
293 "Strong, swift-swimming": Henry C. Bigelow and William W. Welsh, *Fishes of the Gulf of Maine* (Washington, D.C.: Government Printing Office, 1925), p. 48.
293 "The dog": Ibid., p. 50.
295 "You could throw": Seth Rolbern, "One More Tow," *Yankee,* Oct. 1991, p. 68.
297 "We heard men": Ibid., p. 131.
301 "knowledgeable or experienced": Quoted in Dewar, *Industry in Trouble,* p. 143.
302 "the idea of novelty": Tocqueville, *Democracy.*
306 "In the old days": "Stripping the Sea's Life," *Boston Globe,* Apr. 17, 1994, p. 24.
306 "As a result": Charles H. Collins, "Beyond Denial: The Northeast Fisheries Crisis, Causes, Ramifications, and Choices for the Future" (Watertown, Mass.: Northeast Fisheries and Sustainable Communities Project, 1994), p. 6.
309 "But every time": "Regulators Ignored Depletion Warnings," *Maine Sunday Times,* Sept. 20, 1994, p. 14A.
309 "The result": "The Promise of Bounty Goes Bust," *Boston Globe,* Apr. 13, 1994, p. 10.
309 "about as effective": "Critics Says N.E. Fishery Council has Avoided Tough Decisions," *Boston Globe,* Jan. 7, 1989, p. 22A.
311 "unnatural familiarity": Quoted in Finch, *A Place Apart,* p. 57.
312 "One does not": "NMFS' 'Risk Averse' Attitude Threatens Industry," *Commercial Fisheries News,* Feb. 1997, p. 6B.
312 "What we did": William Amaru, "Sharing the Blame for Over-Harvesting Groundfish," *Cape Codder,* Mar. 1995.
314 "Remember Laevinus": Brian Gibbons to RAC, Mar. 1996.

14. Men's Lives

324 "Schooling up": Peter Matthiessen, *Men's Lives: The Surfmen and Baymen of the South Fork* (New York: Random House, 1986), pp. 83, 84.
325 "During the past": Ibid., p. 85.
327 "There are no resource": "MA Commission, OMF Director Split on Petition," *Commercial Fisheries News,* Mar. 1996, p. 22A.
327 "I see this move": "Hearing Tackles Proposed Striper Ban," *Cape Cod Times,* May 21, 1996, p. 12A.

329 "a knife in the heart": "Lobstermen Are Steaming over Judge's Verdict," *Cape Codder,* Jan. 23, 1996, p. 1.

330 "Instead of fighting": Victoria Ogden, "Idea of Fish Farming Worth Cultivating on Cape," *Cape Codder,* Mar. 19, 1996, p. 6.

330 "Jim Harrington": Victoria Ogden, "Tilapia Versus a Cape Cod Clam," *Cape Codder,* Mar. 26, 1996.

338 "The hell with that": Philip K. Dick, *Time Out of Joint* (New York: Carroll & Graf, 1959), p. 142.

340 "there have always been": Paul Kemprecos, "A Cause for Alarm," *Cape Codder,* Nov. 18, 1983, sec. 2, p. 4.

340 "can no longer": Finch, *A Place Apart,* p. 71.

341 "I find that": Gary Colas, "Family Assistance Centers Offer Retraining Opportunities," *Commercial Fisheries News,* Apr. 1996, p. 25A.

342 "It's no fish": Walter Scott, *The Antiquary.*

343 "The good news": "Fishery Act Gets Review by Congress," *Anchorage Daily News,* Mar. 15, 1993.

344 "Nobody gave": Jim O'Malley, telephone interview with RAC, Oct. 17, 1996.

347 "There's been a turn": "Magnuson Bill Posits Big Changes," *National Fisherman,* Dec. 1995, p. 7.

347 "There is, I believe": "Major Fisheries Legislation Signed into Law," *Commercial Fisheries News,* Nov. 1996, p. 15A.

347 "We hope": "Shaky Regional Alliance Praises Senate Passage of Magnuson Act," *Tundra Drums* (Bethel, Alaska), Sept. 26, 1996, p. 15B.

Epilogue

363 "We can and should": Sam Smith, "Where Do You Stand?" *National Fisherman,* July 1997, p. 7.

364 "A little mishap": "Along the Coast," *Commercial Fisheries News,* Aug. 1996, p. 4A.

Bibliography

Acheson, James M. 1988. *The Lobster Gangs of Maine*. Hanover, N.H.: University Press of New England.

Beck, Horace. 1973. *Folklore and the Sea*. Middletown, Conn.: Wesleyan University Press.

Belding, David L. 1931. *The Scallop Fishery of Massachusetts*. Boston: Massachusetts Department of Conservation.

———. 1931. *The Soft-Shelled Clam Fishery of Massachusetts*. Boston: Massachusetts Department of Conservation.

———. 1932. *The Quahaug Fishery of Massachusetts*. Boston: Massachusetts Department of Conservation.

Beston, Henry. 1928. *The Outermost House*. New York: Henry Holt.

Bigelow, Henry B., and William D. Welsh. 1925. *Fishes of the Gulf of Maine*. Washington, D.C.: Government Printing Office.

Bradford, William. 1898 (reprint). *"Of Plimoth Plantation" from the Original Manuscript*. Boston: Wright & Potter.

Bourne, Russell. 1990. *The Red King's Rebellion: Racial Politics in New England 1675–1678*. New York: Atheneum.

Cobb, Clifford, Ted Halstead, and Jonathan Rowe. "If the GDP Is Up, Why Is America Down?" *The Atlantic Monthly*, October 1995.

Collins, Charles H. 1994. "Beyond Denial: The Northeast Fisheries Crisis, Causes, Ramifications and Choices for the Future." Watertown, Mass.: Northeast Fisheries and Sustainable Communities Project.

Conservation and Utilization Division, Northeast Fisheries Science Center. 1995. *Status of the Fishery Resources Off the Northeastern United States for 1994*. Woods Hole, Mass.: Northeast Fisheries Science Center.

Darling, Warren S. n.d. *Quahoging Out of Rock Harbor: 1890–1930*. Orleans, Mass.: Orleans Public Library.

Dewar, Margaret E. 1983. *An Industry in Trouble: The Federal Government and the New England Fisheries*. Philadelphia: Temple University Press.

Dick, Philip K. 1959. *Time Out of Joint.* New York: Carroll and Graf.

Estrella, Bruce T., and Daniel J. McKiernan. 1989. "Catch-per-unit-effort and Biological Parameters from the Massachusetts Coastal Lobster ('Homarus americanus') Resource: Description and Trends." Washington, D.C.: U.S. Department of Commerce.

Finch, Robert. 1981. *Common Ground: A Naturalist's Cape Cod.* New York: W. W. Norton.

———, ed. 1993. *A Place Apart: A Cape Cod Reader.* New York: W. W. Norton.

Garber, Andrew, and Clarke Canfield. "Empty Nets, Sinking Hopes." *Maine Sunday Telegram,* September 18, 1994.

Gibbons, Euell. 1964. *Stalking the Blue-Eyed Scallop.* New York: David McKay.

Gosner, Kenneth L. 1978. *A Field Guide to the Atlantic Seashore from the Bay of Fundy to Cape Hatteras.* Boston: Houghton Mifflin.

Greer, Jed. 1995. "The Big Business Takeover of U.S. Fisheries: Privatizing the Oceans Through Individual Transferable Quotas (ITQs)." Washington, D.C.: Greenpeace.

Griffin, Susan. 1995. *The Eros of Everyday Life: Essays on Ecology, Gender, and Society.* New York: Doubleday.

Hardin, Garrett. "The Tragedy of the Commons." *Bioscience* 162 (1968).

Hay, John. 1963. *The Great Beach.* New York: W. W. Norton.

Herrick, Francis Hobart. 1977 (reprint). *Natural History of the American Lobster.* New York: Arno Press.

Hobsbawm, Eric. 1994. *The Age of Extremes: A History of the World, 1914–1991.* New York: Random House.

Junger, Sebastian. *The Perfect Storm.* 1997. New York: W. W. Norton.

Kettlewell, John J., ed. 1992. *Reed's Nautical Companion.* Boston: Thomas Reed.

Kipling, Rudyard. 1964 (reprint). *Captains Courageous.* New York: New American Library.

Kittredge, Henry C. 1930. *Cape Cod: Its People and Their History.* Orleans, Mass.: Parnassus Imprints.

———. 1937. *Mooncussers of Cape Cod.* Boston: Houghton Mifflin.

Kurlansky, Mark. 1997. *Cod: A Biography of the Fish That Changed the World.* New York: Walker.

Malraux, André. 1934. *Man's Fate.* New York: Random House.

Mander, Jerry, and Edward Goldsmith, eds. 1996. *The Case Against the Global Economy.* San Francisco: Sierra Club.

Martin, Kenneth R., and Nathan R. Lipfert. 1985. *Lobstering and the Maine Coast.* Bath: Maine Maritime Museum.

Matthiessen, Peter. 1986. *Men's Lives: The Surfmen and Baymen of the South Fork*. New York: Random House.

McCay, Bonnie J. "Social and Ecological Implications of ITQs: An Overview." *Ocean and Coastal Management* 28, nos. 1–3 (1995).

McCay, Bonnie J., et al. "Individual Transferable Quotas (ITQs) in Canadian and US Fisheries." *Ocean and Coastal Management* 28, nos. 1–3 (1995).

McKibben, Bill. 1992. *The Age of Missing Information*. New York: Random House.

Mitchell, Joseph. 1944. *The Bottom of the Harbor.* New York: Modern Library.

Morison, Samuel Eliot. 1921. *The Maritime History of Massachusetts, 1783–1860*. Boston: Northeastern University Press.

Nickerson, Colin. "Stripping the Sea's Life." *Boston Globe,* April 17, 1994.

———. "The Promise of Bounty Goes Bust for New England." *Boston Globe,* April 18, 1994.

———. "A 'Regulated Inefficiency' May Be the Solution." *Boston Globe,* April 19, 1994.

Petry, Loren C., and Marcia G. Norman. 1963. *A Beachcomber's Botany.* Chatham, Mass.: Chatham Conservation Foundation.

Piercy, Marge. 1985 *My Mother's Body.* New York: Knopf.

Reynard, Elizabeth. 1934. *The Narrow Land: Folk Chronicles of Old Cape Cod*. Chatham, Mass.: Chatham Historical Society.

Richardson, Wyman. 1947. *The House on Nauset Marsh*. Riverside, Conn.: Chatham Press.

Rolbern, Seth. "One More Tow." *Yankee,* October 1991.

Safina, Carl. "The World's Imperiled Fish." *Scientific American,* November 1995.

Sargent, William. 1981. *Shallow Waters: A Year on Cape Cod's Pleasant Bay.* Hyannis, Mass.: Parnassus Imprints.

Sartre, Jean-Paul. 1964. *Nausea*. New York: New Directions.

Schwarz, Benjamin. "What Jefferson Helps to Explain." *The Atlantic Monthly,* March 1997.

Sterling, Dorothy. 1978. *The Outer Lands*. New York: W. W. Norton.

Stump, Ken, and Dave Batker. 1996. *Sinking Fast: How Factory Trawlers Are Destroying U.S. Fisheries and Marine Ecosystems*. Washington, D.C.: Greenpeace.

Thoreau, Henry David. 1961 (reprint). *Cape Cod*. Hyannis, Mass.: Parnassus Imprints.

Trull, Peter. 1994. *A Guide to the Common Birds of Cape Cod*. Shank Painter.

Van Dorn, William G. 1974. *Oceanography and Seamanship.* Centreville, Md.: Cornell Maritime Press.

Ward, Nathalie. 1995. *Stellwagen Bank.* Camden, Me.: Down East Books.

Warner, William W. 1976. *Beautiful Swimmers: Watermen, Crabs and the Chesapeake Bay.* Boston: Little, Brown.

———. 1977. *Distant Water: The Fate of the North Atlantic Fisherman.* Boston: Little, Brown.

Weber, Michael, and Judith Gradwohl. 1995. *The Wealth of Oceans.* New York: W. W. Norton.

Whynott, Douglas. 1995. *Giant Bluefin.* New York: North Point Press.

Acknowledgments

I owe a debt of gratitude first and foremost to the four men who courageously shared so much of their lives and their time on the water with me, and endured so many questions: Brian Gibbons, Dan Howes, Carl Johnston, and Mike Russo. My thanks are also due to Suzanne Gibbons, Bindy Johnston, and Susan Russo for entertaining me so generously in their homes and lending me their perspectives on a fisherman's life, and to Brian Gibbons in particular for all the wonderful packets of news clippings he mailed to me in New Hampshire.

I am grateful to many others as well: Bruce T. Estrella, senior marine fisheries biologist for Massachusetts' Division of Marine Fisheries, who helped me with the hard questions about lobsters and lobster regulations; H. Perry Jeffries, recently retired from the University of Rhode Island, a biological oceanographer and my guide on the natural history of shellfish; Bonnie McCay, a professor in the Human Ecology Department of Rutgers University, who shared her research on ITQs as management tools in New Jersey and Nova Scotia; Niaz Dorry, of Greenpeace, and Pat Fiorelli, of the New England Fisheries Management Council, both of whom answered many questions and also guided me to others who had answers; Jim O'Malley, of the East Coast Fisheries Federation, who gave me a glimpse behind the scenes of the politics of fisheries management; Dave Lockwood, of the Holderness School and the jazz fusion band Raccoon Beach, who shared with me a few measures of what a jazzman knows; David Sleeper, of *Vermont Magazine,* who urged this sort of thing in the first place and sent me sources I might otherwise have missed; Kevin Jones, Tom Carey, Eric Yeager, Emlyn Stokes, and Ethan Van Veghten, who provided various forms of technical support; and Robin Alden, formerly Maine's commissioner of marine resources, and John and Angela San Filippo, of Gloucester, all of whom were enormously helpful in getting this project going.

My agent, David Black, was also invaluable in that respect, as was his assistant Susan Raihofer. I am thankful to Richard Todd for his early reading of my manuscript, and also to Jim Brewer, formerly of the Holderness School and always a terrific editor and teacher of writing, who generously brought his considerable powers to bear on my behalf. At Houghton Mifflin, Mindy Keskinen and Chris Coffin were unfailingly helpful, and I remain fortunate in my association with Liz Duvall, world's best copy editor. I was fortunate as well — very fortunate — in my editor and guiding spirit, Chris Carduff, who always understood exactly what I wanted to say and so perceptively found ways for me to say it.

Finally, I am grateful to the people whose personal support, patience, and flexibility were also crucial to this endeavor: Pete Woodward, headmaster of the Holderness School, who cheered me on; Lois Carey, who took up slack when I was doing research or fighting a deadline, and who also generously cheered me on, as she always has; Ryan and Kyle Carey, who have uncomplainingly endured the inconveniences of this work; and Sue Pelizza, who tracked down hard-to-find sources, brought me sources I didn't know existed, listened to early drafts of the manuscript, and whose encouragement was always a draft of pure oxygen to me.